Crystals

Growth, Morphology, and Perfection

How do crystals nucleate and grow? Why and how do crystals form such a wide variety of morphologies, from polyhedral to dendritic and spherulitic forms? These are questions that have been posed since the seventeenth century, and are still of vital importance today, both for modern technology, and to understand the Earth's interior and the formation of minerals by living organisms. In this book, Ichiro Sunagawa sets out clearly the atomic processes behind crystal growth, and describes case studies of complex systems from diamond, calcite, and pyrite, to crystals formed through biomineralization, such as the aragonite of shells, and apatite of teeth. It will be essential reading for advanced graduates and researchers in mineralogy and materials science.

Ichiro Sunagawa is the Principal of the Yamanashi Institute of Gemmology and Jewellery Arts, and Emeritus Professor at Tohoku University. He has written or edited a number of books in English or Japanese, including *Morphology of Crystals* (1988) and *Handbook of Crystal Growth* (2000). His contribution to the scientific literature is considerable, with over 200 papers in English and more than 300 in Japanese.

Crystals

Growth, Morphology, and Perfection

Ichiro Sunagawa
Tohoku University

CAMBRIDGE UNIVERSITY PRESS
Cambridge, New York, Melbourne, Madrid, Cape Town, Singapore, São Paulo

Cambridge University Press
The Edinburgh Building, Cambridge CB2 8RU, UK

Published in the United States of America by Cambridge University Press, New York

www.cambridge.org
Information on this title: www.cambridge.org/9780521841894

First published 2005
This digitally printed version 2007

A catalogue record for this publication is available from the British Library

ISBN 978-0-521-84189-4 hardback
ISBN 978-0-521-71479-2 paperback

Contents

Foreword to the English translation

The English version of this book was prepared based on translation by the author of the Japanese version; however, a few modifications have been made to assist the reader. The photographs appearing as plates at the front of the Japanese edition have been moved to the appropriate pages in the text, and references and suggested reading were moved to follow each chapter.

In preparing the English version, several individuals have assisted the author. Professor Andrew Putnis arranged the publication of the English version, and Professor Alan L. Mackay read through the manuscript and gave useful suggestions from a scientific standpoint. Dr. Sally Thomas and Irene Pizzie very carefully checked the manuscript and improved the English. The author sincerely appreciates their contributions. The permissions granted by many authors to reproduce figures from their various books and journals is gratefully acknowledged.

Preface

Crystals are solid materials having regular atomic arrangements characterized by periodicity and anisotropy. These properties are universally present, irrespective of whether the crystal is inorganic or organic, in living systems or in the inanimate world. Crystals exhibit various external forms, as represented by the elaborately varied dendritic forms of snow crystals or the hexagonal prismatic forms of rock-crystal. This variety of shape has stimulated scientific curiosity since the seventeenth century, since when intensive efforts have been made to understand the reasons why and how crystals can take a variety of forms.

The forms that crystals take result from the way in which crystals grow. The mechanism of growth is recorded in various forms in each individual crystal, regardless of size. The same crystal species may show different crystal forms (for example, polyhedral, skeletal, and dendritic), depending on growth conditions. Spiral growth step patterns, which record the growth process at the nanometer scale, have been observed on crystal faces. In single crystals, fluctuations in growth rates during the growth process are recorded as variations in perfection and homogeneity, such as growth sectors, growth banding, and three-dimensional distribution of lattice defects such as dislocations. The texture and structure of minute polycrystalline aggregate also record the growth history. These fluctuations are observed not only in crystals formed by inorganic processes, but also in those formed in living organs like bones, teeth or shells, or in calculus formed in various organs through the excretion of unnecessary components. To understand the phenomena occurring in complicated and complex systems, in which the growth process is unobservable *in situ*, we must regard the characteristics exhibited by crystal forms (morphology of crystals) as very important sources of information.

Our understanding of the atomic process and the mechanism of crystal growth progressed greatly during the latter half of the twentieth century. Various techniques have been developed that will produce bulk single crystals and thin films with high perfection and homogeneity by strict control of the growth parameters.

Various crystalline materials with desired properties have been synthesized, and this has driven the utilization of single crystals in the production of semiconductor, opto-electronic, piezoelectric, and pyroelectric materials.

Understanding crystal growth mechanisms is just as important in the field of industrial crystallization, where a large quantity of minute crystals with well controlled sizes and forms are required, such as in pharmaceuticals, chemical seasonings or photographic emulsion, and even proteins. By coupling intellectual curiosity with the demands from industry, we have reached the point at which we can understand, at the atomic level, the mechanisms that determine crystal form, perfection, and homogeneity, at least in simple systems. It will be the task of researchers in the twenty-first century to deepen our understanding of phenomena that occur in more complicated and complex systems.

The main purpose of this book is to present the route that we need to take in order to decipher the phenomena and history occurring in complicated and complex systems, based on our present day understanding of simple and single systems. It is also hoped to present the root of the science and technology of crystal growth to those who are already actively involved in growing bulk single crystals and thin films using industrially established growth techniques.

The findings summarized in this book have been achieved through research activities by the author during half a century's work at the Geological Survey of Japan, Tohoku University, and the Yamanashi Institute of Gemmology and Jewellery Arts. Throughout this scientific career, a deeper understanding has been achieved thanks to my joint efforts with colleagues, and postgraduate and undergraduate students. I have unforgettable memories from individual research works, and, although no names are mentioned, I wish to express my sincere thanks to all those that have been involved. My thanks are also due to Ikuo Hirayama and Hideya Fukase of Kyoritsu Shuppan Co., for their overall support in publishing this book.

This book is dedicated to my wife, Michiko, and those unnamed others who have supported my research activities behind the scenes.

PART I FUNDAMENTAL CONCEPTS

In Part I we summarize the fundamental concepts relating to the growth and morphology of crystals, such as atomic processes and mechanisms of crystal growth, and the principles governing the morphology, perfection, and homogeneity of crystals. We base this description on a historical review of the development of the subject. Such fundamental concepts relating to atomic processes and the mechanisms of crystal growth have been acquired throughout the twentieth century through investigations on simple systems and the single phase, and have formed a base for the development of industries, such as semiconductors, in which single crystals are used. Our understanding of the phenomena occurring in complicated and complex systems, such as in the formation of solid earth materials or biomineralization, will be deepened based on the fundamental concepts explained in Part I.

1

Introduction

The crystal, with its regular atomic construction, is the most commonly encountered state of solid materials. The three properties of a crystal, external form, perfection, and homogeneity, are directly related to how the crystal grows. Individual crystals of the same species and atomic construction may have different properties. The mechanism of crystal growth has long been understood at an atomic level, at least for simple systems. Our understanding of the factors determining external form, perfection, and homogeneity provides us with the information necessary for the development of industries which utilize the physical properties of single crystals in forms of bulk and thin film, such as semiconductors. In the twenty-first century, we expect to clarify the relevant phenomena occurring in complex and complicated systems.

1.1 Historical review

On hearing the word "crystal," most of us call up images of regular, symmetric forms, perhaps the prismatic form of rock-crystal, or the dendritic form of snow crystals. We use the term "crystal clear" to imply something transparent and pure. The ancient Greeks used the term crystal (κρνσταλλο) to imply clear, transparent, and hard ice; rock-crystal was so named as it was thought to be unmeltable ice that existed in fissures of rocks. As in the present day, the regular polyhedral forms were the ones that roused the most interest. Pythagoras, for example, used the word crystal to imply perfection, harmony, and beauty; and Plato listed his famous five polyhedra, related to fire, earth, air, water, and the universe.

Theophrastos (372–287 BC), the author of the oldest book on minerals, *On Stones*, referred briefly to regular polyhedral forms exhibited by mineral crystals [1]. In later books on minerals, polyhedral forms bounded by mirror-flat faces are mentioned.

For example, in the lengthy *Natural History* [2], by the famous Roman natural historian Gaius Plinius Secundus, there are descriptions of polyhedral forms of beryl, rock-crystal, diamond, etc.

It was not until the seventeenth century, however, that special attention was paid to the *forms* of crystals. Kepler was deeply attracted by the highly varied dendritic forms of snow crystals, and he believed that the units constituting snow crystals were equally sized spheres [3]. Kepler described the observed elaborately varied external forms as simply the result of the combination of these spherical units. This marked the beginning of the concept of crystal structure. Starting from Kepler's idea of equally sized spheres as the constituent units, the concept was developed to include equally sized ellipsoidal units by Huygens, and further to polyhedral units by Haüy. These were systematized into seven crystal systems based on the axial lengths and axial angles, and were further subdivided into thirty-two crystal groups (point groups) by the combination of symmetry elements compatible with the seven crystal systems. In the Appendixes we present: the crystal axes; the fourteen Bravais lattices and the seven crystal systems; the indexing of crystal faces and zones; the symmetry elements and their respective symbols; and the stereographic projections of the thirty-two crystal groups. These are mathematical systems, but numerous observations on the real crystals of minerals formed the basis for this ordering. Based on this mathematical structure, the forms exhibited by polyhedral crystals were investigated, and the relationships between these forms and the lattice types were discussed. This type of investigation is called crystal morphology, and was of great interest in the nineteenth and early twentieth centuries. A vast amount of data of descriptive and classification type were accumulated during this period. The results were complied in books such as Goldschmidt's *Atlas der Kristalformen* and Dana's *The System of Mineralogy* (see refs. [1] and [4], Chapter 9).

In 1912, Von Laue demonstrated using X-rays that a crystal is constructed of unit cells, and that it is possible to determine atomic positions by means of diffraction phenomena. Investigations analyzing crystal structure spread quickly, and within a short period the crystal structures of most simple inorganic crystals had been analyzed, deepening the knowledge of chemical bonding and atomic and ionic radius, and leading to the emergence and rapid progress of a new scientific discipline, crystal chemistry. It was during this period that the importance of the 230 space groups, which had been mathematically introduced by Federov, Schoenflies, and Barlow, was pointed out by a Japanese researcher, S. Nishikawa, in analyzing crystal structures. In addition to X-rays, electron and neutron beams are now also used in structural analyses; nowadays, stronger X-ray beams are used, and there have been great developments in the apparatus used for structural analyses, which have now become almost routine, except for crystals of macromolecular proteins.

1.2 The birth of the concept of crystal growth

In the seventeeth century, the Danish anatomist Nicolaus Steno (1638–1687) collected many samples of rock-crystals, and he measured the interfacial angles of hexagonal prisms. He found that the interfacial angles of corresponding faces were constant, irrespective of external form. This was the beginning of the now accepted "law of the constancy of interfacial angles." In the treatise describing his finding, Steno put forward two important concepts. He argued that rock-crystals were not of organic origin formed by the activity of bacteria in the soil, as was widely believed, but that they grew through the agglutination of tiny particles formed through inorganic processes taking place in high-temperature aqueous solution. He explained that the reason why rock-crystals show hexagonal prismatic forms bounded by six triangular faces at the tip and six rectangular faces, and sometimes show tapering prismatic or platy forms, is that the growth rates are different depending on crystallographic direction, i.e. on the different crystal faces [4]. These concepts of crystal growth and growth rate anisotropy provide a basis for the present-day science of crystal growth.

Interest in crystal form continued into the eighteenth century. Hooke noted that alum crystals took regular octahedral forms when they grew on the tip of a string immersed in aqueous solution, whereas their forms changed into platy triangular or hexagonal platy, though bounded by the same faces, when they grew at the bottom of a beaker (see ref. [11], Chapter 4). De l'Isle reported that when NaCl crystals grew in aqueous solution containing a small amount of urea, they took a simple octahedral form, in contrast to the simple cubic form grown in pure solution (see ref. [26], Chapter 4).

In the nineteenth century, when crystal morphology was systematized to fourteen types of unit cells, seven crystal systems and thirty-two crystal groups, the following two macroscopic treatments on the morphology of crystals emerged.

(1) The Bravais empirical rule, which states that there is a close correlation between the polyhedral forms of a crystal and the lattice type.
(2) Theoretical treatments by Gibbs, Curie, and Wulff on equilibrium form showed that the equilibrium form of a crystal having anisotropy should be polyhedral, not spherical, as would be expected for the equilibrium form of an isotropic liquid droplet. (See refs. [6]–[8], Chapter 4.)

It became necessary to understand how crystals grow at the atomic level so as to form a deeper understanding of why crystals can take a variety of forms. This was achieved through the layer growth theory put forward in the 1930s by Volmer, Kossel, and Stranski on the structure and implication of the solid–liquid interface, the spiral growth theory by Frank in 1949, and the theory of morphological

Figure 1.1. (a) This hexagonal prismatic form of rock-crystal appears due to spiral growth on smooth interfaces and growth rate anisotropy. (b) Twinned rock-crystals from the Otome Mine in Japan. According to the Japan Law (see Section 10.6), these are larger and show a more flattened form than the coexisting single crystals.

instability by Mullins and Sekerka in 1963 (see refs. [7]–[11] and [20], Chapter 3). The understanding of the mechanism of crystal growth progressed experimentally in the twentieth century. Before the 1930s, the main work was performed on the relationship between the macroscopic ambient phase and macroscopic crystal forms. Later, advancements in optical microscopy and electron microscopy made it possible to observe and measure growth spirals with step heights of nanometer order. It is now possible to observe the behavior of spiral step advancement of nanometer order in height *in situ*.

1.3 Morphology, perfection, and homogeneity

Since the growth mechanisms of crystals have become understood at the atomic level, at least in single and simple systems, it has become clear how the micromorphology (the morphology of growth spirals and etch pits) and the macromorphology (polyhedral, skeletal, and dendritic forms) of crystals are determined. Since the morphology of crystals is the result of crystal growth, this is related to how lattice defects (point, linear, and planar) are induced and distributed in single crystals, and to how impurity elements participate. This leads to the development of the method of controlling perfection and homogeneity of bulk and thin film single crystals and to the exploration of semiconductor and opto-electronic materials with desirable properties. It is with these developments that today's industries are concerned.

1.4 Complicated and complex systems

Knowledge of the growth mechanism, morphology, perfection, and homogeneity of single crystals creates a base from which we may understand the morphology exhibited by polycrystalline aggregates, or the mechanism of formation of textures and structures that appear in polycrystalline aggregates of multiple phases, such as those shown by ceramics, metals, and rocks.

Many examples are known to exist that show that polycrystalline aggregates exhibit properties different from those of single crystals. For example, on the Mohs scale the hardness of diamond is given as 10, the highest for any known mineral, yet a diamond single crystal is weak to shock due to the cleavage. It is due to the textures formed by polycrystalline aggregation that carbonado and ballas, polycrystalline aggregates of diamonds, have higher toughness and can be used as boring heads, unlike the relatively weak single crystals. Similarly, the toughness of bamboo results from the textures formed by aggregation of soft cells. The outer and inner skins have different textures and different functions.

What determines the physical forms of animals and plants is a subject of much debate within the biological sciences. We know that the basic unit of living bodies is not crystalline; however, should we entirely disregard the units that comprise living organisms as having no link to crystals? Does DNA alone uniquely control all living phenomena?

There are inorganic and organic crystals formed in living bodies as a result of being alive, and these are the link between living phenomena and crystals.

Tooth, bone, shell, and the exoskeleton are formed to maintain life, whereas gallstones, bladder stones, and sodium uric acid (a cause of gout) are the results of excretion processes of excess components. Tooth, bone, and shell are composed of inorganic crystals of apatite, aragonite, calcite, and protein. It is assumed that these crystals grow through the cooperation of both crystals and proteins. Teeth consist of enamel and dentine portions, both constituting aggregates of apatite crystals of different forms and sizes, leading to different textures and functions. In contrast to well controlled forms, sizes, and textures of these polycrystalline aggregates formed by cooperation with protein, crystals constituting calculi in organs, such as gallstones, usually exhibit uncontrolled radial polycrystalline aggregations, suggesting no cooperation between protein and crystals.

It is conjectured that there can be two types of cooperation with proteins for crystals to grow in living bodies: in one, a protein film acts as an envelope; and in the other it acts as a substrate in epitaxial growth (see Section 7.4). When crystals of the same or different species grow in a definite crystallographic relation (orientation) on a crystal face of the substrate, the phenomenon is called *epitaxy*. This phenomenon was originally known to be present in mineral crystals, but it has now been developed as an important technique of crystal growth used to grow thin films of single crystals for semiconductor devices.

Growth mechanisms in epitaxy are now understood at an atomic level in relation to host–guest crystals (interface energy) and driving force. Investigations on host–guest relations have been expanded from those between inorganic and inorganic, to inorganic/van der Waals, inorganic/polymers, and inorganic/protein crystals. The growth of inorganic crystals in living bodies corresponds to the relation between protein crystals as the host and inorganic crystals as the guest. The opportunity may arise for us to unlock a new relationship between inorganic and organic, or inanimate and animate, forms using morphology and textures exhibited by crystals formed through physiological activities.

The aim of this book is to analyze phenomena in complicated and complex systems, such as crystallization in minerals and in the living world, using the morphology of crystals as the key.

Crystal form is the direct result of crystal growth, and we will therefore develop our arguments based on the mechanism of crystal growth. The book consists of two parts: Part I presents the fundamental concepts, and Part II deals with the application of these concepts to complicated and complex systems (by looking at case studies).

In Part I, after systematically summarizing the hitherto used terms in the morphology of crystals, we summarize the developments in the atomic process of crystal growth and morphology achieved in the twentieth century.

In Part II, factual examples concerning minerals and physiology are presented to demonstrate how complicated phenomena may be analyzed based on the understanding explained in Part I.

References

1 E. R. Caley and J. F. C. Richards, *Theophrastus On Stones*, Colombia, Ohio State University, 1956

2 Pliny the Elder, *Natural History*, trans. J. F. Heales, Harmondsworth, Penguin, 1991

3 J. Kepler, *Strena Seu de Nive Sexangula*, Frankfurt, Godfrey Tampach, 1611, trans. C. Hardie, *The Six-Cornered Snowflake*, Oxford, Clarendon Press, 1966

4 N. Steno, *De Solido Intra Solidum Naturaliter Contento Dissertationis Prodromus*, Florence, 1669, trans. J. G. Winter, *The Prodromus of Nicolous Steno's Dissertation Concerning a Solid Body Enclosed by Process of Nature Within a Solid*, New York, Hafner, 1968

Suggested reading

History of crystallography

J. H. Burke, *Origin of the Science of Crystals*, Berkeley, University of California Press, 1966

Collected classic papers on forms of crystals and crystal growth

N. Kato (ed.), *Crystal Growth*, New Edition of Classical Papers in Physics, vol. 44, Tokyo, Physical Society of Japan

C. J. Schneer, *Crystal Form and Structure*, Benchmark Papers in Geology, vol. 34, Stroudsburg, Pa., Dowden, Hutchinson and Ross, Inc., 1977

D. T. J. Hurle (ed.), *A Perspective on Crystal Growth*, Amsterdam, North-Holland, 1992

2

Crystal forms

Crystals are solid materials having regular arrangements of atoms, ions, or molecules. Crystal forms are determined not only by structure but also by the factors involved in growth. The same crystal species may therefore appear in various forms. In this chapter, the external forms of real crystals are systematically classified.

2.1 Morphology of crystals – the problems

The morphology of crystals is the central theme of this book. Our intention is to present systematically the fundamental concepts that allow us to analyze the factors that determine the various forms of crystals. We may then deduce and analyze the phenomena and processes that we cannot observe *in situ*, such as those occurring in the depths of the Earth or in the animate world.

When there is no distortion in the structure or no change in orientation throughout a crystal, we refer to the structure as a single crystal. A solid consisting of many single crystals with different orientations is called a polycrystalline aggregate. There are also polycrystalline aggregates of multiple phases, as in metals, rocks, and ceramics.

Since real crystals may contain various defects of lattice order, we may regard these crystals as nearly perfect single crystals, and we accept the defects as part of the original system. Assuming this to be the case, individual crystals may be classified as single crystals, no matter how big or small they are, and a combination of individual crystals is termed a polycrystalline aggregate. However, when crystalline materials are used in specific industrial purposes, a single crystal of a particular size may be required. Single crystals of silicon of centimeter to meter size are necessary in semiconductor devices, or specific sizes of quartz or ADP (ammonium dihydrogen phosphate, $NH_4H_2PO_4$) may be required for use in piezoelectrics.

Clear, transparent single crystals of centimeter size are used for facet cutting in jewelry making.

This centimeter to meter order of size is generally assumed when we refer to single crystals. Individual crystals of clay minerals, however, are single crystals, although their sizes are of micrometer order. On the other hand, metals and ceramics are generally chosen for the bulk physical properties exhibited by polycrystalline aggregates; single crystalline materials of these solids are in general not required (exceptionally, large single crystals of metals are used). However, to understand the physical properties of metals, it is necessary to grow single crystals of appreciable size and to investigate their properties. Investigations performed on single metal crystals have promoted the rapid development of dislocation theory and physical metallurgy.

The morphology of single crystals is determined by the crystal structure (the internal factors), the crystal growth conditions, and the process of that growth (the external factors). First, it is necessary to understand how the morphology of a single crystal is determined through the interconnection between the external and internal factors. In the case of polycrystalline aggregates of a single phase or multiple phases, the texture and the structure* are determined by the time and density of nucleation of the multiple phases and the morphologies of the respective crystals. Similarly, just as the toughness of bamboo is determined by the texture, which consists of soft cells, characteristic morphologies and properties that are not seen in single crystals may be evident in polycrystalline aggregates. So, we see that understanding the morphology of single crystals assists us in understanding more complicated systems. We shall use this method to extend our discussions to include complex structures.

Crystals having different crystal structures usually take geometrical external forms following the symmetries involved in their respective structures, but they may also take very different external forms if the growth conditions are dissimilar. Snow crystals are a representative example: they exhibit the polyhedral forms of the hexagonal prism and hexagonal plate, but they also may exist in the familiar dendritic form, with branches in six directions, which in Japan is referred to, rather poetically, as the "six petal flower." It is the purpose of this book to demonstrate why elaborately varied forms of crystals appear, and we are going to base our discussion on the understanding of the atomic process of crystal growth. First, therefore, we will summarize and systematize the problems involved in determining the morphology of crystals.

* "Texture" and "structure" are not explicitly defined terms. *Structure* is usually applied to macroscopic heterogeneity due to macroscopic movement such as flow structure or foliated structure, whereas *texture* usually refers to microscopic heterogeneity, such as holocrystalline texture, porphyritic texture, and lamellar texture due to exsolution.

To do this, we must first investigate the origin of the variations in morphology exhibited by the same single crystal. For example, a single crystal might exhibit a polyhedral form bounded by flat crystallographic faces, a hopper or skeletal form bounded by faces with a step-wise depressed center at a face, or a dendritic form (see Fig. 2.1).

On crystal faces bounding a polyhedral crystal, step patterns resembling the contour lines on a topographic map or striations in one direction are observable depending on the nature of the face. These show the process of crystal growth or dissolution at an atomic level, and are referred to as the surface morphology or surface microtopography.

If an appropriate detection method is applied, one may visualize growth sectors formed by the growth of different faces, growth banding induced by fluctuation of growth rates and impurity contents, and the distribution of lattice defects contributed by the growth. These observations represent an important record of the variations in the morphology during the growth process of the crystal.

Crystals formed under small driving force conditions (see Section 3.2) in a dilute ambient phase, such as the vapor phase or solution phase, will generally exhibit polyhedral forms, irrespective of their size. Even crystals of micrometer size, such as clay minerals, show polyhedral forms. However, there are crystals that show elongated needle forms that resemble whiskers, coils, hollow tubes, and even ice cream cones (see Figs. 2.2 (a), (b)); others exhibit tree-like polycrystalline aggregates of dendrites (see Fig. 2.2 (c)).

When a crystal is synthesized using a seed crystal or by a controlled growth technique such as one-directional solidification, it takes on a very different morphology from that of a freely grown crystal; for example, it may be a column bounded by a curved surface, or in an entirely different polyhedral form from that exhibited by a freely grown crystal (see, for example, Figs. 2.1 (f), (g)). Although crystals created in this way are bounded by non-flat, curved surfaces, they are still single crystals.

Twinning and epitaxial growth are also methods of controlled growth, and therefore crystals grown by these methods often exhibit a very different characteristic morphology from that of coexisting single crystals.

2.2 *Tracht* and *Habitus*

Polyhedral crystals bounded by flat crystal faces usually take characteristic forms controlled by the symmetry elements of the crystal (point) group to which the crystal belongs and the form and size of the unit cell (see Appendix A.5). When a unit cell is of equal or nearly equal size along the three axes, crystals usually take an isometric form, such as a tetrahedron, cube, octahedron, or dodecahedron. Crystals having a unit cell size of $a \approx b < c$ take a platy form, and those having $a \approx b > c$ take a prismatic form (see Fig. 2.3). The characteristic forms exhibited by different crystal species are called "crystal habit" in English. In German, this is called *Habitus*, distinguished from *Tracht* to be described later. The term "crystal habit" is sometimes used in a broader sense to describe the characteristic forms shown by polycrystalline aggregates, such as spherulitic, botryoidal, or reniform.

A crystal having a unit cell size of $a \approx b \approx c$ does not always take an isometric *Habitus*. When NaCl crystals grow in an aqueous solution, they usually take on a cubic habit, but after the addition of Pb or Mn ions as impurities, the *Habitus* changes to octahedral. If NaCl aqueous solution is kept in a wineskin in dry shade, numerous needle crystals of NaCl spring out of the skin's surface. Such a highly anisotropic *Habitus* is called a whisker. In the bulk aqueous solution, thin platy or needle-like NaCl crystals bounded by {100} faces rarely grow together with the cubic *Habitus*. Figure 2.4 shows the unusual *Habitus* of chalcopyrite ($CuFeS_2$) crystals. Under specific conditions, an upper and lower set of four crystallographically equivalent {111}faces of tetrahedral *Habitus* develop anisotropically and result in triangular plate or needle forms [1]. Chalcopyrite crystals showing such unusual forms are called "triangular chalcopyrite," and are found only in vein-type hydrothermal deposits in the Tertiary formation in North-Eastern Japan. *Habitus* variation like this shown by the same crystal species also belongs to the category of variation of crystal habit.

As well as the *Habitus* variation described above, another term, *Tracht*, is applied to describe the variation of forms due to the combination and degree of development of faces within the same category of *Habitus*. There is no distinction in English between *Habitus* and *Tracht*, both being called crystal habit, but in German *Tracht* is distinguished from *Habitus*.

Figure 2.5 shows a representative example showing changes in pyrite (FeS_2) crystals within an isometric *Habitus*, due to a combination and degree of development of three faces, {100}, {111}, and {210} [2]. Variations seen in *Tracht* and *Habitus* are determined by the relative normal growth rates of the respective faces. The face with the slowest normal growth rate is the largest, and this determines the crystal habit. The face with the highest normal growth rate disappears from the crystal. If the degree of development and the frequency of appearance of crystal faces observed on many crystals of the same species grown in different ways are treated statistically, the order of the morphological importance of crystal faces may be determined. It became empirically known that the order of morphological importance is often correlated to the order of reticular densities of faces (density of lattice points per unit area), and this became known as the Bravais empirical rule (see Section 4.2).

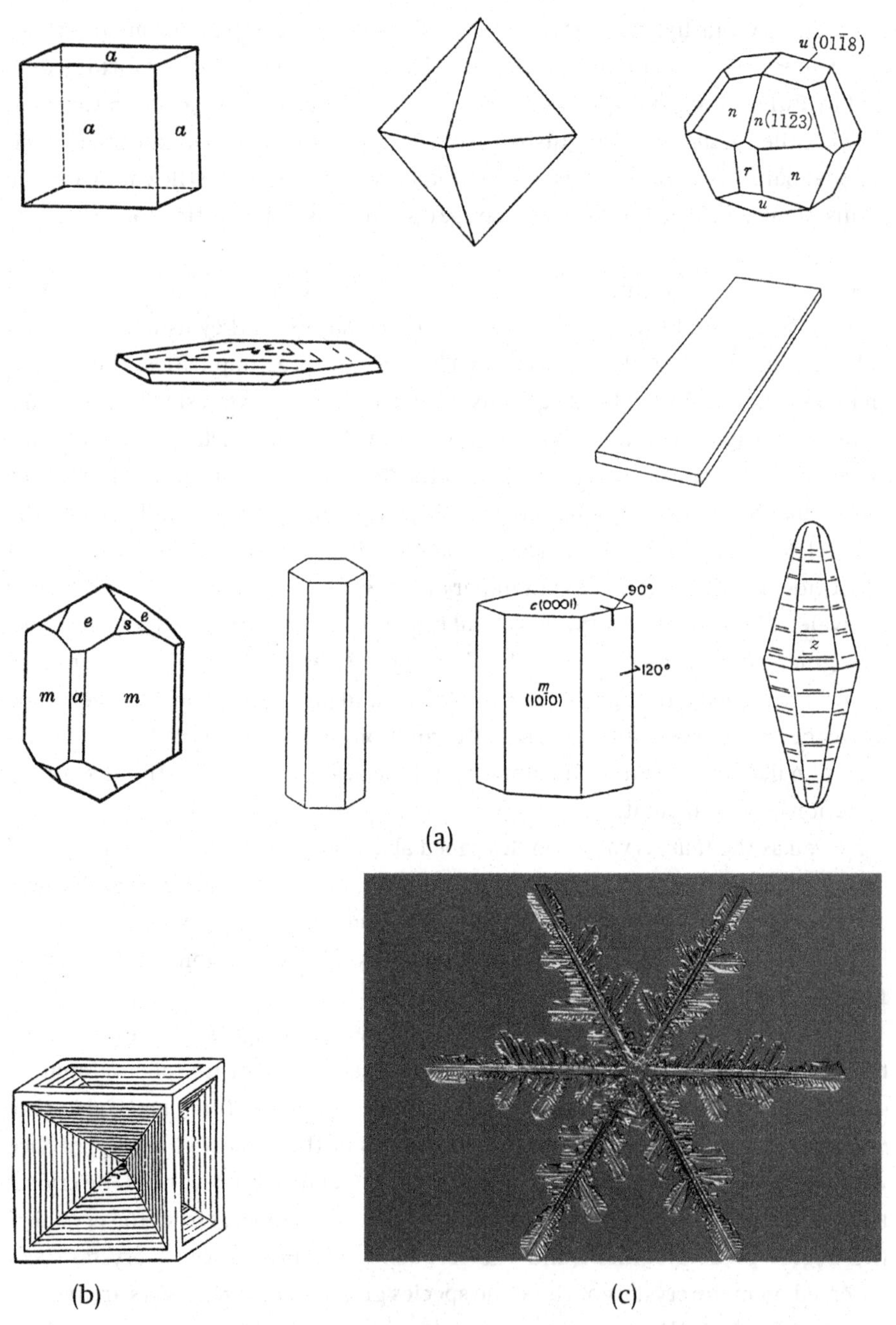

Figure 2.1. Various forms exhibited by crystals. (a) Polyhedral crystals; (b) hopper crystal; (c) dendritic crystal (snow crystal, photographed by the late T. Kobayashi); (d) step pattern observed on a hematite crystal (0001) face; (e) internal texture of a single crystal (diamond-cut stone, X-ray topograph taken by T. Yasuda); (f) synthetic single crystal boule, Si grown by the Czochralski method; (g) synthetic corundum grown by the Verneuil method.

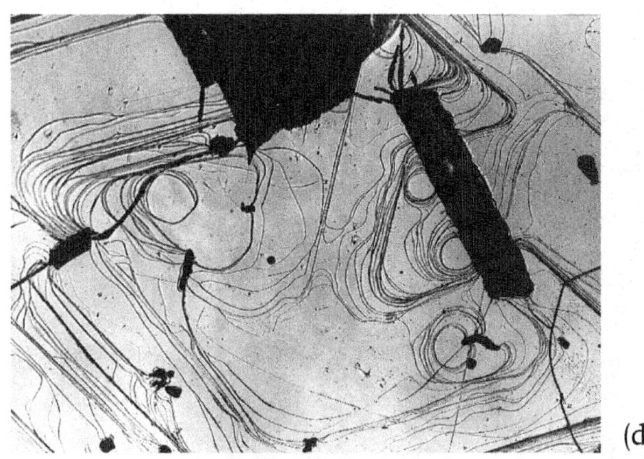

(d)

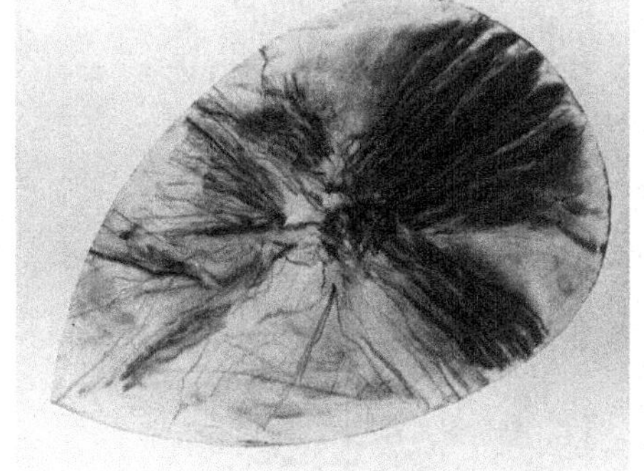

(e)

(f) (g)

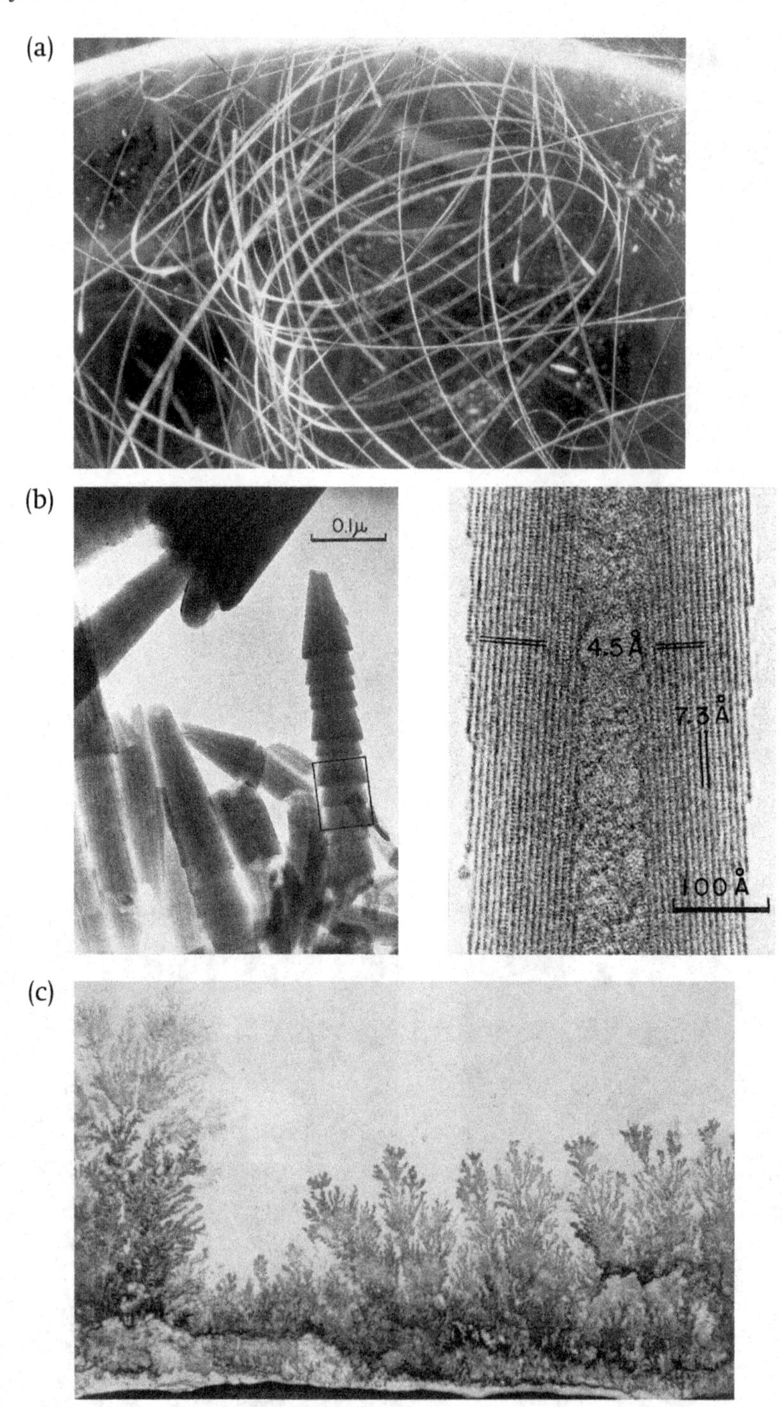

Figure 2.2. (a) Coiled whisker of rutile in cabochon-cut quartz (courtesy of E. A. Jobbins). (b) Crysotile crystals resembling stacked ice cream cones. (Transmission electron microphotograph taken by K. Yada.) (c) Pyrolusite polycrystalline dendritic pattern formed along a bedding plane of sedimentary rock.

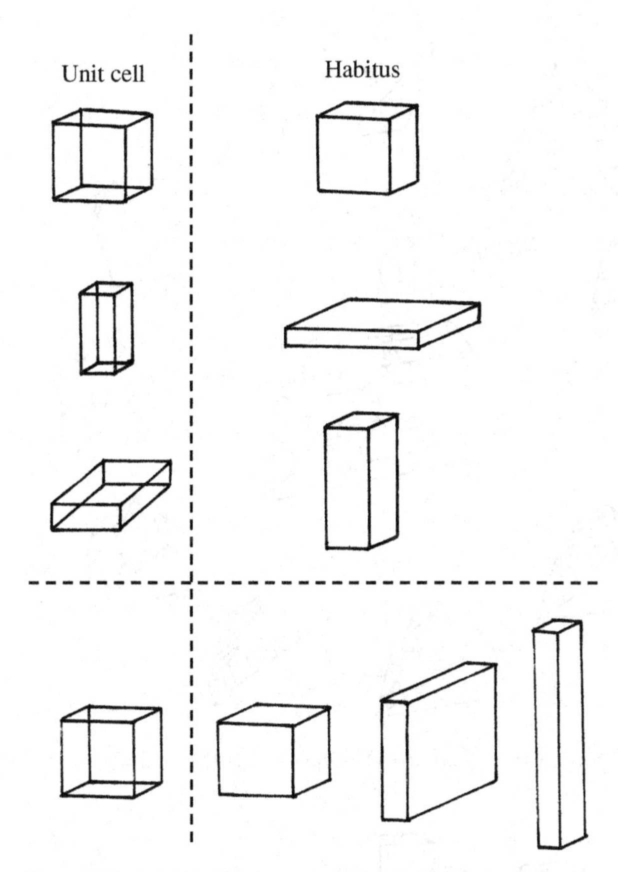

Figure 2.3. Relation between crystal habit (*Habitus*) and the forms of unit cells. The figures in the upper two quadrants indicate the different *Habitus* expected for different unit cells, and those in the lower two quadrants indicate the different *Habitus* that the same unit cell can take.

When two polyhedral single crystals unite in a twinned relation, a new interface is introduced. Since the presence of an interface between the two individuals affects the growth, the morphology of the twinned crystal may sometimes differ from that of the coexisting single crystals (see, for example, Fig. 1.1). Similar morphological changes may also be expected to occur in epitaxial relationships, in which crystals of different species grow in a fixed crystallographic relation to the face of the substrate crystal, as well as between crystals freely grown in solution and those growing on the bottom surface of a beaker.

When numerous crystals grow together in aggregation, they show characteristic forms or textures that differ from those exhibited by single crystals. In the radial growth of numerous crystals around a nucleus, divergent, sheaf, bow-tie, and spherulitic morphologies appear. Gathering these forms together results in the formation of oolitic, botryoidal, and reniform aggregations or agate banding. Textures due to polycrystalline and polyphase aggregation, which are observed in

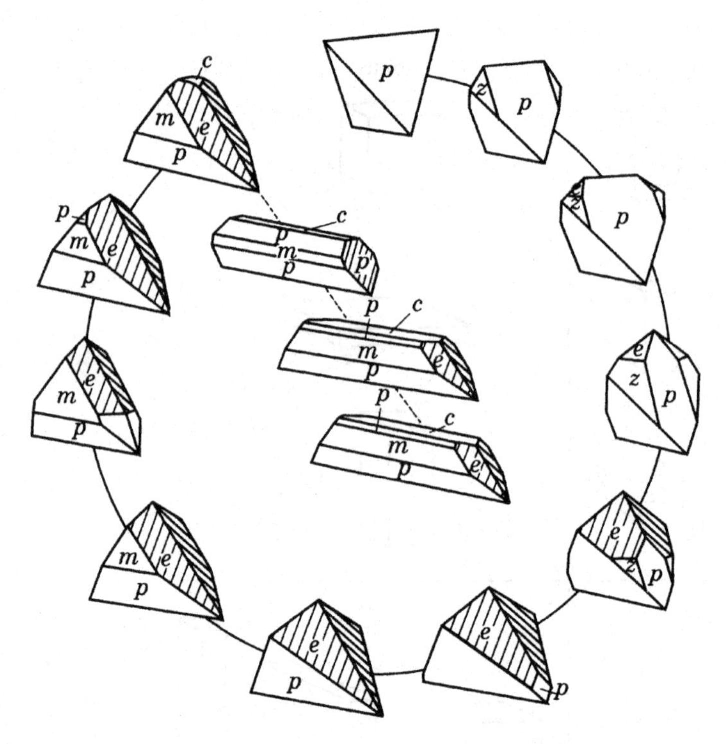

Figure 2.4. Habit (*Habitus*) variation of chalcopyrite: *c* (001), *p* (111), e (101), *m* (110), z (201) [1].

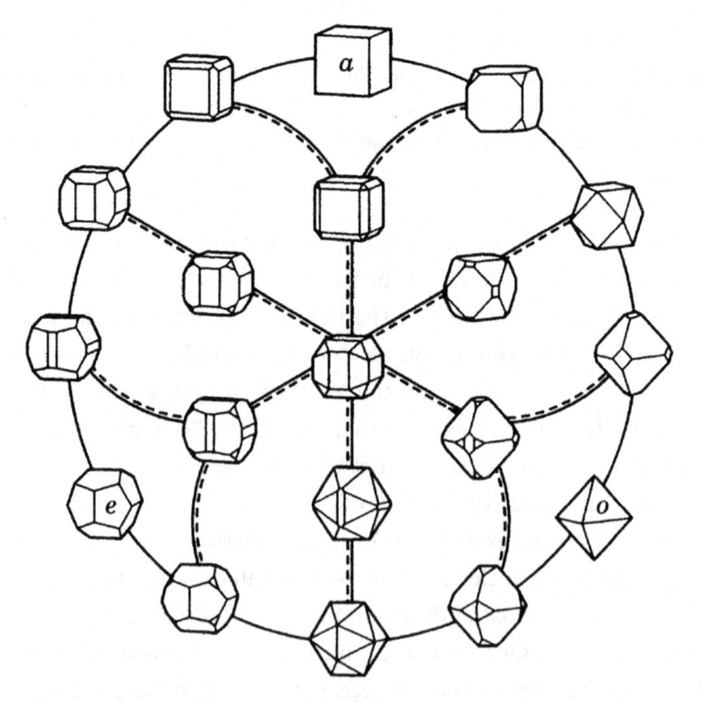

Figure 2.5. *Tracht* variation seen in pyrite crystals: *a* (100), *o* (111), *e* (210) [2].

rocks and ores, are also determined by the nucleation, growth, and morphology of crystals. All these are directly related to how crystals grow. Therefore, we will now explain the fundamental concepts pertaining to the mechanism of crystal growth.

References

1 I. Sunagawa, On the so-called triangular chalcopyrite, *Bull. Geol. Surv., Japan* **2**, 1951, 253–67

2 I. Sunagawa, Variation in crystal habit of pyrite, *Rep. Geol. Surv., Japan*, no. 175, 1957, 41

Suggested reading

N. Morimoto, I. Sunagawa, and A. Miyashiro, *Mineralogy*, Tokyo, Iwanami Pub. Co. (in Japanese), 1975

J. Sinkankas, *Mineralogy, A First Course*, Princeton, D. Van Nostrand Co., 1964

3

Crystal growth

It is only when a driving force causes a system to depart from its equilibrium condition that a nucleus of a crystal is formed and growth begins. Solid, melt, solution, and vapor phases may be distinguished as ambient phases in which crystals can grow. These can be systematically classified depending on whether they are condensed or dilute phases, and whether solute–solvent interaction is involved or not. In a driving force, heat and mass transfers are coupled, but the degree of their respective contributions depends on the type of ambient phase involved. The interface between the crystal and the ambient phase is the unique place where growth or dissolution takes place.

We may consider the interface problem by classifying its structure into atomically rough and smooth interfaces. An adhesive-type growth mechanism is expected for an atomically rough interface, whereas on atomically smooth interfaces crystals grow either by two-dimensional nucleation and layer growth or by a spiral growth mechanism. Based on the above analyses, mutual relations among polyhedral, hopper, and dendritic morphologies can be systematized.

3.1 Equilibrium thermodynamics versus kinetic thermodynamics

Crystal growth is the process of the birth and development of a solid phase with a regular structure out of a disordered and irregular state, and thus it can be regarded as a first-order phase transition.

The science dealing with phase transitions is thermodynamics. Using thermodynamics, we may discuss which phase will eventually be formed when a material (of composition c and phase p) is maintained under the same conditions for an infinite time, and the phase reaches the minimum energy state (equilibrium state) under given thermodynamic conditions (temperature and pressure). Experimentally, a phase diagram (equilibrium phase diagram) is prepared, and we may use the data from this diagram to extract information about the system. If the system remains for a long period under the same conditions, such as in geological processes, it is assumed to have reached equilibrium, and can be a subject of thermodynamic discussion.

In dealing with systems of metals, alloys, ceramics, and silicates, as well as growing single crystals, phase diagrams are the prerequisite in understanding the basic phase relations. There are many forms of phase diagram. When trying to grow crystals whose phase diagrams are not known, we must first prepare a preliminary phase diagram.

In contrast to equilibrium thermodynamics, the main concerns of this book are to show how the phase transitions involved in growth and dissolution proceed at an atomic level, and to explain how the morphology, perfection, and homogeneity of single crystals and the textures of polycrystalline aggregates are determined through these processes. Phase transition and the formation of a crystal, the final product, do not occur instantaneously. Along with our curiosity as to how crystal growth proceeds at an atomic level, the science and technology of this subject have also made great progress during the twentieth century, due to the strong demands from within the semiconductor and opto-electronic industries, in which single crystals and thin films are used as basic materials. Real crystals contain various lattice defects, and cannot be used for industrial purposes unless these defects, which are generally induced during the growth process, are controlled. A good example is the ruby laser. If a single crystal of ruby containing uncontrolled dislocations (see Section 3.7) is used, the laser beam will be emitted only once. Only after the successful synthesis of ruby single crystals with a controlled dislocation density was achieved did the ruby laser become usable in practice. The reason why silicon has come to play the principal role in the semiconductor industry is because of the synthesis of dislocation-free silicon single crystals. Although the control of lattice defects in single crystals was achieved in the twentieth century for single-component systems like silicon, we are still a long way from a similar situation in the synthesis of single crystals of compound semiconductors or quartz. If we could understand the phenomena relating to growth and dissolution occurring in complicated and complex systems, such as mineral crystals constituting terrestrial and extra-terrestrial solid materials, or those formed as a result of biological activities, such as teeth, bones, and shells, we would be able to offer new insights to these disciplines. These phenomena will be a target of research for the science of crystal growth in the twenty-first century.

3.2 Driving force

The equilibrium state corresponds to the state when the exchange of heat and mass between the starting material and the product becomes zero at the

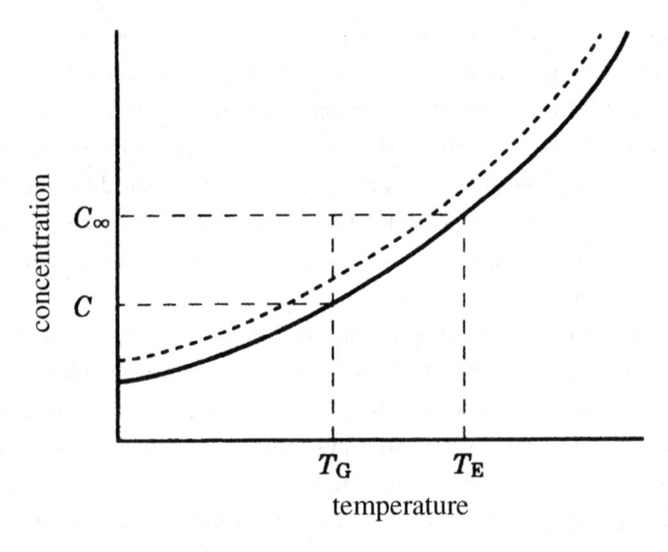

Figure 3.1. Solubility curve (solid line, equilibrium concentration curve) and the region in which nucleation and growth hardly occur (the Miers region, in between the dotted and solid lines). T_E and C_∞ are the equilibrium temperature and concentration, respectively; and T_G and C are the growth temperature and concentration, respectively.

interface of the two phases. This corresponds to the state where neither temperature nor degree of concentration (through cooling or evaporation) changes at a given point on a solubility curve on the phase diagram of a solution system. In this situation, neither growth nor dissolution of a crystal occurs. To achieve growth or dissolution of a crystal, it is necessary to attain a situation away from the equilibrium state, either by lowering the temperature or by concentrating the component by evaporation. The degree of departure from the equilibrium state corresponds to the driving force for growth or dissolution to take place. This relates to the difference between the dotted line and the solubility curve indicated by the solid line on the phase diagram shown in Fig. 3.1.

In the growth of crystals from the vapor phase, the degree of the driving force is expressed by the ratio S, i.e. the ratio between the equilibrium vapor pressure p_∞ and the pressure p at growth:

$$S = p/p_\infty = J/J_\infty,$$

where J and J_∞ are incoming flows from the vapor phase to the crystal at growth and equilibrium, respectively.

In the case of solution growth, the driving force corresponds to the difference between the concentration C_∞ at the equilibrium temperature T_E of the saturated solution and the concentration C at the growth temperature T_G and it is expressed as follows:

$$\Delta C = C_\infty - C,$$

$$S = C_\infty / C, \text{ (solute supersaturation ratio)}$$

$$\sigma = (C_\infty - C)/C, \text{ (relative supersaturation)}$$

$$\sigma = S - 1.$$

In melt growth, the driving force may be evaluated as the difference between the equilibrium and growth temperatures:

$$\Delta T = T - T_\infty \text{ (supercooling)}.$$

The degree of driving force can be generalized by the difference of the chemical potentials between the two phases:

$$\Delta\mu = \mu^{(g)} - \mu_\infty^{(g)} = \mu^{(g)} - \mu_\infty^{(c)}.$$

Here, $\mu^{(g)}$, $\mu_\infty^{(g)}$, and $\mu_\infty^{(c)}$ are, respectively, the chemical potentials of the supersaturated vapor (or the ambient phase such as the solution phase), the saturated phase, and the solid phase. From this, a generalized driving force may be expressed as

$$\Delta\mu = kT_B \ln(p/p_\infty) = kT_B \ln S,$$

where k is the Boltzmann constant, and T_B is the absolute temperature. We shall express the driving force in terms of the generalized driving force, $\Delta\mu/kT$, throughout this book.

3.3 Heat and mass transfer

As can be seen from the expression for the driving force in terms of the chemical potential differences, which are related to the differences in temperature and concentration, the two transporting processes, heat transfer and mass transfer, are coupled in crystal growth. The degree of contribution from the respective transport process is determined by the degree of condensation of the environmental (ambient) phase. To grow crystals in a diluted ambient phase, a condensation process is required, and so mass transfer plays an essential role. The contribution of heat generated by crystallization in this case is small compared with that of the mass transfer. However, for crystallization in a condensed phase, such as a melt phase, heat transfer plays the essential role, and the contribution from the mass transfer will be very small, because the difference in concentration (density) between the solid and liquid phases is very small, smaller, say, than 1 or 2%. It is therefore necessary to classify the types of ambient phases and to be familiar with their respective characteristics from this standpoint.

Solid state crystallization, or recrystallization,* is the process whereby both the starting materials and the final product are aggregations of particles with the same crystal structure, and the coarsening of grain sizes is the only process that occurs by crystallization. In this case, grain boundaries become movable, and coarsening of grains occurs due to the release of strain energy that is stored in grain boundaries by the thermal energy resulting from heating. Primary recrystallization is the name given to the process that describes the situation preceding the state in which a state with texture consisting of particles of almost equal size is achieved, whereas secondary recrystallization describes the situation where particular grains rapidly grow much larger due to the diffusion of impurities in grain boundaries by further heating the sample. However, the situation will be different from this when the precipitation of the B phase occurs in the solid state in a solid solution of an (A, B) component.

In contrast to solid state crystallization, crystallization from vapor, solution, and melt phases, which correspond to ambient phases having random structures, may be further classified into condensed and dilute phases. Vapor and solution phases are dilute phases, in which the condensation process of mass transfer plays an essential role in crystal growth. In the condensed melt phase, however, heat transfer plays the essential role. In addition to heat and mass transfer, an additional factor, solute–solvent interaction, should be taken into account.

Within the three types of ambient phases, crystal growth from pure melt of a congruent melting material (i.e. the solid and liquid phases have the same composition) is called "melt growth." In this situation, interactions with other components are not involved. A similar concept is applied to physical vapor deposition (transport) (PVD, PVT) in which crystallization occurs through simple evaporation by heating and condensation by cooling. In the solution phase, however, there are solute and solvent components, which mutually interact to form a solution. This may be compared with chemical vapor deposition (transport) (CVD, CVT), which involves chemical reactions. The strength of the solute–solvent interaction energy is closely related to the solubility of the solute component, the nucleation, and the growth rates of crystals.

The basic criteria to consider in crystal growth from vapor, solution, and melt phases are therefore whether the phase is condensed or dilute, and whether the phase involves a solute–solvent interaction or not.

In the condensed phase, mass transfer is not significantly involved, and crystal growth occurs by the rearrangement of the atoms at the solid–liquid interface through the removal of heat. Heat transfer is the driving force required for atomic

* In chemistry, the term recrystallization has a different meaning; it implies the purification process by solution and recrystallization.

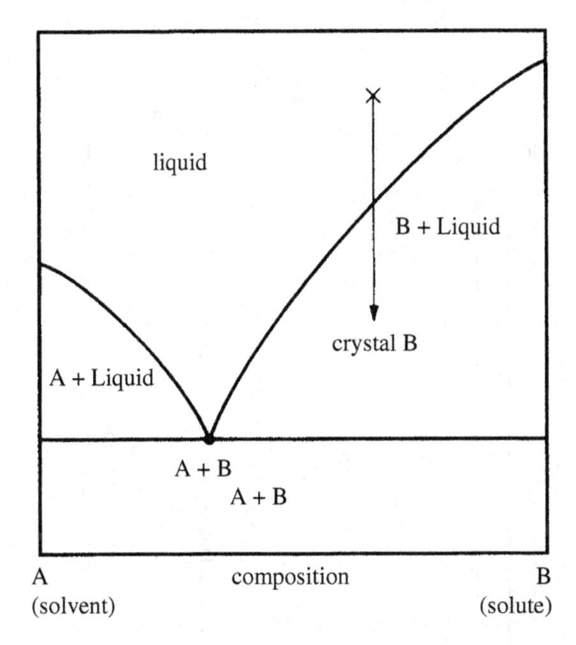

Figure 3.2. An imaginary phase diagram of a solution with solvent component A and solute component B. Due to the solute–solvent interaction energy, the melting point of the solution is lowered from that of the pure solute and solvent components to the eutectic point. This phase diagram shows that a crystal may be grown at a much lower temperature than in melt phase growth by preparing a solution, and that the resulting crystal is determined not by which component has the higher melting point, but by the composition of the solution. These are the essential characteristics of solution growth.

rearrangement to occur; the growth temperature is high, and, as a result, the solid–liquid interface tends to be rougher than that formed in a dilute ambient phase (see Section 3.6).

In a dilute phase, atoms or ions are transported from far away, and, on arrival at the crystal surface, they are incorporated into the crystal. There is a desolvation process involved in crystallization in the solution phase or in CVT. (For an explanation of desolvation, please see Section 3.4.) The essential role of the driving force is mass transfer; growth temperature is much lower than that in melt growth, and the solid–liquid interface tends to be smoother than that in the condensed phase.

Table 3.1 summarizes the essential differences in crystal growth for the various ambient phases. Crystal growth taking place in geological processes and in biological mechanisms is also indicated for reference. Since the main topic of this book is to understand phenomena occurring in complicated and complex systems, we shall focus our discussion on solution and CVT growth rather than crystal growth from much simpler melt or PVT growth.

Table 3.1 *Characteristics of crystal growth in melt, solution, and vapor phases*

	Melt phase	*Solution phase*			*Vapor phase*	
		High-temperature solution	*Hydrothermal solution*	*Ordinary temperature solution*	*CVT*	*PVT*
State	condensed	←			→	diluted
Driving force	heat transfer	←			→	mass transfer
Growth temperature	high	high–medium		low	medium–low	
Solute–solvent interaction	none	strong		strongest	strong	none
α factor	small	medium	medium	medium	large	largest
Interface roughness	rough	smooth	smooth	smooth	smooth	smooth
Growth mechanism	adhesive-type			two-dimensional nucleation growth or spiral growth		
Terrestrial and extra-terrestrial	none	igneous rock	vein	Earth's surface	fumaroles	cosmic dust
Living activity	none	none	rare	biomineralization		

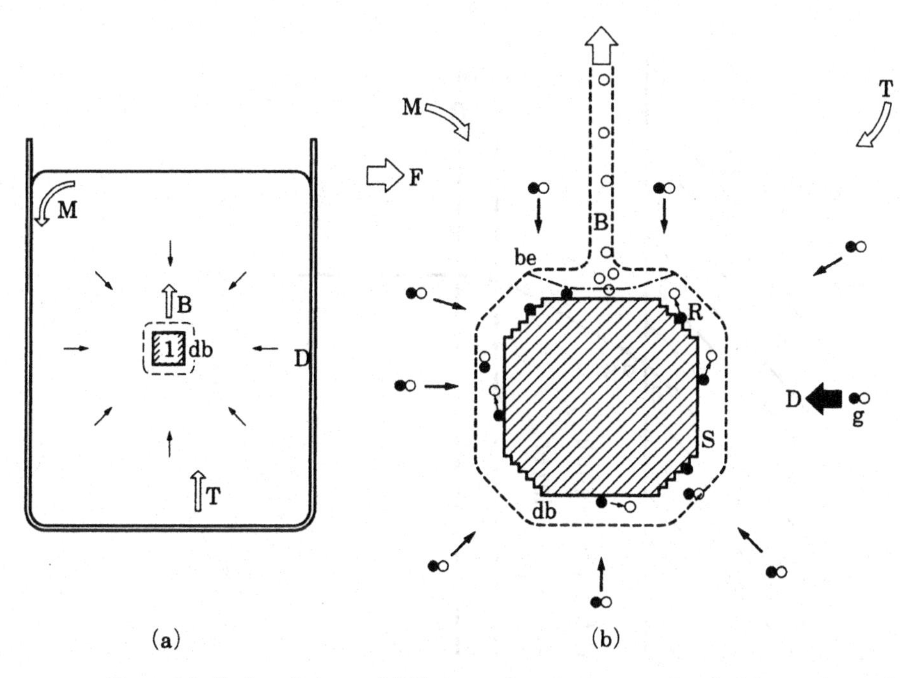

Figure 3.3. Various features of diffusion and convection associated with crystal growth in solution (a) in a beaker and (b) around a crystal. The crystal is denoted by the shaded area. Shown are: the diffusion boundary layer (db); the bulk diffusion (D); the convection due to thermal or gravity difference (T); Marangoni convection (M); buoyancy-driven convection (B); laminar flow, turbulent flow (F); Berg effect (be); smooth interface (S); rough interface (R); growth unit (g). The attachment and detachment of the solute (solid line) and the solvent (open line) are illustrated in (b).

3.4 Examples of mass transfer

We will consider in this section how mass transfer proceeds, using crystal growth from a solution phase as a representative example of crystal growth in which heat and mass transfer are coupled. We will use the growth of ionic crystals in aqueous solution in a beaker as an example (Fig. 3.3).

For NaCl aqueous solution, which is a dilute ambient phase, all the solute components of Na^+, Cl^-, or NaCl, $(NaCl)_n$ are bonded with the solvent component H_2O. Whether the solute component is present in an isolated ionic state, or whether NaCl molecules or clusters of $(NaCl)_n$ are present, will depend on the driving force. Although the situation is not fully understood, it is expected that, as the supersaturation increases, the proportion of clusters will increase. Although there can be various bonding states between these clusters and H_2O, it is generally considered that there are two regions surrounding the solvent component: one structure forming a region with strong bonding, and the other structure forming a breakable region with much weaker bonding between the solvent molecules.

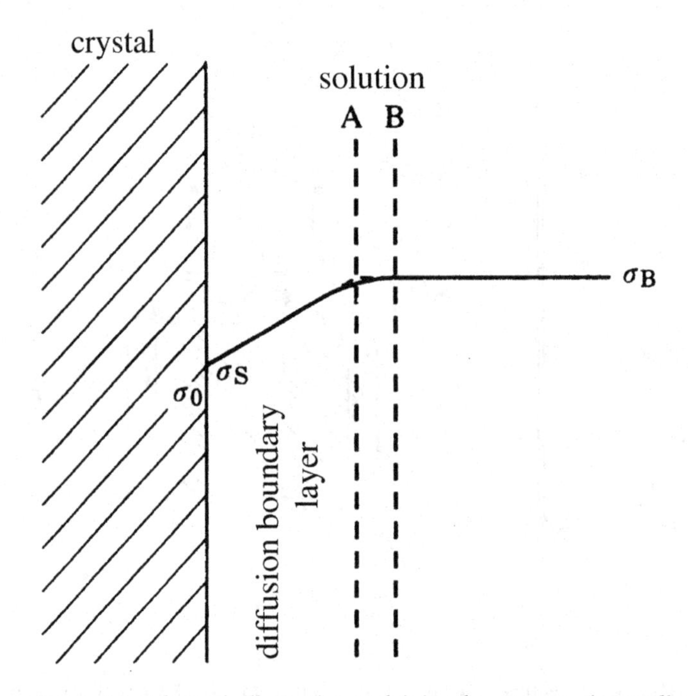

Figure 3.4. Schematic illustration explaining the concentration gradient around a growing crystal and the presence of the diffusion boundary layer. Bulk diffusion indicated by diffusion boundary layer (for A or B). The concentration gradient is (σ_B–σ_S), where σ_B is the bulk supersaturation and σ_S is the surface supersaturation. σ_0 is the surface equilibrium saturation.

When the temperature of the aqueous solution is lowered, or evaporation occurs, a supersaturated state results. If the driving force of the system exceeds the energy barrier necessary for nucleation to take place (for a discussion, see Section 3.6), nucleation of the solute component occurs in the system. Accompanying the nucleation, a more dilute region will appear surrounding the nucleus, and diffusion from the bulk ambient phase will occur due to the concentration difference. This is called *volume* or *bulk diffusion*. The concentration gradient is not linear from the bulk ambient phase, away from the nuclei and the surface of the nuclei or crystal. In the region close to the surface of the nuclei or the crystal, the concentration gradient is sharp, whereas that between the bulk ambient solution away from the surface shows almost no gradient. This was already clear from investigations conducted before the 1930s (see Fig. 3.4). The region with a sharp gradient is called the *diffusion boundary layer*, and the concentration gradient in this layer plays an essential role in crystal growth. For the growth of ionic crystals in an aqueous solution, the thickness of this layer is of the order of 100 μm.

The concentration at the surface of a crystal (a nucleus) is not the same as the

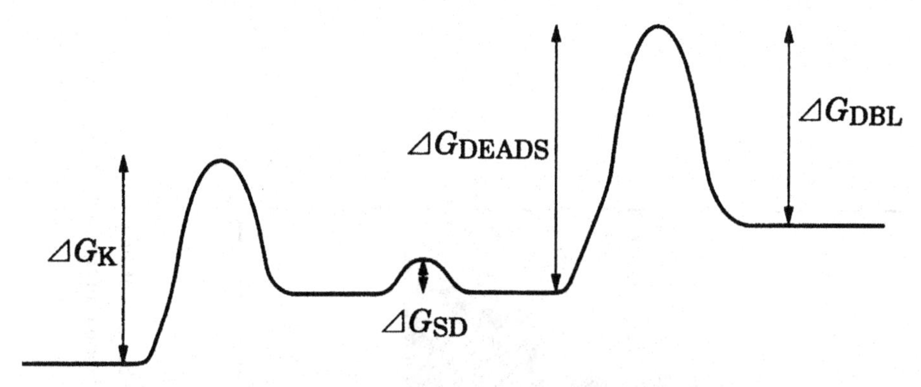

Figure 3.5. The energy barrier at various steps in solution growth. K = kink; SD = surface diffusion; DEADS = desolvation; DBL = diffusion boundary layer.

equilibrium concentration. In the days before we could measure the surface concentration directly, the driving force of a bulk phase was evaluated for simplicity assuming the equilibrium concentration, but now it is assumed that the surface concentration is slightly higher than this value. Therefore, the driving force for crystal growth is now considered to be the difference and its gradient between the concentration at the outermost region of the diffusion boundary layer and the concentration at the surface (surface supersaturation).

Crystallizing particles arriving at the surface will diffuse onto the surface (surface diffusion). As this occurs, some may return to the ambient phase, while some will be caught at kinks or steps (see Section 3.6) on the surface and will be incorporated into the crystal. When these particles are incorporated into the crystal, the solvent component will be dissociated. This process is called *desolvation*. In solution growth, this process will determine the growth rate. At certain points in these processes, it is necessary to overcome the energy barriers required to climb the respective steps (Fig. 3.5).

The solvent component detached by the desolvation process returns to the ambient phase. As a result, the concentration in the diffusion boundary layer will be lowered. Through competition between the supply of solute component from the bulk phase and that of the solvent component by the desolvation process, the thickness of the diffusion boundary layer will vary according to the change in concentration, or buoyancy-driven convection will occur when crystals are growing in the gravity field of the Earth. This change is related to the bulk supersaturation. In Fig. 3.6, changes in behavior of the diffusion boundary layer around a growing $Ba(NO_3)_2$ crystal in aqueous solution are illustrated in relation to the bulk supersaturation σ. At $\sigma < 0.5\%$, the thickness of the diffusion boundary layer simply increases, whereas at $0.5\% > \sigma > 3.0\%$, unstable uplifting convection appears, and at $\sigma > 3.0\%$ a steady buoyancy-driven convection plume rising perpendicularly

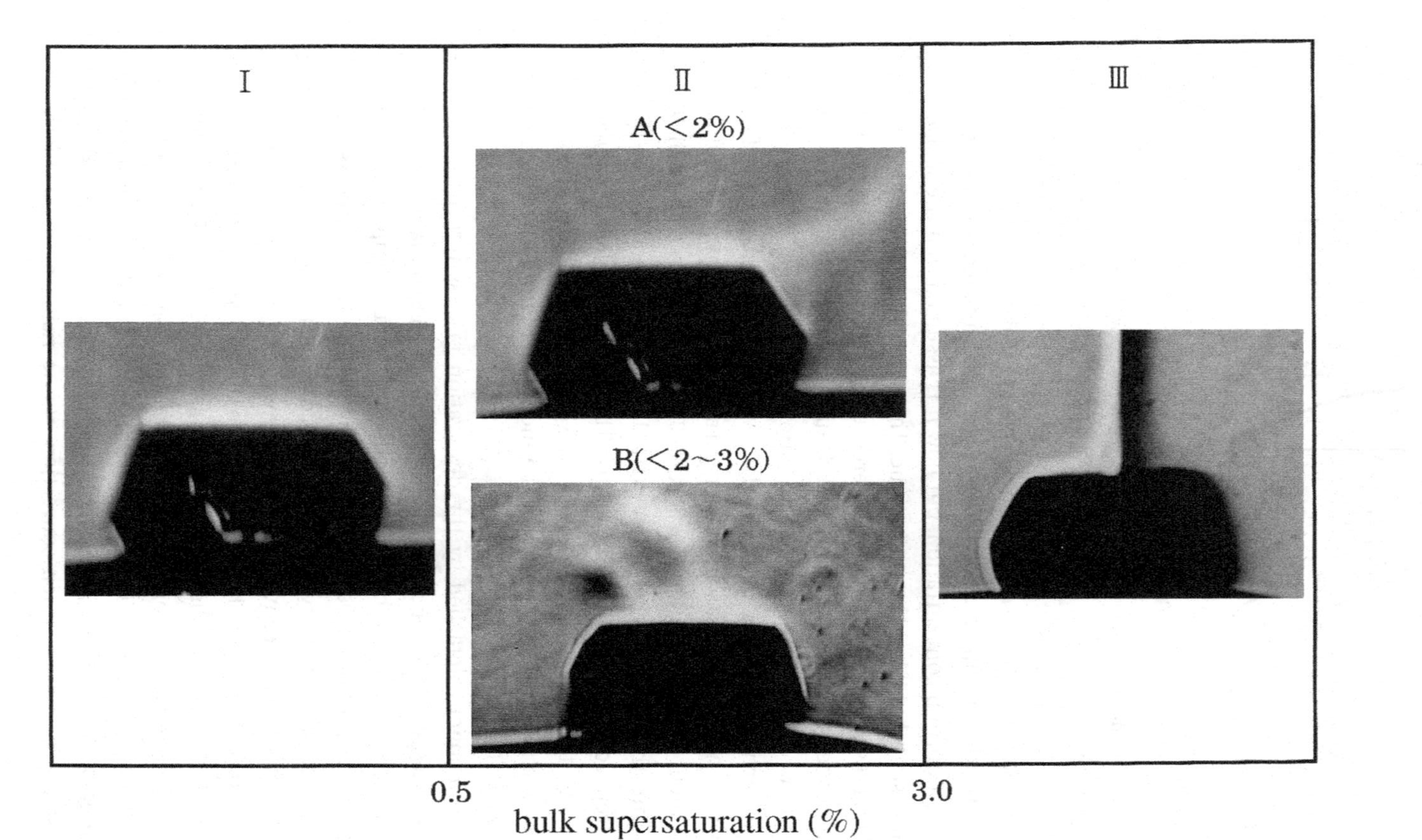

Figure 3.6. Schlieren photographs showing the changes in thickness of the diffusion boundary layer and the behavior of buoyancy-driven convection shown in relation to bulk supersaturation [1], [2]. The figure shows the (111) face of a $Ba(NO_3)_2$ crystal from an aqueous solution. In region I, only the thickness of the diffusion boundary layer increases; in region II, we see unstable lateral convection (IIA) and intermittently rising plumes (IIB); and in region III we see steady buoyancy-driven convection.

from the center of the top face appears. At $\sigma > 8.0\%$, rising convection plumes appear, creating a considerable depletion at the root, which greatly affects the perfection of the crystal.

Concentration variation in the bulk phase is isotropic around a crystal. Consider a crystal growing in a concentric, equi-concentration distribution field. In a polyhedral crystal bounded by corners, edges, and faces, we expect the concentration difference at a corner to be larger than that at an edge, which is in turn larger than that at a face, and thus corners and edges are at a higher concentration state than is the center of a face. This creates surface diffusion. The concentration differences at the corners, edges, and face centers of a growing crystal will differ depending on the bulk supersaturation or flow rate along the surface. The difference in concentration at different sites on a face was demonstrated by Berg [3] by means of an interferometric technique, which is now referred to as the Berg effect or Berg phenomenon. This is an important phenomenon in analyzing the morphology of crystals.

3.5 Laminar flow and convection

Like diffusion, convection is also an important factor for mass or heat transfer. Convection is a mode of flow in a fluid that creates rotation or circulation of the fluid, and is distinguished from laminar flow and turbulent flow. If there is a temperature difference between the fluid at the top and bottom of the beaker shown in Fig. 3.3, convection occurs due to the concentration difference arising from the temperature difference. Convection will also occur when there is a concentration gradient, or it can arise due to buoyancy. Convection due to the surface tension of a solution is called Marangoni convection. The difference in surface tension may be due to a temperature difference or to the concentration of chemical species on the surface. Convection may be taken to mean thermal convection, solutal convection, or a combination of the two. In contrast to these natural processes of convection, various methods, such as rotating the seed crystals or the crucible itself, are adopted when growing single crystals in order to control convection and to effect the homogenization of the liquid phase. Since it is desirable to achieve a homogeneous liquid phase in order to grow homogeneous single crystals, convection occurring in the liquid phase has been extensively investigated by means of computer simulation or various other visualization techniques.

Laminar flow, which is a directional flow, changes into turbulent flow when the critical Reynolds number (*Re*)[†] exceeds 200. When there is a flow of solution around

[†] The most representative non-dimensional value in fluid dynamics is *Re*, the ratio of the magnitude of inertia force and viscosity, $Re = \rho \nu l/\mu = \mu l/v$, where ρ is the density, μ is the viscosity, v is the dynamic viscosity, l is the representative length of matter in the flow, and ν is the representative flow rate.

a growing crystal, the thickness of the diffusion boundary layer, the Berg effect, and how the solute component is supplied will be affected depending on the flow rate of the laminar flow or the nature of the turbulent flow, all of which influence the growth rate, perfection, and morphology of the crystal.

3.6 Nucleation

The processes of crystal growth are divided into the following three stages.

(1) A driving force is applied, which causes the process to proceed by the formation of a supersaturated or supercooled state.

(2) Particles smaller than a critical size, which may be referred to as clusters, are formed in the system. Some of these clusters, out of the many that repeatedly come together and part again, may by chance grow larger than the critical size. The process that describes the situation up to the point where the particles reach critical size may be called the nucleation stage.

(3) Once it exceeds the critical size, the particle can grow larger. This process corresponds to the growth stage in its strictest sense.

Among the clusters that are formed, those exceeding the critical size do not dissociate and can grow larger. As shown schematically in Fig. 3.7, this is because the density of dangling bonds per unit area on the surface decreases as the size increases, and it eventually reaches a critical value. Figure 3.7(b) shows this relation in terms of energy changes. The nucleation energy is the sum of the energy spent in forming a particle by coagulating atoms, $-\Delta G_v$, which is proportional to r^3, and the energy gained by creating the surface, $+\Delta G_s$, which is proportional to r^2, namely $\Delta G = -\Delta G_v + \Delta G_s$. As can be seen in Fig. 3.7(b), the energy required for nucleation to occur increases as r increases, reaches a maximum at r_c and decreases thereafter. The nucleus which attains r_c is called the critical nucleus, and r_c is referred to as the radius of critical nucleus.

Since a critical value exists for nucleation to occur, nucleation or growth hardly occur, except in a narrow region away from the solubility curve, which corresponds to the equilibrium situation. This narrow region along the solubility curve is called the Miers region (see Fig. 3.1).

Nucleation can be classified as (i) "spontaneous nucleation" or "homogeneous nucleation," in which no intervention by other factors is necessary for nucleation to take place, and (ii) "heterogeneous nucleation," which occurs due to the presence of a surface wall of a vessel, or foreign particles, or an impurity component. A much higher driving force is required for the former than the latter. There is a difference of a few tens of degrees centigrade in terms of supercooling between the

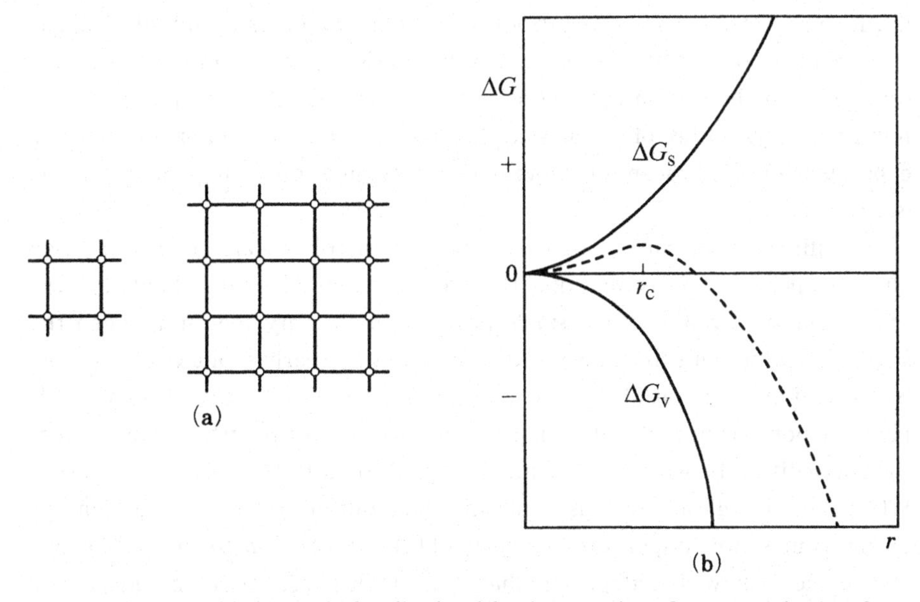

Figure 3.7. (a) Changes in dangling bond density per unit surface area as the size of a particle increases. When the number of units constituting the nucleus increases from four to sixteen, the dangling bond density per unit area decreases from 8 to 16/9. (b) Changes in nucleation energy ΔG as the radius of a nucleus r changes. ΔG reaches a maximum at r_c. The energy required to attain volume ΔG_v is proportional to r^3, whereas the energy gained in creating the surface, ΔG_s, is proportional to r^2.

two in melt growth. Most nucleation phenomena in real systems may be regarded as heterogeneous. In contrast to this, it is accepted that homogeneous nucleation plays the essential role in nucleation under microgravity conditions.

The critical energy for nucleation is determined by the free energy of the interface (the surface of a nucleus) and the driving force. The smaller the critical energy, the easier it is for nucleation to take place, so impurities with crystal chemical similarity, unresolved tiny particles, steps, and unevenness or scratches on the surface of the vessel wall may all act as heterogeneous nucleation sites. The decoration method in electron microscopy (see Section 5.2), which enables us to observe steps on a nanometer scale, and the epitaxial method for growing single crystalline thin films (see Section 7.4), are techniques that utilize heterogeneous nucleation phenomena by taking account of the energy at steps and interfaces. The use of seed crystals allows growth to start immediately, bypassing the nucleation stage. Secondary nucleation, a method widely used in the field of industrial crystallization, is essentially the same as crystal growth using seeds, although traditionally a term of nucleation has been used.

It is the main purpose of industrial crystallization to obtain numerous tiny crystallites of equal size and form. To achieve this, a practical method has been adopted in which large seed crystals are introduced into a reaction vessel and agitated. The surfaces of the seed crystals are detached by agitation, so providing numerous tiny crystallites in the system that act as seed crystals. Hence, the aim of obtaining numerous crystallites of equal size is achieved. Similar phenomena may be expected to occur in ascending movement of a magma containing phenocrysts [4] (see Section 8.5).

In equilibrium thermodynamics, the final stable phases are determined by a given temperature–pressure range. However, it is not always possible to achieve nucleation and growth of the stable phase from the very beginning when the system is kept under this condition. In many cases, the earliest phase that appears by nucleation is a metastable phase, which is different from the stable phase. This phenomenon is called Ostwald's step rule [5]. This can be understood as a result of the competition between the driving force and the interface free energy terms. When the driving force term is small, the contribution from the surface energy term becomes more important. As a result of this competition, the nucleation of a metastable phase will occur prior to that of the stable phase. Before the appearance of the stable phase, the earlier formed metastable phase can develop in a stable manner, and behaves as if it is the stable phase. However, once the stable phase appears in the system, the earlier formed metastable phase starts to dissolve or transform, and the stable phase grows. If conditions are maintained that suppress the appearance of a stable phase, the metastable phase can become stable [6]. The formation of aragonite (a high-pressure phase of $CaCO_3$) existing in living bodies, or of CVD diamonds (a high-pressure phase of carbon) under 1 atm pressure are such examples.

In this section we have explained the relation between a bulk ambient phase and a bulk crystal (where the crystal is assumed to have an ideal, regular structure). Since the 1930s, theoretical and experimental investigations have indicated that a real crystal contains defects of lattice order, and that these defects have a significant effect on the physical properties of the crystal. This understanding has led to the present day semiconductor industry.

3.7 Lattice defects

Physical properties of solid materials which are greatly influenced by the presence of defects of lattice order in real single crystals are called "structural-sensitive properties," and are distinguished from "intrinsic properties," which are determined by the elements constituting the crystal, for example the chemical bonds, the structure, etc. Color, plasticity, glide, and semiconductor properties are structural-sensitive properties, whereas density, hardness, elasticity, and optical, thermal, and magnetic properties are the intrinsic properties. Structural-sensitive

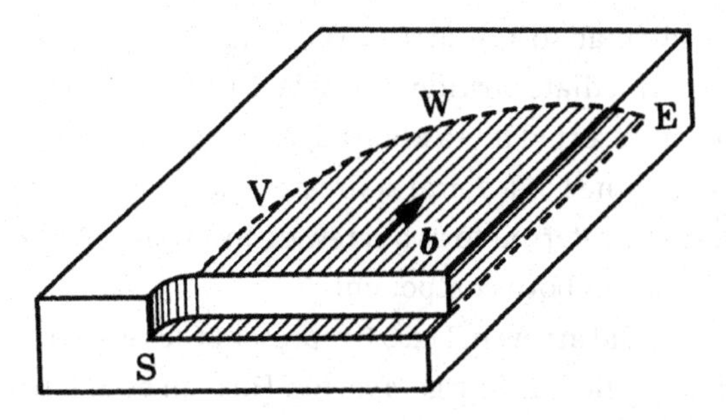

Figure 3.8. Explanation of dislocations in relation to glide. The solid arrow, b, corresponds to the Burgers vector of the dislocation. SV is the screw dislocation, WE is the edge dislocation, and VW is a mixed dislocation. The shaded area represents a glide plane.

properties will be different for each individual sample. There are cases where even the intrinsic properties are influenced by lattice-order defects.

The lattice defects are classified as (i) "point defects," such as vacancies, interstitial atoms, substitutional impurity atoms, and interstitial impurity atoms, (ii) "line defects," such as edge, screw, and mixed dislocations, and (iii) "planar defects," such as stacking faults, twin planes, and grain boundaries.

Within these three categories, the earliest attention was paid to point defects; this arose from intellectual curiosity about the origin of the deep blue color seen in large colorless single crystals of NaCl occurring in salt deposits. From this starting point, the idea of color centers was developed. Since the color centers are point defects, they modify the electronic states around the defects and affect the heat and electronic conductivity; consequently, they have become directly connected to the development of the semiconductor industry.

The concept of line defects was theoretically introduced to provide an answer to the discrepancy between the theoretical and experimental values of shear stress to create glide (slip). It took nearly twenty years to prove the presence of dislocations in crystals. Glide is a phenomenon in which part of a crystal is displaced (without losing adhesive force on a glide plane) by a distance (either a lattice spacing or a multiple of it) in the glide direction. The theoretical value of the shear modulus, assuming that glide occurs instantaneously in a perfect ideal crystal, was calculated to be up to 1000 times the experimentally measured value. From this emerged a theory that considered the glide phenomenon to be linked to a movement of dislocations. The theory assumed that glide occurs because dislocations move on the glide plane (see Fig. 3.8).

If we assume that gliding occurs on the glide plane in the direction of and by the amount indicated by the solid arrow, we see that the lattice is distorted along the front of the glide, SVWE. The mode of distortion of the lattice plane is different between the portion perpendicular to the arrow, WE, and that parallel to the arrow, SV, and again at the intermediate portion VW. At WE, the lattice plane is distorted as if a half lattice plane is inserted like a knife-edge, and at SV it continues to the next lattice plane in the manner of a spiral staircase. The former is called an edge dislocation, and the latter is referred to as a screw dislocation. VW is called a mixed dislocation, since it possesses both components.

The direction and size of the solid arrow in Fig. 3.8 represent the vector that characterizes the dislocation; this is called the Burgers vector. This can be of lattice size, or an integer fraction or integer multiple of the lattice size. Since the lattice plane is distorted along the narrow region around a dislocation line, a strain field is concentrated there. This region is called the dislocation core, and it can be a preferential site for dissolution or adsorption to take place. Indeed, the earliest experimental verification of the presence of dislocations in single crystals was achieved by producing point-bottomed etch pits, corresponding to the outcrops of dislocations, and subsequently etching the specimen after applying stress. A change in form of the etch pit from point-bottomed to flat-bottomed was observed at the original outcrop of the dislocation, and point-bottomed etch pits were noted at the position to which the dislocation had traveled. Dislocations may now be directly observed by various techniques, such as X-ray topography or transmission electron microscopy.

The concept of dislocations was theoretically introduced in the 1930s by E. Orowan and G. I. Taylor, and it immediately played an essential role in the understanding of the plastic properties of crystalline materials, but it took a further twenty years to understand fully the importance of dislocations in crystal growth. As will be described in Section 3.9, it was only in 1949 that the spiral growth theory, in which the growth of a smooth interface is assumed to proceed in a spiral step manner, with the step serving as a self-perpetuating step source, was put forward [7].

Dislocations themselves play a role in promoting growth, but they may also be induced into existing, growing crystals at various stages. Lattice mismatch can result where the side branches of a dendritic crystal unite, when an inclusion trapped during growth is enclosed, or along the boundary of two crystals with slightly different orientations. Lattice parameter mismatches between host and guest crystals in epitaxial growth are also origins of growth-induced dislocations. When a real crystal grows in an uncontrolled system, it will usually have a dislocation density of the order of $10^{8\sim10}$ lines/cm^2. In utilizing single crystals for devices, it is necessary to grow single crystals containing a reduced dislocation density; however, the distribution of the dislocations in single crystals can serve as a reliable record of the growth process.

Growth-induced dislocations may multiply during the post-growth process. If stress is applied to dislocations fixed by impurities, the induced dislocations multiply to form loop-like dislocation rings. This is an important mechanism in the multiplication of dislocations, and is referred to as a Frank–Read source. In this way, growth-induced dislocations move during the post-growth period and form energetically more stable accumulations of dislocations. Secondary structures, such as dislocation tangles, networks or honeycomb structures of dislocations, small-angle grain boundaries, twist boundaries, and punched-out dislocations, are typical examples. On the other hand, growth-induced dislocations in crystals growing from the solution phase are generally distributed in bundles originating from the center of the crystal or from inclusions trapped during growth. They also originate on the surface of a seed crystal or from growth bands. Growth-induced dislocations usually run nearly perpendicularly to the growing surface.

In addition to lattice-order defects, much larger defects may often be induced into a growing crystal during or after growth; for example, the inclusion of solid, liquid, or vapor phases. Inclusions trapped simultaneously during the growth process are called syngenetic inclusions, whereas those induced after growth is complete are called epigenetic inclusions. Solid state syngenetic inclusions provide evidence of the compositions and thermodynamic conditions of the ambient phases at the time of crystal growth. If a liquid inclusion were trapped as a single phase, the dissolved component in the liquid would be dissociated in the form of vapor or solid phases due to the lowering of temperature and pressure, and would result in vapor–liquid two-phase inclusions or multi-phase inclusions in which vapor–liquid–solid phases coexist. Based on the temperature at which two phases change to a single phase on heating, the temperature at the time the inclusion was trapped may be evaluated, after first calibrating the effect of the pressure.

A phase is often encountered among silicate or sulfide minerals in spotted, star, lamellar, or lattice textures in the host phase. The phase was originally mixed as a homogeneous single solid solution with the host phase at the time of growth, and later precipitated or exsolved the phase of one solid solution component within the host crystalline phase as the temperature decreased. This phenomenon is called precipitation in physical or metallurgical fields and exsolution in geological disciplines. The exsolved phase has a definite crystallographic relation with the host phase to maintain minimum interface energy between the two phases.

3.8 Interfaces

After clusters attain a critical size in the system and the nucleation stage is complete, the growth stage commences in a narrower sense. Since the structure is already formed, the solute component will be incorporated into the crystal at the expense of a much smaller energy than that necessary for nucleation. Here, the interface structure between the solid and liquid phases that appeared as a result of

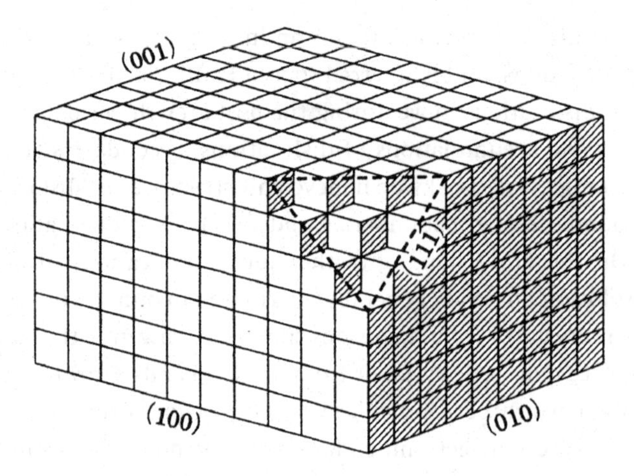

Figure 3.9. Kossel crystal.

nucleation becomes important. This is because the growth mechanism will be different depending on whether the interface is atomically rough or smooth. The rough interface consists entirely of kinks, and the smooth interface comprises a flat surface (terrace), a small number of steps, and a site where a step is bent, i.e. a kink.

Kossel [8] and Stranski [9] were the first to focus attention on the interface structure, after considering the experimental results obtained by Volmer [10], who demonstrated the existence of surface diffusion. It thus became possible to discuss the mechanism of crystal growth at an atomic level, starting from these analyses.

Let us assume that the constituent units of both a crystal and that of growth are simple cubes. This kind of model crystal is called a Kossel crystal, and is shown in Fig. 3.9. The {100} face is completely paved by the unit, and the surface is atomically flat. This face is called the "complete plane." The {111} face, however, consists of kinks, as can be seen in Fig. 3.9, and has an uneven surface, and so it is called an "incomplete plane." In contrast to {111}, {110} corresponds to a face consisting entirely of steps. Kossel did not give a particular name to this type of crystal face.

There are three different sites having different attachment or detachment energies in a Kossel crystal. A constituent unit of simple cubic form on the surface of a {100} face is connected to the crystal at five faces out of six; on the edge of a {100} face or on a {110} face, it is connected to four faces out of six; whereas at the corner of a {100} face or on the surface of an incomplete {111} face, it is connected to three out of six faces. These sites may be called a smooth face (terrace face), a step, and a kink, respectively. We can immediately understand that the energy of attachment for the growth unit to the crystal decreases in this order. Since at a kink site the attachment energy is equal to one-half of the bonding energy between the neighboring constituent units, a kink site is called a half-crystal site.

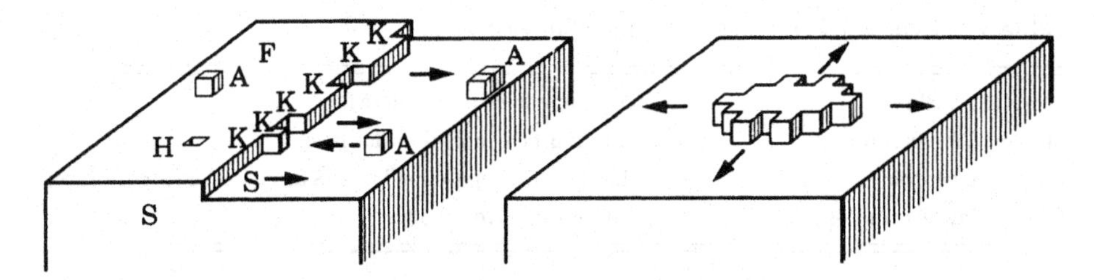

Figure 3.10. Schematic to show the layer-by-layer growth model due to two-dimensional nucleation. This figure assumes the mode of nucleation to be the mono-nuclear model. Other models, such as the poly-nuclear or birth and spread models, as explained in the text, may also be considered.

Considering the attachment energy, we see that at an incomplete face an impinging growth unit will be immediately incorporated into the crystal. This indicates (i) that the growth mechanism of an incomplete face will be adhesive-type growth, (ii) that the interface will advance homogeneously, and (iii) that the rate of advancement of the interface (the normal growth rate) will be linearly related to the driving force.

In contrast, a growth unit arriving at a flat terrace surface on a complete interface cannot immediately be incorporated into the crystal. It may diffuse on the surface or leave the interface, returning to the initial phase. If there are steps or kinks, which are folded steps, on the terrace surface, an arriving growth unit will be incorporated into the crystal. As a result, the face will grow by two-dimensional spreading of the step. If the step reaches the edge of the face and the surface becomes completely flat, a new step source is required to continue growth. Kossel [8] and Stranski [9] assumed two-dimensional nucleation to take place. This mechanism is called layer-by-layer growth. Figure 3.10 shows this model schematically.

In this theory, since the growth rate of a face is controlled by the rate of two-dimensional nucleation, we should expect the presence of a critical driving force, only above which growth can take place. Below this value, there will be no growth. As possible modes of two-dimensional nucleation, three different models may be considered.

(1) A mono-nuclear model, which allows only one nucleation, the next nucleation taking place only after the entire surface is covered by growth layers originating from the earlier nucleus.

(2) A poly-nuclear model, which allows many nucleations to take place on one surface.

(3) A birth and spread model, which allows nucleation and advancement of one growth layer at a time on one surface.

In all these models, we assume two-dimensional nucleation.

Table 3.2 *Terms used to express the states of interfaces*

Kossel, Stranski	complete face, incomplete face
Burton, Cabrera, Frank	singular face, non-singular face
Hartman, Perdok	flat face (F face), stepped face (S face), kinked face (K face)
Present-day terminology	smooth face, rough face

Many theories on interface structures have been expounded, and consequently many different terms have been introduced; we summarize these in Table 3.2. Instead of referring to complete and incomplete faces, Burton, Cabrera, and Frank [11] used the terms singular and non-singular faces, in the sense of one face showing a sharp minimum (cusp) and one showing no minimum, respectively, on a polar diagram of surface energy. This classification is based on the calculation of statistical mechanics, and is closely related to the discussion of the equilibrium form (see Section 4.3). Hartman and Perdok [12] classified crystal faces into F (for flat face), S (stepped), and K (kinked face), depending on the numbers of periodic bond chains (PBCs) that were found in a crystal structure by connecting strong bonds. It is clear that these faces correspond to the {100}, {110}, and {111} faces of a Kossel crystal, respectively (Fig. 3.9). Today, the terms smooth and rough interface are commonly used alongside perfect and imperfect faces, and singular and non-singular faces.

As can be seen from the Kossel crystal shown in Fig. 3.9, the roughness of an interface depends on which faces we are considering, i.e. on the crystallographic directions. It also varies depending on the crystal species, the temperature, and the driving force. Increasing the temperature or driving force will cause a smooth interface to change into a rough interface. The transition point that exists in going from a smooth to a rough interface as the temperature increases is called a roughening transition. The existence of roughening transitions has been demonstrated by computer simulation. A roughening transition resulting from increasing driving force is called a kinetic roughening transition. Depending on the degree of roughness of an interface, the growth mechanism or growth rate versus the driving force relation changes, and so the interface roughness is an essentially important concept in analyzing the morphology of crystals and element partitioning (see Section 3.14).

The earliest theoretical predictions on the state of interfaces were made by Burton, Cabrera, and Frank [11], who demonstrated that the interface would, in most cases, be rough at the growth temperature for metal crystals. Jackson [13] suggested a system in which the solid and liquid phases are separated by an interface with one-layer thickness, and he calculated the energy changes as a function of the ratio of site occupancy of the constituent unit on the interface. When the site occupancy ratio was 50%, the interface was rough, whereas 0 or 100% site occupancy

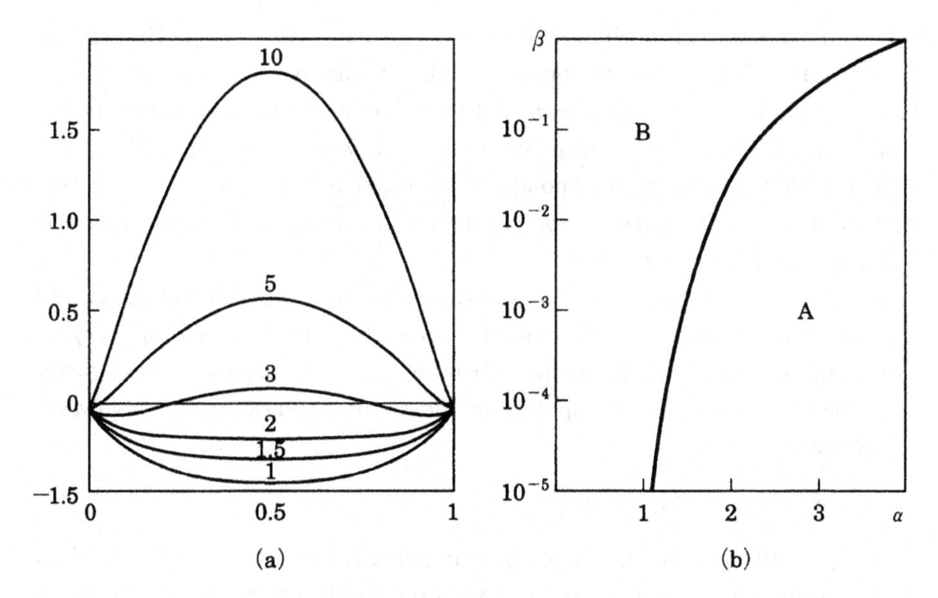

Figure 3.11. Prediction of interface roughness obtained by (a) Jackson [13] and (b) Temkin [14], [15]. (a) The horizontal axis shows the surface site occupancy ratio, and the vertical axis depicts the change in free energy. Numerals are α factors. (b) The vertical axis indicates the driving force, and the horizontal axis shows the α factor. A is the stable area for a smooth interface; B is the stable area for a rough interface.

corresponded to a smooth interface. In this calculation, he used the so-called Jackson α factor defined by

$$\alpha = \xi(L/kT_M),$$

where ξ is the orientation factor, L is the heat of fusion, k is the Boltzmann constant, and T_M is the melting point. Figure 3.11 (a) shows the results of Jackson's calculation, and indicates that there are two types of curves depending on the α factor. For materials with $\alpha<2$, there is only one minimum at 50% of surface site occupancy, indicating that the interface energetically prefers to be rough. On the other hand, in a material with $\alpha>3$, there are two energy minima at site occupancy 0 and 100%, indicating that the interface will be smooth.

The α factors are different for different materials. Metal crystals generally have $\alpha<2$; semiconductor crystals have $2<\alpha<3$; and oxides, silicates, and polymer crystals, which have complicated crystal structures, have $\alpha>3$. Due to the presence of ξ (corresponding to the ratio between two- and three-dimensional coordination numbers z_s/z), the α factors will be different in different crystallographic directions in the same crystal. For ice crystals, the (0001) face has $\alpha>3$, whereas for the $(10\bar{1}0)$ face it is $2<\alpha<3$, indicating that $(10\bar{1}0)$ is a rougher face than (0001).

In contrast to the equilibrium thermodynamic treatment performed by Jackson, Temkin [14], [15] considered the problem of the roughening transition of an interface in relation to the driving force and the α factor. His result is shown in Fig. 3.11(b), in which area A corresponds to an area where a smooth interface is expected, whereas area B corresponds to that expected for a rough interface. On increasing the driving force, a smooth interface transforms into a rough interface: a kinetic roughening transition.

Jackson's α factor is the value measured at the melting point; the solute–solvent interaction is not taken into account. This corresponds to the situation of crystal growth in the melt phase. Bennema and Gilmer [16] generalized Jackson's α factor as follows so that it can be applied to the solution phase, which involves a solute–solvent interaction:

$$\alpha_G = \xi\{\varphi_{sf} - 1/2(\varphi_{ss} + \varphi_{ff})\}/kT_G.$$

α_G is a generalized α factor, φ_{ss} and φ_{ff} are bonding energies in the solid and fluid phases, respectively, φ_{sf} is the solute–solvent interaction energy, k is the Boltzmann constant, and T_G is the growth temperature, which is lower than the melting point. In crystal growth from the melt phase, α_G becomes the same as that in Jackson's expression. In this generalization, however, α_G is given in terms of the growth temperature and the solute–solvent interaction energy, so it is different from Jackson's α factor.

Extensive computer experiments have been made since the 1980s to investigate how interface structure changes as $\Delta\mu/kT$ changes, to determine whether the roughening transition takes place, and to establish how the growth rate versus the driving force relation (and thus the growth mechanism) change depending on changes in the α factor, α_G factor, or $\Delta\mu/kT$. The results were as expected; Fig. 3.12 shows two examples of the results of computer experiments.

The above analyses assume that the interface structure does not change under any growth conditions. It has been experimentally demonstrated using Si crystals that the interface structure may be reconstructed in some ambient phases. Atoms in the interior of the crystal are bonded three-dimensionally and symmetrically to neighboring atoms, whereas those on the surface are not. The surface atoms are bonded with atoms in the structure, but not with those in the ambient phase. Consequently, they are in a higher energy state than the atoms in the crystal. Under certain conditions, "surface reconstruction" is required to relax the high-energy state; this is achieved by the formation of partial bonding, for example the formation of a 3×3 structure instead of the original 1×1 structure. It has been shown that reconstructed structures appear in Si 2×1 or 7×7, and in Ir (100) a 1×1 structure changes to a 1×5 structure. Crystal faces having such a reconstructed structure will be modified accordingly, and they exhibit a different surface morphology from ordinary non-reconstructed faces (see Section 4.4).

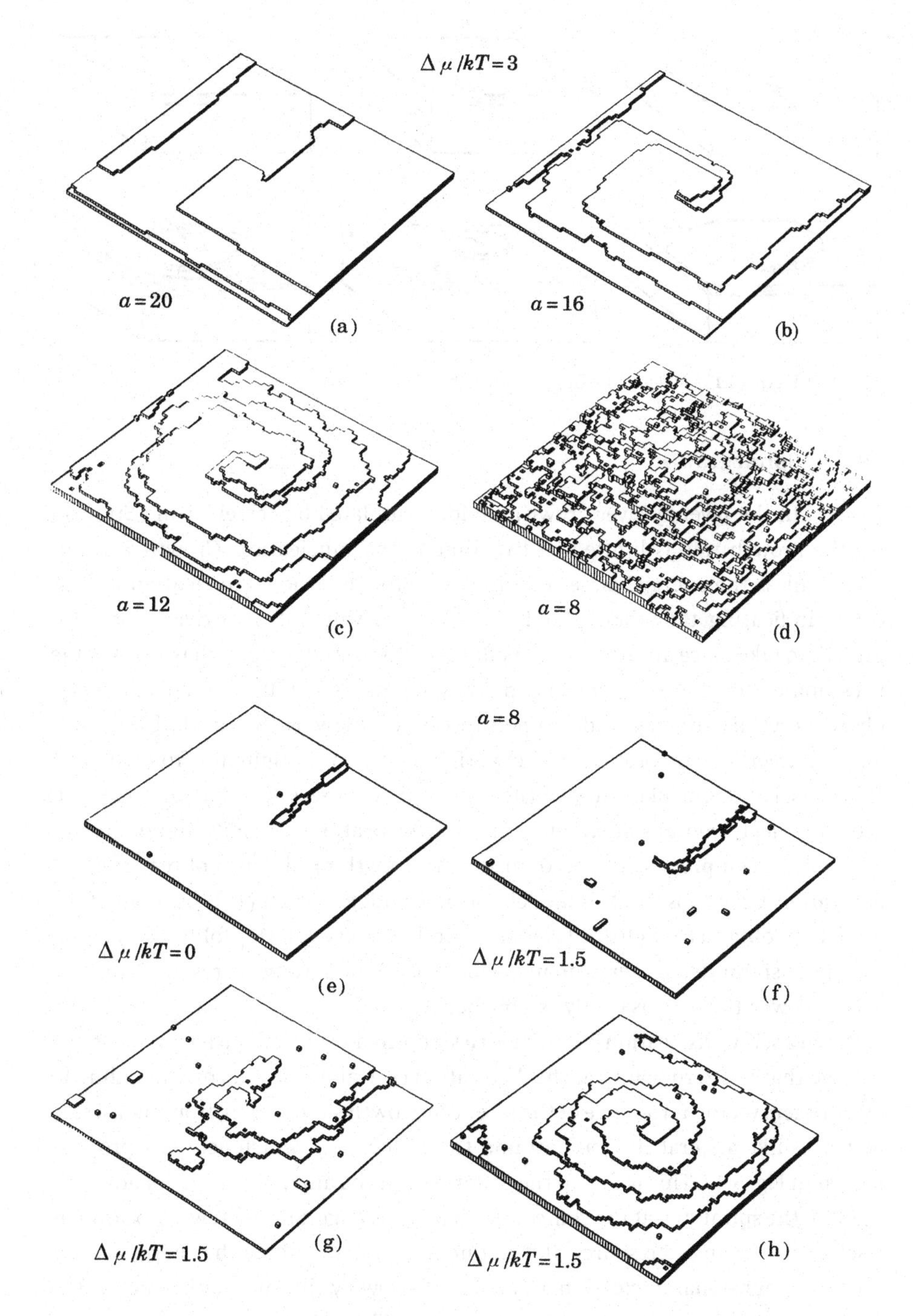

Figure 3.12. Changes in interface roughness as a result of changing α and $\Delta\mu$, taken from the results of a computer experiment [17]. (a)–(d) indicate changes in interface state for varying α while keeping $\Delta\mu/kT$ constant; (e) shows a step created by a screw dislocation at equilibirum, $\Delta\mu/kT = 0$, and (f)–(h) show how it advances under $\Delta\mu/kT = 1.5$, with α constant.

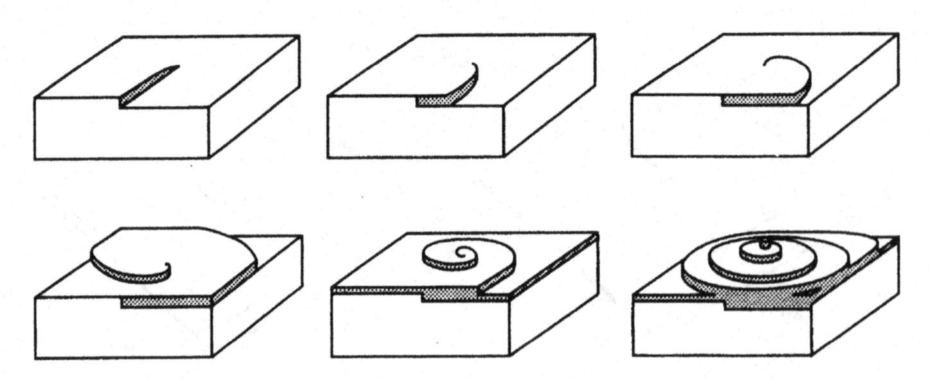

Figure 3.13. Mechanism of spiral growth.

3.9 Spiral growth

In the layer-by-layer growth model formulated by Kossel [8] and Stranski [9], the growth rate is limited by two-dimensional nucleation. Growth will not occur unless the energy barrier required for two-dimensional nucleation is overcome, indicating that there should be a critical value of the driving force for growth to take place. In the case of crystal growth from the vapor phase, this value is estimated to be around 25–50% in terms of supersaturation. In real cases, we observe crystals growing under supersaturation as low as 1%. This large discrepancy between the theoretical and experimental values originates from the fact that Kossel and Stranski assumed the crystal to be perfect. Real crystals, however, are imperfect, containing impurities and dislocations. Frank's spiral growth model has been proposed to account for the growth mechanism of real crystals. Soon after the proposal of this model, the first evidence to support the proof of the model was obtained on $\{10\bar{1}0\}$ faces of natural beryl crystals [18], followed by observations of spiral step patterns on many faces of a wide variety of crystals. Thus, the spiral growth theory was firmly established [19].

When a screw dislocation outcrops on a smooth interface, a step is created on the surface that has zero height at the dislocation core and a height corresponding to one Burgers vector at the edge of the face. The growth step starting from such a step advances like a spiral staircase around the dislocation, which acts as a prop, as shown in Fig. 3.13. This is due to the difference in the angular velocity of advancement of the spiral step at the center and that at the edge. Since screw dislocation is a self-perpetuating step source, it is not necessary to overcome the energy barrier for two-dimensional nucleation. Crystals can grow by this mechanism below the critical driving force for layer-by-layer growth. The growth rate and the driving force are related as follows:

$$R = A(\Delta\mu/kT)^2,$$

where A is a constant, and $\Delta\mu/kT$ is the driving force.

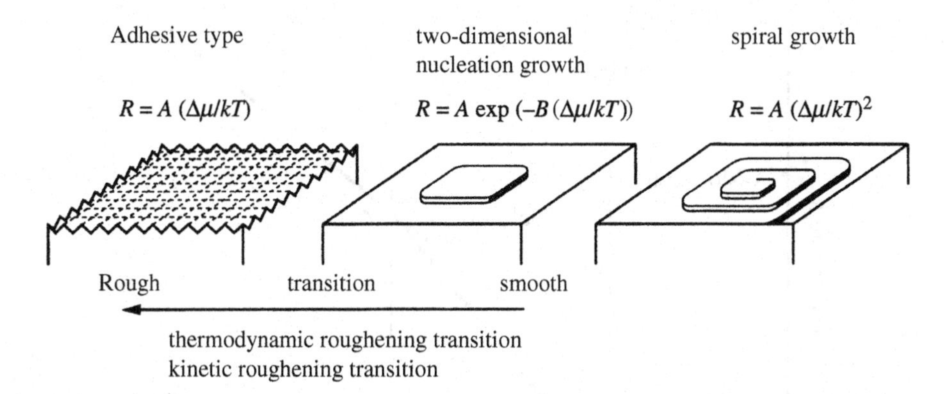

Figure 3.14. Differences between growth mechanisms on rough and smooth interfaces, showing the relation between the growth rate and the driving force in the adhesive-type growth mechanism, the two-dimensional nucleation growth mechanism, and the spiral growth mechanism. The roughening transition from a smooth to a rough interface is also shown. R is the growth rate, $\Delta\mu/kT$ is the driving force, and A and B are constants.

Spiral growth is a mechanism that is expected only on smooth interfaces. The assistance provided by screw dislocations is not necessary in the growth of rough interfaces, where an adhesive-type growth operates.

As explained above, three fundamental models of crystal growth mechanism were established in relation to the roughness of interfaces; these are illustrated in Fig. 3.14. At present, there is no other known growth mechanism that is essentially different from these three. Therefore, we shall analyze the morphology of crystals, the main topic of this book, based on these three growth mechanisms.

3.10 Growth mechanism and morphology of crystals

Figure 3.15 shows the growth rate R versus the driving force $\Delta\mu/kT$ for the three models of growth mechanism. This figure illustrates the following two points.

(1) As the driving force increases, an interface becomes rougher.

(2) Two bending points appear at $\Delta\mu/kT^*$ and $\Delta\mu/kT^{**}$, since the curves of R versus the $\Delta\mu/kT$ relations expected for the three models of growth mechanisms are different. An interface becomes rough and the growth mechanism will be of adhesive-type above $\Delta\mu/kT^{**}$, whereas the interface will be smooth and growth will be principally controlled by the spiral growth mechanism below $\Delta\mu/kT^*$. In between $\Delta\mu/kT^*$ and $\Delta\mu/kT^{**}$, the interface will be smooth, but the growth mechanism will be principally by two-dimensional nucleation.

Figure 3.15 is a schematic diagram of R versus the $\Delta\mu/kT$ relation expected in one crystallographic direction in an imaginary ambient phase.

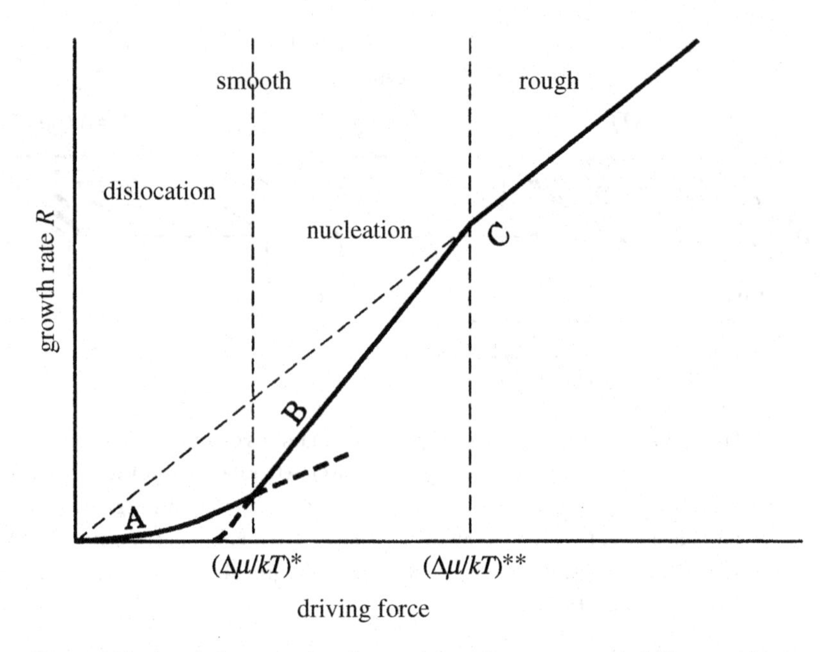

Figure 3.15. Areas where rough and smooth interfaces are expected. The growth rate versus the driving force relations expected for the three models of growth are indicated on the growth rate, R (vertical), axis versus the driving force ($\Delta\mu/kT$) diagram. Curve A shows the spiral growth mechanism; B represents the two-dimensional nucleation growth mechanism; C denotes the adhesive-type mechanism.

The positions of $\Delta\mu/kT^{*}$ and $\Delta\mu/kT^{**}$ are different, even in the same crystallographic direction of the same crystal species, depending on the difference in ambient phase. In the condensed melt phase, the position of $\Delta\mu/kT^{**}$ of a material with $\alpha<2$ is expected to be very close to the origin (namely, the interface will be rough), whereas the position will be at a much larger value in the case of growth from vapor phase, which is the most dilute ambient phase. In solution growth, the position of $\Delta\mu/kT^{**}$ will be in between the two positions (Fig. 3.16). Depending on the difference of solute–solvent interaction energy, the positions of $\Delta\mu/kT^{**}$ will be different from material to material growing from the solution phase with the same solvent component. This implies that when the crystal grows from the melt phase the interface will be rough, whereas a crystal from the same species will have a smooth interface and spiral growth is expected when the crystal grows from the vapor phase. In other words, this implies that a polyhedral crystal bounded by flat faces will be more generally expected when a crystal grows from the vapor phase than when it grows from the melt phase. Also, depending on crystal species, the implication is that the morphology will be different, even if the crystals grow from the same aqueous solution, because the $\Delta\mu/kT^{**}$ values are different for different crystal species due to solute–solvent interactions.

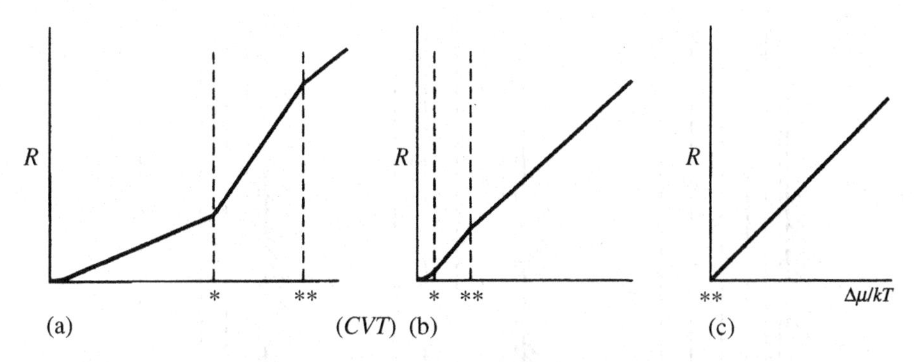

Figure 3.16. The different positions of $\Delta\mu/kT^*$ and $\Delta\mu/kT^{**}$ depending on the ambient phases. (a) Vapor phase; (b) solution phase; (c) melt phase.

Figure 3.15 shows R versus $\Delta\mu/kT$ for one crystallographic direction. The positions of $\Delta\mu/kT^*$ and $\Delta\mu/kT^{**}$ are different in different crystallographic directions of the same crystal. The crystal face ranked as the most important face in order of morphological importance (i.e. that which shows the sharpest cusp on the polar diagram of the γ (interface energy) plot, and belongs to a category of the most important F face in the PBC analysis, and has the highest reticular density in the BFDH (Bravais–Friedel–Donnay–Harker) law; see Chapter 4) is assumed to have the largest values of $\Delta\mu/kT^*$ and $\Delta\mu/kT^{**}$. As the order of morphological importance decreases, these values will be lowered among the faces categorized as F faces. The position of $\Delta\mu/kT^{**}$ of K or S faces in the PBC analysis is expected to be closer to the origin.

3.11 Morphological instability

The interface is the unique place where crystal growth (and dissolution) takes place. We have classified the interface into rough and smooth in the preceding section, and we have presented the expected growth mechanisms. In crystal growth from the melt phase, a rough interface will take a curved form that follows a curved plane of equal temperature, whereas in growth from the vapor or solution phases, it will follow a curved surface following lines of equal concentration. In contrast, a smooth interface takes a straight form truncating equi-thermal or equi-concentration lines. We shall discuss in this section whether the interface form can be maintained throughout the growth or not, what sort of instability will take place when the interface form is not maintained, and how these instabilities affect the morphology of crystals.

Let us assume that a small bulge appears on a rough, curved interface, and that for some reason the interface morphology is altered. Intervals between the lines of equal temperature or concentration become narrower at the bulge; hence, the

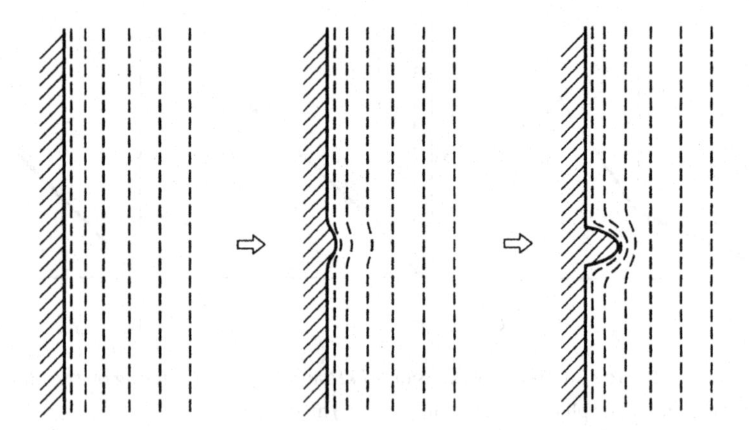

Figure 3.17. Relation between interface and lines of equal concentration in the ambient phase when morphological instability occurs.

temperature or concentration gradient becomes sharper, resulting in an increasing growth rate (Fig. 3.17). The situations are the same for melt growth, in which heat transfer plays the essential role, and for diluted vapor or solution growth, in which mass transfer plays the essential role. When changes occur in the gradient of the temperature and concentration, the interface morphology loses its stability, and the bulge will be enhanced. Morphological instability is stronger on rough, curved interfaces than on smooth, flat interfaces.

Unless cooperation of forces results in the suppression of interface instability, the instability will be enhanced as growth proceeds. As a result, periodic cellular structure, arrays of cusps, rod or lamellar structure, and further dendritic morphology will appear. Various patterns arising from morphological instability are indicated in Fig. 3.18. It is now understood that the dendritic morphology of snow crystals, which attracted the attention of early observers of crystals, results from morphological changes arising from morphological instability. The six symmetrical branches are due to the structure.

The problem of morphological instability was solved theoretically by Mullins and Sekerka [20], who proposed a linear theory demonstrating that the morphology of a spherical crystal growing in supercooled melt is destabilized due to thermal diffusion; the theory dealt quantitatively with and gave linear analysis of the interface instability in one-directional solidification.

The interface morphology is unstable for a variety of reasons; we should consider first constitutional supercooling, i.e. the change of melting point due to a change in impurity contents. In the solidification process of an alloy, even if crystallization starts from a flat solid–liquid interface with homogeneous impurity content, impurity concentrations in the solid and the liquid will change as growth proceeds (as the interface advances) because of the partitioning of impurity com-

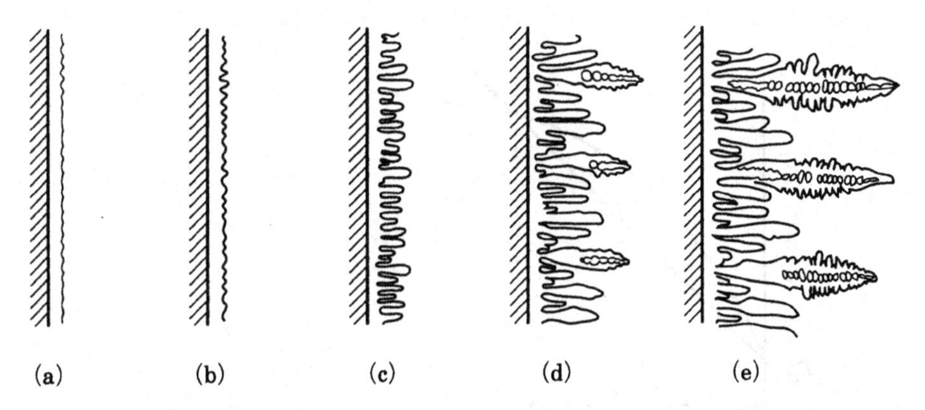

Figure 3.18. The process of changes in morphology due to interface instability. (a)–(e) Starting from a smooth interface, it is seen that the morphology changes to a cusp array, to a rod structure, to a dendritic structure.

ponents between the solid and liquid phases. Impurity components with a partitioning (distribution) coefficient smaller than unity will concentrate on the interface. The rate of concentration will be determined by the diffusion constants of the impurity component, the growth rate of the crystal, and the distribution coefficient of the impurity component. This results in an apparent supercooling state due to the relation between the temperature gradient between the crystal and the melt and the concentration gradient, as schematically illustrated in Fig. 3.19. This is referred to as constitutional supercooling, and provides an explanation for crystal growth from a melt phase. As a result, the causes of rod, cellular, or lamellar structure observed as the solidification structure in alloys are explained. However, the same concept may be applicable to vapor or solution growth in the presence of impurity components.

Why polyhedral forms bounded by smooth interfaces can grow, whilst maintaining their polyhedral forms, was not properly accounted for until the layer-by-layer growth theory (which considers atomic process of crystal growth) formulated by Kossel and Stranski appeared.

If the growth is entirely controlled by diffusion, the crystal should take a spherical form, with no flat faces. The appearance of flat crystal faces was accounted for by the introduction of the concept of the layer-by-layer growth mechanism on a smooth interface. As an example of the violation of morphological instability, we may mention the Berg effect. As Berg demonstrated (see Section 3.4), the driving force over a crystal face is higher at the corners or on the edges than on the face, and it is lowest at the center of the face. There is a critical value for two-dimensional nucleation to occur. If the driving force at the corners and edges is lower than this critical value, the growth of this face will be controlled by the spiral growth mechanism (see Fig. 3.15), and the growth layer originating from screw dislocations in

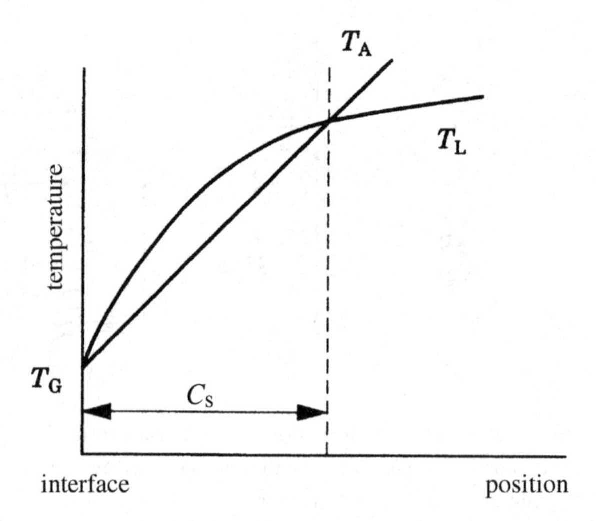

Figure 3.19. Schematic illustration of constitutional supercooling (C_S). T_L is the liquid temperature gradient, T_G is the growth temperature gradient, and T_A is the actual temperature gradient.

the central area of the face will advance outward, and so the polyhedral stability of the morphology can be maintained. If the bulk driving force increases above the value critical for two-dimensional nucleation to take place at the corners or edges (the area between $\Delta\mu/kT^*$ and $\Delta\mu/kT^{**}$ in Fig. 3.15), dislocations outcropping near the edges or corners, or two-dimensional nucleation in these places, will act as sources for growth layers, which then advance inwards to the central area of the face. This corresponds to the formation of a hopper face. If the driving force increases further, three-dimensional nucleation will take place at sites where the driving force is maximum, namely at the corners of the crystal. It is expected that this protrusion will appear in the direction of the corners, and that it will possibly develop into a dendritic crystal for the same reason that morphological instability occurs on a rough interface.

Taking the Berg effect into account, Kuroda *et al.* [21] demonstrated that the boundary between the area where the polyhedral morphology remains stable and the region where the stability is violated, and hopper or dendritic forms appear, will change depending on crystal size. Figure 3.20 shows their results.

3.12 Driving force and morphology of crystals

If we predict morphological changes based on the analyses of interface states, on the R versus $\Delta\mu/kT$ relations for the three models of growth mechanism, and on the morphological stability of interfaces, as discussed in Sections 3.2–3.9,

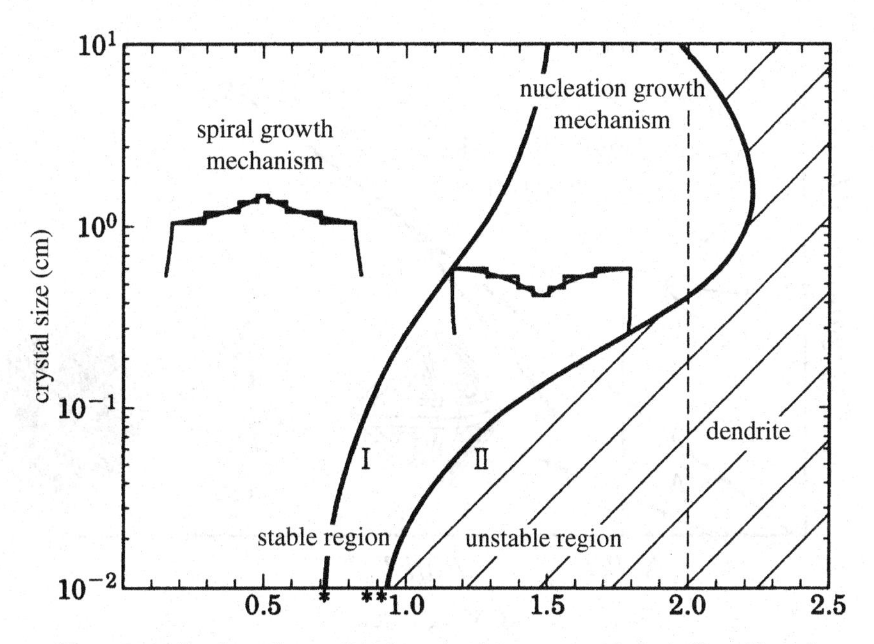

Figure 3.20. Size dependence of the boundary between morphologically stable and unstable regions when the Berg effect is taken into consideration [21].

and then plot the results on an R versus $\Delta\mu/kT$ diagram as given in Fig. 3.15, we create Fig. 3. 21.

Figure 3.21 is a schematic diagram showing only one crystallographic direction. Below the driving force condition $\Delta\mu/kT^{*}$, the spiral growth mechanism is the controlling mechanism, and a polyhedral crystal bounded by flat faces is expected. Above $\Delta\mu/kT^{**}$, the interface will be rough, the growth mechanism will be of adhesive-type, and a dendritic morphology is expected. In the region between $\Delta\mu/kT^{*}$ and $\Delta\mu/kT^{**}$, a hopper morphology is expected due to the two-dimensional nucleation mechanism. In this way, coherent logic has been established between dendritic and polyhedral morphologies of crystals, which were regarded as entirely independent problems at the early stages of study of the morphology of crystals. This range of morphology, from polyhedral to dendritic, is exhibited by single crystals. If $\Delta\mu/kT$ increases further, the rate of nucleation increases, resulting in the aggregation of many crystals. Depending on the conditions, spherical polycrystalline aggregates will appear due to radiating growth from a center. In the process of formation of a spherulitic, divergent, bow-tie, or semi-spherulitic morphology, either polycrystalline aggregates will be formed, or a random polycrystalline aggregation will appear.

The positions of $\Delta\mu/kT^{*}$ and $\Delta\mu/kT^{**}$ in the diagram will be different for different

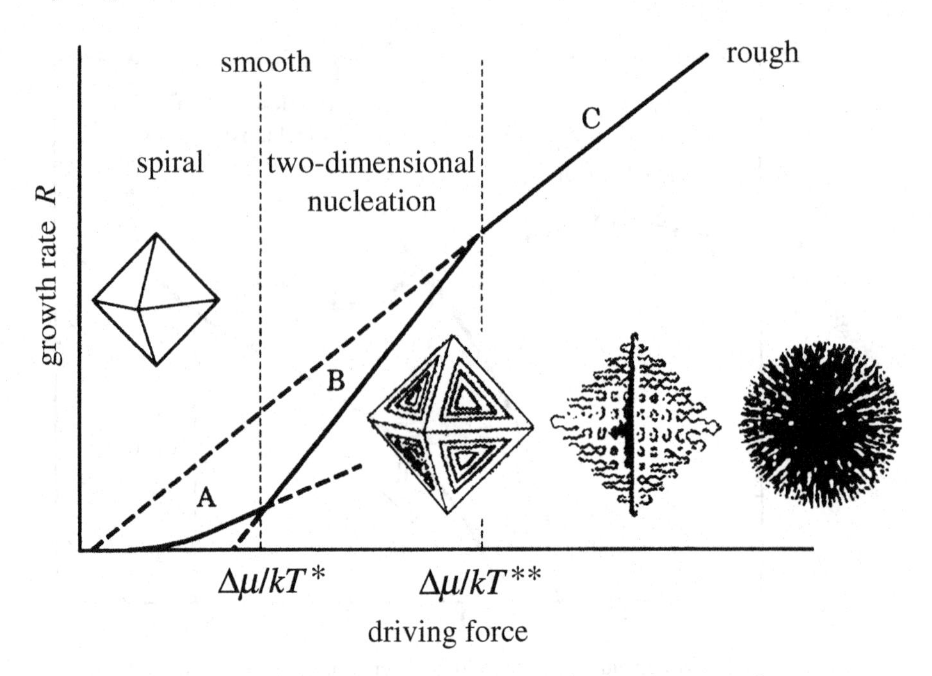

Figure 3.21. Changes in the morphology of crystals, shown on a growth rate versus driving force diagram (see Fig. 3.15), assuming a crystal bounded by the {111} face only.

crystallographic directions. The higher the morphological importance, the larger these values will be. In other words, a growing crystal is bounded by faces having different driving forces, and so each face exhibits a different interface roughness.

In the case of a crystal bounded by both smooth and rough interfaces, the rough interface will disappear as the growth proceeds, and the crystal will eventually take a polyhedral form bounded by a smooth interface alone, unless the growth rate of the rough interface is suppressed due to environmental conditions. Since the growth of a polyhedral crystal is controlled by the spiral growth mechanism, the normal growth rate of the face is determined by the height, the spacing, and the advancing rate of the steps of the growth spiral. The factors which determine these three features therefore control the *Tracht* and the *Habitus* of polyhedral crystals. This will be analyzed in Chapter 4.

When a crystal grows under uncontrolled conditions, as in the crystal growth of natural minerals, nucleation first occurs under high driving force conditions, which will diminish as growth proceeds. This implies that a crystal which is originally dendritic will eventually take a polyhedral form bounded by a flat face. If a polyhedral crystal bounded by flat faces is bisected, a skeleton showing dendritic form will be observed in the interior. Abrupt changes in driving force conditions occur frequently in natural crystallization, such as in the uplifting of magma, and

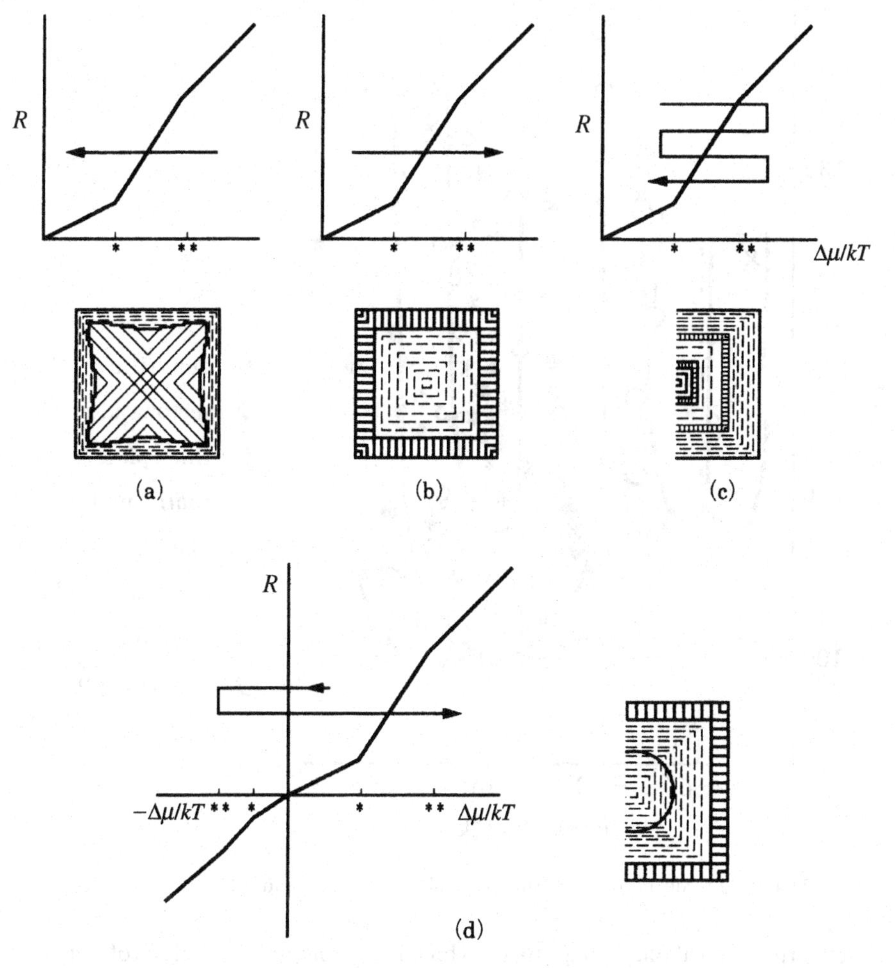

Figure 3.22. Textures recorded within a single crystal due to changes in the growth conditions.

we therefore expect to find polyhedral crystals coated in a fibrous texture due to dendritic growth, or crystals showing multiple changes in cross-section. Figure 3.22 presents a summary of these internal textures.

3.13 Morphodroms

We discussed in Section 3.12 that there is a mutual relation among spherulitic, dendritic, hopper, and polyhedral crystals, with respect to the driving force. We will see how these mutual relations appear in real systems, using, as representative examples, low-temperature snow crystals (vapor phase growth) and high-temperature silicate crystals growing in silicate solution phases.

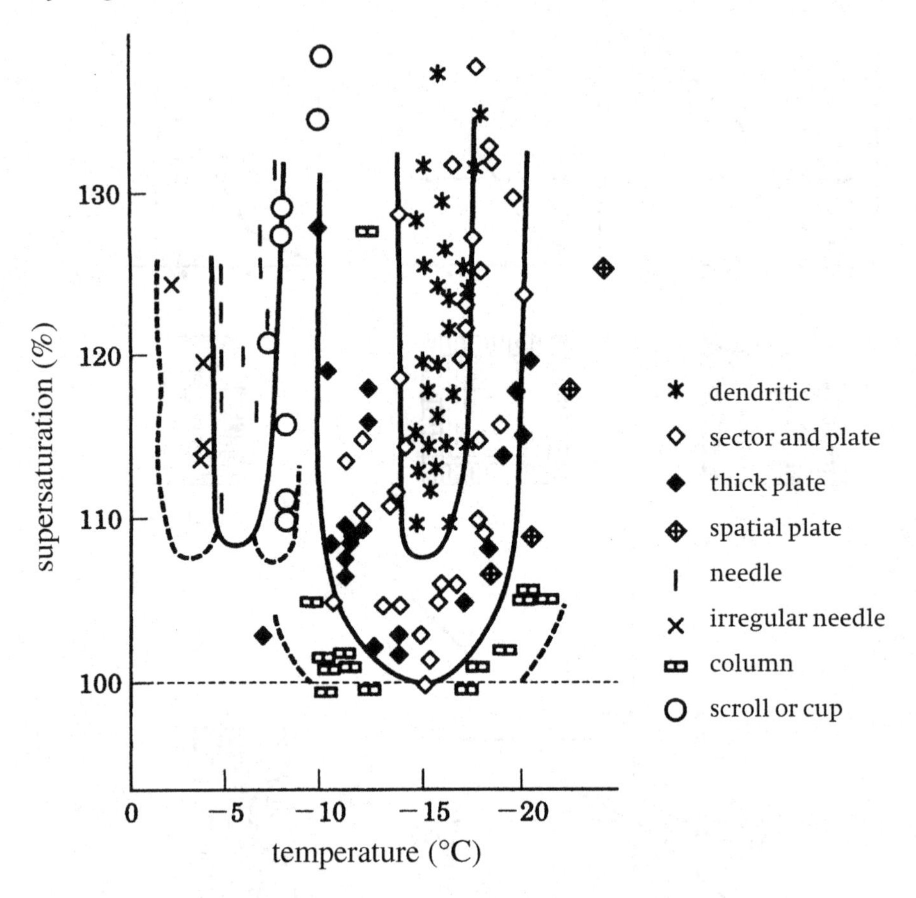

Figure 3.23. Morphodrom of snow crystals: Nakaya's diagram [22].

The morphological phase diagram on which morphological variations of a crystal are illustrated in relation to growth conditions is called a "morphodrom." This word was originally used by Kern *et al.* to illustrate variations of *Tracht* and *Habitus* of polyhedral ionic crystals growing in aqueous solution (see ref. [29], Chapter 4), but in fact there exist much older examples showing the morphological variation in the form of a morphological phase diagram. The earliest example is Nakaya's famous diagram of snow crystals [22]; Nakaya depicted the morphologies of artificial snow crystals in relation to supercooling and supersaturation of H_2O. Nakaya's diagram is shown in Fig. 3.23. It is clearly seen from this diagram that snow crystals take on a polyhedral morphology in the region of low supersaturation of H_2O, whereas they assume a dendritic morphology in the region of high driving force. An analysis of the reason why polyhedral snow crystals change their *Habitus* alternately from platy to prismatic depending on the degree of supercooling will be given in Chapter 4. The Nakaya diagram clearly shows that the morphology of snow crystals changes from polyhedral to dendritic as the supersaturation (driving force) increases.

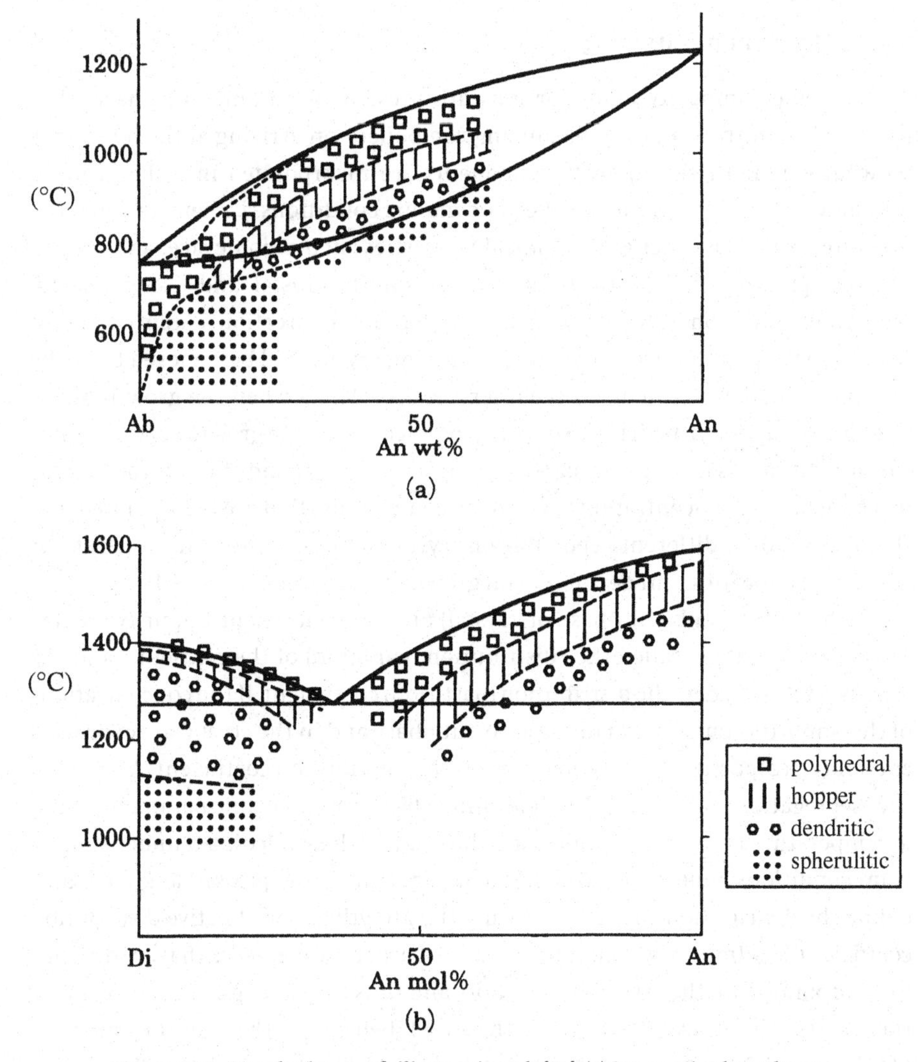

Figure 3.24. Morphodroms of silicate minerals [23]. (a) Summarized results on quenched samples: anorthite (An)–albite (Ab) series. (b) Summarized observation obtained by *in situ* method: diopside (Di)–anorthite (An) system.

Figures 3.24(a) and (b) show morphodroms of silicate crystals growing from silicate solutions: Fig. 3.24(a) shows the results of observations on quenched products, and Fig. 3.24(b) summarizes the results obtained by a high-temperature *in situ* observation method of growth [23]. In these cases also, as apart from the liquidus of solid–solution component, it is seen that the morphology of crystals changes from polyhedral, through hopper, to dendritic, then spherulitic. The predictions described in Section 3.12 are thus confirmed by experiment.

3.14 Element partitioning

When impurity atoms or ions are present in an ambient phase, they behave differently from the crystallizing component on arriving at the interfaces. To what extent these impurity components are incorporated into the growing crystal will differ depending on thermodynamic conditions, i.e. temperature and pressure, as well as on crystal chemical properties such as ionic size, and properties and spatial size of substitutional positions in the crystal, and also will be different depending on growth rate and crystallographic directions. Atoms and ions which are difficult to incorporate into a growing crystal will be removed from the crystal and will accumulate on the interface completely when the growth rate is slow, but at higher growth rates some will be enclosed by the growing crystal. Since the growth rate is dependent on the interface structure, and thus on the growth mechanism, the concentration of impurity components that are incorporated into the crystal will be different depending on crystallographic direction, namely their contents will be different depending on growth sectors (see Section 6.1).

To which side (ambient phase or crystal) and to what extent impurity components concentrate is called the "distribution (partition) of the element," and the ratio is represented by the distribution (partition) coefficient. If the concentrations of the impurity component i in the ambient phase and in the crystal are expressed as c_L and c_S, respectively, the ratio $c_L/c_S = k$ is the distribution coefficient. Also called the "segregation coefficient," k is determined by thermodynamic parameters such as temperature and pressure, but is not the same as the k calculated under equilibrium conditions, where the growth rate is zero (this k is expressed as k_0). We may define the distribution coefficient during the growth as the effective distribution coefficient, k_{eff}. Impurity elements with $k_{eff} < 1$ are those elements that are difficult to incorporate into the crystal in this ratio, and thus they sweep out to and concentrate at the interface; elements with $k_{eff} > 1$ behave in the opposite manner. Elements with $k_{eff} < 1$ that are swept out to the interface form a concentrated layer with a gradient that depends on the growth rate, R, if we assume a diffusion layer with thickness δ. Under the equilibrium condition where $R = 0$, there is no gradient, but at rapid quenching, with $R = \infty$, the gradient is the steepest, and at a finite R the gradient is intermediate. At $R = \infty$, impurity concentrations are equal in the ambient phase and in the crystal, and $k_{eff} = 1$, whereas under equilibrium conditions, where $R = 0$, $k_{eff} = k_0$. Assuming the diffusion constant of an impurity to be D, Burton, Prim, and Slichter [24] expressed k_{eff} by the following formula:

$$k_{eff} = k_0/[k_0 + (1 - k_0)\exp(-R\delta/D)].$$

This is the famous BPS equation; it shows that k_{eff} depends on R, but it does not show that it depends on orientation.

The first example which demonstrated that element partitioning has orientation dependence was the observation of the distribution of impurity Bi atoms in a Si single crystal grown from the melt. The Bi atoms are mainly distributed in the smooth interface growth sector. This was explained as being due to a much larger advancing rate of growth steps on a smooth interface than the normal growth rate R of a rough interface. If the advancing rate is rapid, impurities arriving at the interface will be captured by the advancing growth steps before they are swept to the ambient phase. If we extend this concept further, it is predicted that impurity partitioning will depend on crystallographic orientation, and that the growth sectors of a face with the highest morphological importance will contain more impurities with $k_{eff} < 1$. Impurity distribution in a single crystal cannot be homogeneous and will be dependent on growth sectors.

Impurity elements with $k_{eff} < 1$ will concentrate on the interface. This affects the morphological stability of the interface through growth rate and constitutional supercooling phenomena, and may modify the smoothness of the interface. As a result, fluctuations in impurity concentration will appear in the form of growth banding.

3.15 Inclusions

Tiny crystallites of the same or different species may precipitate on the growing surface of a crystal. Whether these are expelled from or occluded into the growing crystal is determined by the relation between attractive and repulsive forces with the growing crystal, similarly as partitioning of impurity elements. At higher growth rates, it is more likely that the precipitates will be occluded into the crystal. They become solid inclusions, which serve as good indicators of the chemical composition of the ambient phase and growth conditions, temperature, and pressure. Solid inclusions in natural diamond crystals have been extensively investigated, from which two types of ambient phases, ultramafic and eclogitic, have been considered as the probable ambient phases in which the diamonds were formed.

When the ambient phase is a solution phase with low viscosity, such as a hydrothermal solution, occlusion of the mother liquid (the liquid in which the crystal grows) into a growing crystal occurs at places where dendritic branches conjugate, or through overhanging of macro-growth steps, or at places where advancing macro-steps meet. When growth proceeds further on the wall of a mother–host liquid inclusion, it will dissociate into arrayed smaller inclusions by necking, or will become a negative crystal bounded by crystallographic faces.

References

1 K. Onuma, K. Tsukamoto, and I. Sunagawa, Effect of buoyancy driven convection upon the surface microtopographs of $Ba(NO_3)_2$ and CdI_2 crystals, *J. Crystal Growth*, **98**, 1989, 384–90

2 I. Sunagawa, K. Tsukamoto, K. Maiwa, and K. Onuma, Growth and perfection of crystals grown from aqueous solution: case studies on barium nitrate and K-alum, *Prog. Crystal Growth and Charact.*, **30**, 1995, 153–90

3 W. F. Berg, Crystal growth from solutions, *Proc. Roy. Soc. (London)*, **A164**, 1938, 79–95

4 A. Kouchi, A. Tsuchiyama, and I. Sunagawa, Effect of stirring on crystallization kinetics of basalt; texture and element partitioning, *Contrib. Mineral. Petrol.*, **93**, 1986, 429–38

5 W. Ostwald, *Z. Physik. Chem.*, **34**, 1900, 493–503

6 T. Abe, K. Tsukamoto, and I. Sunagawa, Nucleation, growth and stability of $CaAl_2Si_2O_8$ polymorphs, *Phys. Chem. Minerals*, **17**, 1991, 473–84

7 F. C. Frank, The influence of dislocations on crystal growth, *Disc. Faraday Soc.*, no. 5, 1949, 48–54

8 W. Kossel, Zur Theorie der Kristallwachstums, *Nachur. Ges. Gottingen*, **2**, 1927, 135–45

9 I. N. Stranski, Zur Theorie der Kristallwachstums, *Z. Phys. Chem.*, **136**, 1928, 259–78

10 M. Volmer, Zum Problem des Kristallwachstums, *Z. Phys. Chem.*, **102**, 1922, 267–75

11 W. K. Burton, N. Cabrera, and F. C. Frank, The growth of crystals and the equilibrium structure of their surfaces, *Phil. Trans.*, **A243**, 1951, 299–358

12 P. Hartman and W. G. Perdok, On the relation between structure and morphology of crystals, *Acta Crist.*, **8**, 1955, 49–52, 521–4, 525–9

13 K. A. Jackson, *Liquid Metals and Solidification*, Metals Park, Ohio, American Society of Metals, p. 174

14 O. E. Temkin, Phenomenological kinetics of the motion of a phase boundary, *Sov. Phys. Cryst.*, **15**, 1971, 767–72

15 O. E. Temkin, Kinetic phase transition during a phase conversion in a binary alloy, *Sov. Phys. Cryst.*, **15**, 1971, 773–80

16 P. Bennema and G. H. Gilmer, Kinetics of crystal growth, in *Crystal Growth, An Introduction*, ed. P. Hartman, Amsterdam, North-Holland, 1973, pp. 263–327

17 G. H. Gilmer and K. A. Jackson, Computer simulation of crystal growth, in *Current Topics in Materials Science*, eds. E. Kaldis and H. J. Scheel, Amsterdam, North-Holland, 1977, pp. 80–114

18 L. J. Griffin, Observation of uni-molecular growth steps on crystal surfaces, *Phil. Mag.*, **41**, 1950, 196–9

19 I. Sunagawa and P. Bennema, Morphology of growth spirals, theoretical and experimental, in *Preparation and Properties of Solid State Materials*, vol. 7, ed. W. R. Wilcox, New York, Marcel Dekker Inc., 1982, pp. 1–129

20 W. W. Mullins and R. F. Sekerka, Morphological stability of a particle growing by diffusion or heat flow, *J. Appl. Phys.*, **34**, 1963, 323–9

21 T. Kuroda, T. Irisawa, and A. Ookawa, Growth of a polyhedral crystal and its morphological stability, *J. Crystal Growth*, **42**, 1977, 41–6

22 U. Nakaya, *Snow Crystals, Natural and Artificial*, Cambridge, Mass., Harvard University Press, 1954

23 I. Sunagawa, Morphology of minerals, in *Morphology of Crystals*, Part B, ed. I. Sunagawa, Dordrecht, D. Reidel, 1988, pp. 509–87

24 J. A. Burton, R. C. Prim, and W. P. Slichter, The distribution of solute in crystals grown from the melt, Part 1, *Theor. J. Chem. Phys.*, **21**, 1953, 1987–91

Suggested reading

Handbooks on crystal growth and synthesis

Editorial Committee, *Handbook of Crystal Technology*, Tokyo, Kyoritsu Pub. Co., 1971 (in Japanese)

D. T. J. Hurle (ed.), *Handbook of Crystal Growth*, 1A, 1B, 2A, 2B, 3A, 3B, Amsterdam, North-Holland, 1994

Editorial Committee, *Handbook of Crystal Growth*, Tokyo, Kyoritsu Pub. Co., 1995 (in Japanese)

Editorial Committee, *Dictionary of Crystal Growth*, Tokyo, Kyoritsu Pub. Co., 2001 (in Japanese); see also references therein

Fundamentals of crystal growth

P. Hartman (ed.), *Crystal Growth, An Introduction*, North-Holland, Amsterdam, 1973

A. Ookawa, *Crystal Growth*, Shokabo, 1977 (in Japanese)

A. A. Chernov, *Modern Crystallography III, Crystal Growth*, Berlin, Springer-Verlag, 1984

T. Kuroda, *Crystals are Alive*, Science Co., 1984 (in Japanese)

T. Nishinaga, S. Miyazawa, and K. Sato (eds.), *Series on Dynamics of Crystal Growth*, 7 vols, Tokyo, Kyoritsu Pub. Co., 2002 (in Japanese)

Technology of single crystal synthesis

S. Takasu, *Basic Techniques of Growing Crystals*, Tokyo, Tokyo University Press, 1980 (in Japanese)

Morphology of crystals

I. Sunagawa (ed.), *Morphology of Crystals*, Parts A, B, Dordrecht, D. Reidel, 1987

W. A. Tiller, *The Science of Crystallization, Macroscopic Phenomena and Defect Generation*, Cambridge, Cambridge University Press, 1991

I. Sunagawa (ed.), *Morphology of Crystals*, Part C, Dordrecht, D. Reidel, 1994

Interfaces

D. P. Woodruff, *The Solid-Liquid Interface*, Cambridge, Cambridge University Press, 1973

W. A. Tiller, *The Science of Crystallization, Microscopic Interfacial Phenomena*, Cambridge, Cambridge University Press, 1991

4

Factors determining the morphology of polyhedral crystals

Polyhedral crystals bounded by flat faces can exhibit various *Tracht* and *Habitus* because they result from the interconnection of internal structural factors and the external factors involved during crystal growth. The concepts of the structural form and the equilibrium form were first suggested in an attempt to provide a greater understanding of the origin of the morphology of polyhedral crystals. Furthermore, extensive experiments on and observations of natural minerals were carried out in an attempt to clarify the origin of morphological variation of growth forms. The results obtained from the many investigations performed in the twentieth century are summarized in this chapter.

4.1 Forms of polyhedral crystals

The various morphologies, such as polyhedral, hopper, dendritic, and spherulitic, that are exhibited by single crystals and polycrystalline aggregates, have been discussed in relation to the driving force in the preceding chapters.

Among all of the various morphologies exhibited by crystals, it is the problem of variations in *Tracht* and *Habitus* exhibited by polyhedral single crystals that has attracted the deepest concern. Polyhedral single crystals are expected when crystals grow by the spiral growth mechanism under a driving force of $\Delta\mu/kT^*$. There are four logical routes that we can take to understand this problem.

The first is the prediction of the *Habitus* made from the characteristics of the crystal structure, entirely neglecting the effect of growth conditions. We will call this the "structural form" or "abstract form." The second logical approach is to predict the *Habitus* thermodynamically when the crystal reaches the equilibrium state. This may be called the "equilibrium form." The third is a method of analyzing the factors that may have an effect by correlating the *Habitus* and *Tracht* shown by real crystals and their growth conditions. Investigations of this type may be referred to as "growth forms." It should be noted that, whereas structural and equilibrium forms may be described as singular, growth forms are plural. The fourth method is to analyze the growth forms based on the observation of surface microtopographs of crystal faces, possible now that molecular information relating to growth forms is available.

We discuss the structural form in Section 4.2 and the equilibrium form in Section 4.3. Section 4.4 presents a summary of the results and investigations of the growth forms. The subject of microtopographs will be discussed in Chapter 5.

4.2 Structural form

The morphology of real crystals is determined by two factors: the internal elements, due to the periodicity and anisotropy of the bonding that constitutes the crystal structure, and the external elements involved in the growth parameters. If a crystal grows in an isotropic ambient phase, crystal species having different structures will show characteristic *Habitus* corresponding to the respective structures. However, the same crystal species may show various *Habitus*, including dendritic form, depending on the growth conditions. It is known empirically that the order of morphological importance of crystal faces, deduced from statistical treatments of the development and frequency of appearance of faces on polyhedral crystals of a particular crystal species, is closely related in many cases to the lattice type of the crystal.

In Fig. 4.1 we depict three lattice types of the cubic system and the crystal faces with the highest reticular density (the density of lattice points per unit area) in each type.

The crystal face with the highest reticular density (and thus with the widest reticular spacing) is $\{100\}$ in the simple cubic lattice (P), $\{111\}$ in the face-centered cubic lattice (F), and $\{110\}$ in the body-centered cubic lattice (I). The order of morphological importance obtained statistically from mineral crystals agrees well with this in many cases. This is known as the Bravais empirical law [1], the earliest law presented to describe the morphology of polyhedral crystals. There are, however, cases in which there is no agreement. For example, the polyhedral form of quartz predicted from the consideration of its lattice type should be bounded by three equally developed faces, $\{0001\} \approx \{10\bar{1}0\} \approx \{10\bar{1}1\}$ (Fig. 4.2(a)); however, the $\{0001\}$ face appears only exceptionally rarely in real quartz crystals. In pyrite crystals, $\{110\}$ should develop as a large face, but in reality pyrite is mainly bounded by $\{210\}$, and the order of morphological importance of $\{110\}$ is very low.

Lattice type is the basis of calculation of reticular density in the Bravais empirical law. In lattice types, only the symmetry elements with no translation, i.e. the

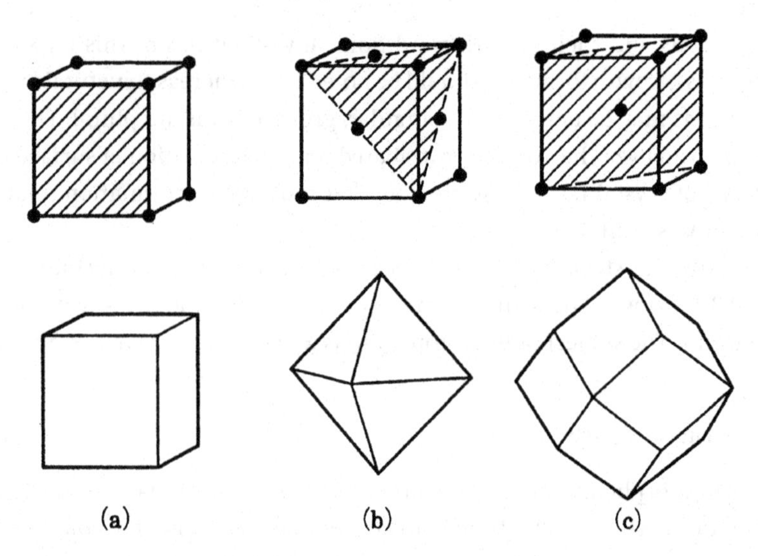

Figure 4.1. Crystal faces with the highest rank in the order of morphological importance (with the highest reticular density) for P, F, and I lattice types of the cubic system: (a) P lattice, {100}; (b) F lattice, {111}; (c) I lattice, {110}.

center of symmetry, the symmetry plane, and the symmetry axes, are included. This means that the geometry of the fourteen types of Bravais lattice and thirty-two crystal groups form the basis of the analysis. In the structural analysis, the 230 space groups containing translational symmetry elements, i.e. glide planes and screw axes, are required. In quartz crystal containing a three-fold screw axis, the reticular density of {0001} should be calculated as {0003} (see Fig. 4.2 (b)), and in pyrite {110} should be treated as {220} (due to the presence of a glide plane), and so the reticular density becomes half that calculated on {110}. If the reticular densities of crystal faces are recalculated by expanding the symmetry elements involved in the thirty-two point groups to those in the 230 space groups, most of the discrepancies between the predicted forms determined from the Bravais empirical law and those that are actually observed disappear. This is an extension of the Bravais (and Friedel) empirical law by Donnay and Harker [2], which we shall abbreviate as the BFDH (Bravais–(Friedel)–Donnay–Harker) law. We see that, in the BFDH law, crystal forms are analyzed by considering the crystal faces.

Polyhedral crystal forms are defined by faces (planes) and edges (zones). One treatment that focuses mainly on the zones is the periodic bond chain (PBC) theory by Hartman and Perdok [3], and see ref. [12], Chapter 3. If we connect atoms and ions having strong bonding in a crystal structure, we find a few strongly bound chains in the structure. Since these chains are arranged periodically, they are called periodic bond chains (PBCs), which are vector quantities. We may define a PBC by its stoichiometric composition. PBCs containing only an integer fraction of composition

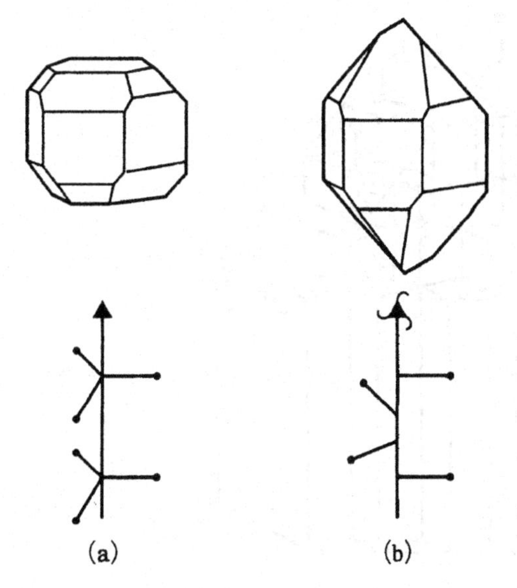

Figure 4.2. The morphology of quartz crystal predicted by (a) the Bravais empirical law and (b) Donnay–Harker's law.

are partial PBCs, but these are not considered in HP (Hartman–Perdok) theory. Crystal faces are classified into three types, depending on the number of PBCs included:

(i) K face (kinked face), containing no PBCs;
(ii) S face (stepped face), containing only one PBC;
(iii) F face (flat face), containing two or more PBCs.

It is clear that K, S, and F faces correspond to the (111), (110), and (001) faces, respectively, of the Kossel crystal shown in Fig. 3.9 (Fig. 4.3). A K face corresponds to a rough interface, an F face to a smooth interface, and an S face to a face having an intermediate nature between that of the K and F faces. Further, a K face grows by the adhesive-type growth mechanism, an F face grows either by a layer-by-layer or a spiral growth mechanism, and an S face appears by the piling up of growth layers advancing on the neighboring F face. Therefore, an F face develops to a large size in order to control *Habitus* and *Tracht* in a real crystal, the K face will disappear from the crystal surface, and the S face will be characterized by striations only, if it appears on a crystal.

Another advantage of the PBC theory is its ability to predict not only the bulk morphology of a polyhedral crystal, but also the morphology of growth layers developing on F faces. When growth layers are polygonal, the step direction is assumed to be defined by the PBC theory. These predictions cannot be made by the BFDH law.

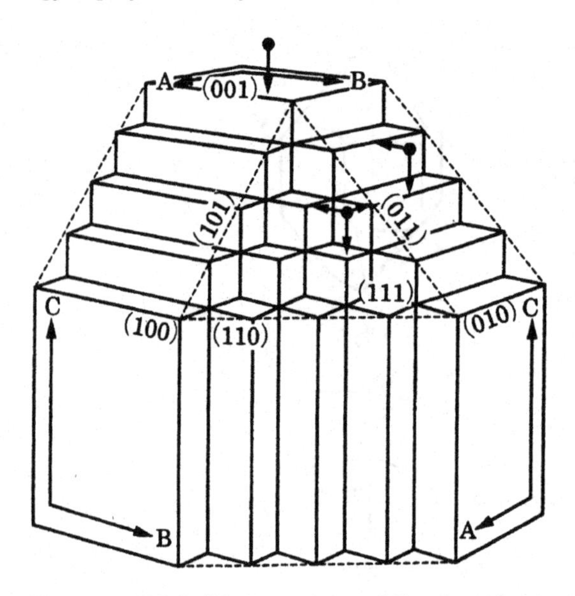

Figure 4.3. PBCs (solid arrows, A, B, and C) and an F face {100}, an S face {110}, and a K face {111} in PBC (HP) theory (ref. [12], Chapter 3).

There are two downsides to the PBC theory: the first is that a certain arbitrariness is unavoidable in finding PBCs in real crystal structures, and the second is that PBC analysis is difficult in complicated structures. In answer to the first criticism, Hartman [3] calculated the attachment energy, E_{att}, and correlated this to Jackson's α factor. To answer the second criticism, a connected net model was proposed by Bennema and van der Eerden [4]. Since PBC analysis and interface roughness can be correlated through these two suggested solutions, a brief explanation will be given.

The lattice energy at a half-crystal position (kink site) is defined as the attachment energy E_{att}, and the energy released in forming a slice containing more than two PBCs is denoted by E_{sl}. These are related to the lattice energy E_{cr} as follows:

$$E_{sl} + E_{att} = E_{cr},$$

where the energy is given per mole. The normal growth rate R of a crystal face is proportional to E_{att}; the smaller the value of E_{att}, the smaller the value of R, and the larger the face becomes. The normal growth rate of an *hkl* face, R_{hkl}, is related in an S-figure form to E_{att}.

By generalizing E_{att}, the following formula may be expressed:

$$\varepsilon = gE_{cr}/RT,$$

$$g = (E_{ss} + E_{ff} - 2E_{sf})/E_{ss},$$

where ss, ff, and sf represent solid–solid, liquid–liquid, and liquid–solid interactions, respectively.

By putting

$$P = E_{att}/E_{cr},$$

it is possible to express the following relationship with Jackson's α factor:

$$\alpha = (1-P)\,\varepsilon.$$

When E_{sl} is small and the interaction energy between liquid and solid is small (i.e. g is small), the roughening transition temperature becomes lower. Through this treatment, PBC analysis, which was originally criticized as being an arbitrary qualitative theory, is now correlated energetically to interface roughness.

Considerable difficulties arise during the analysis of the PBCs of crystals with complicated crystal structure, such as garnet or organic compounds. Bennema and van der Eerden [4] suggested a connected net model to analyze complicated structures and to provide a correlation with the roughening transition of an interface. Using this model, it became possible for the first time to construct a unified relation between PBC analysis and the roughening transition, and the structural form, the growth forms, and the effect of growth conditions predicted from computer calculations [5]. In the model, the centers of gravity of the atoms and molecules that constitute the crystal are connected, and nets are obtained. Connected nets correspond to two-dimensional crystals, and therefore depict the same thing as an F face in PBC analysis. Therefore, the net has a corresponding roughening transition temperature.

If we define a non-dimensional temperature $\theta = 2kT/\varphi$, the connected net becomes a crystal face (an F face) on which two-dimensional layer growth takes place when

$\theta^{R}\theta^{c} > 0$ and $\theta^{R} < \theta^{c}$ (from the relationship between θ^{R}, corresponding to the α^{R} value, and θ^{c} of the connected net). An S face consists of only one PBC with zero edge free energy, and $\theta^{R} = \theta^{c} = 0$; and a K face has zero edge free energy in all directions, and again $\theta^{R} = \theta^{c} = 0$. In other words, an S face is a net with no connection, and a K face is not a connected net (Fig. 4.4). An anomaly with PBC theory exists here, in that direct PBC theory (D-PBC theory), which satisfies stoichiometric composition, is presumed, whereas the connected net model allows a partial PBC (P-PBC) model. The importance of the connected net model is that it allows both D-PBC and P-PBC theories, by connecting networks.

4.3 Equilibrium form

A droplet of liquid, which has a random structure and thus isotropic properties, takes a true spherical form at equilibrium. This is because a sphere is the form with minimum surface energy in the case of an isotropic material. Gibbs considered that a crystal, which has a regular structure for which anisotropy is the

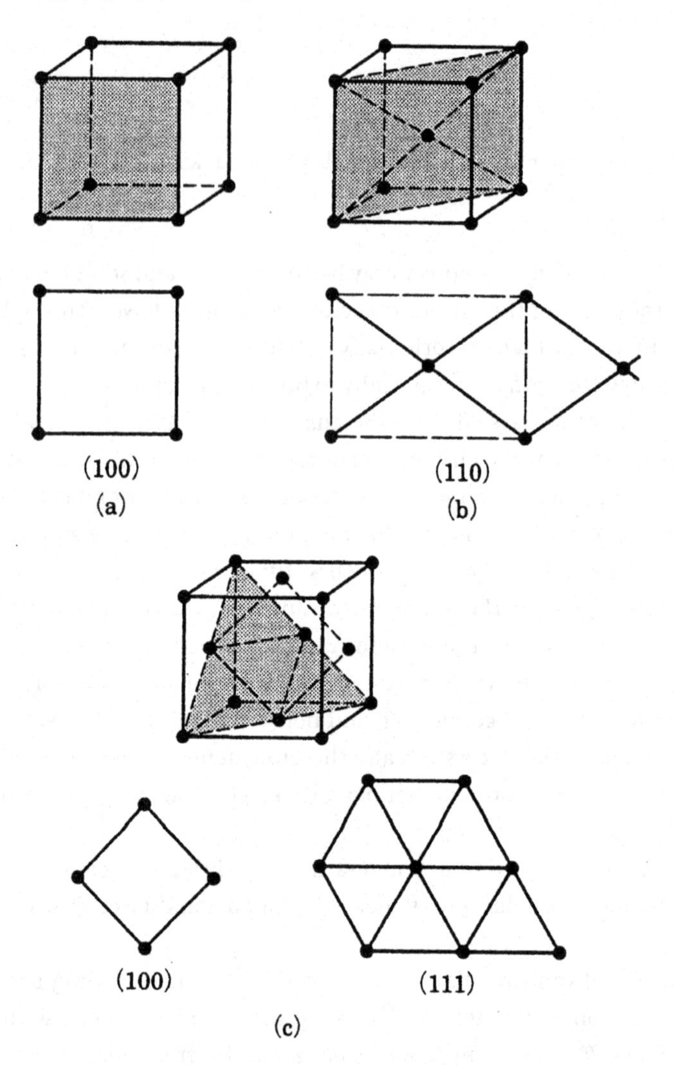

Figure 4.4. The net in the connected net model [4]. (a) In a P lattice the connected net is the square, (100). (b) In a bcc lattice the connected net is (110), which is fundamentally the same as the connected net in (a). (c) An fcc lattice. The connected nets are (100), square, and (111), hexagonal.

essential property, should, at equilibrium, take a form such that the total surface area times the surface free energy is at a minimum. A crystal in this state is called the equilibrium form, which is unique under given temperature and pressure conditions.

Starting from Gibbs' concept [6] that the total surface energy times the surface area is at a minimum, the following concepts emerged:

(1) Curie's concept [7], which considered that the normal growth rates of crystal faces are proportional to the surface free energies; and

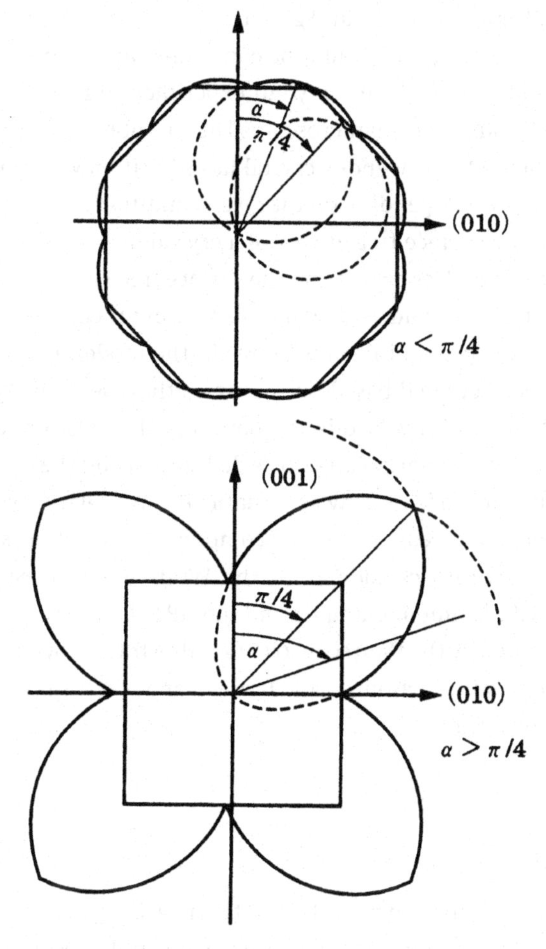

Figure 4.5. Wulff's polar diagram based on ref. [8]. The equilibrium form is obtained by drawing inscribed lines at the cusps.

(2) Wulff's plot, indicating that the equilibrium form is obtained by connecting inscribed lines drawn at cusps on a raspberry-shaped polar diagram, as shown in Fig. 4.5, on which points at distances proportional to the surface free energy from the center of a crystal are plotted.

A closed system is required in order to investigate the equilibrium form experimentally. In the experiment, droplets of a supersaturated solution in gelatin are dispersed using a spray, and the morphological changes of the crystals formed in the droplets enclosed in the gelatin are recorded. The final morphology, i.e. that after which no further morphological changes occur, can be assumed to represent the equilibrium form. Taking NaCl as an example, the initial morphology is of dendritic form, but after transformation it eventually takes a simple cubic form, which is regarded as the equilibrium form under the given conditions. When a mother liquid is trapped in a crystal as an inclusion and growth proceeds on its inner wall, forming a negative crystal, the morphology may be considered to be representing the equilibrium form. It is also possible to carry out experimental investigations under higher temperatures in an electron microscope. The equilibrium form is not just polyhedral in type, bounded by low-index crystal faces [9], it may take a form bounded by flat and curved faces, depending on the given conditions.

It is difficult to measure the surface free energy of a crystal directly. A suitable method that is often adopted involves evaluating the surface free energy by calculating the energy required to cut a bond. Following this type of procedure, Wolff's "broken bond model" [10] predicts equilibrium forms. In the model, the surface free energy of a face, σ_{hkl}, is calculated based on the strength, type, ionicity, and coordination of the dangling bond involved in the bonding. The most important PBC vector, <*uvw*>, in surface free energy can also be deduced, using the σ_{hkl} value for several faces. Using this method, we may systematically and relatively easily deduce the equilibrium form of a real crystal, for complicated as well as simple structures. The method takes into consideration the Wulff plot, the Stranski concept of attachment and detachment energies, and the HP PBC theory.

In Fig. 4.6, the result obtained by the Wolff construction illustrates how the equilibrium form of a crystal consisting of atoms A and B will change depending on the changing ionicity of the A–B bond.

4.4 Growth forms

4.4.1 *Logical route for analysis*

The *Tracht* and *Habitus* exhibited by real crystals vary greatly depending on the perfection of the crystals and the growth environments and conditions. We refer to these "growth forms" as plural (as opposed to the singular structural and equilibrium forms), since they may vary during the growth process of a particular crystal, or they might be different among crystals of the same species formed under different conditions. It is possible to analyze what sorts of growth environment and conditions may influence crystal forms using the structure form and the equilibrium form as criteria.

The growth forms of polyhedral crystals appear as a result of different normal growth rates R of different crystal faces or among different, crystallographically equivalent, faces. Crystal faces with large R will disappear; only those with small R will survive. When a crystal reaches an equilibrium state, the crystal will be bounded by crystal faces with the smallest surface free energy γ, namely with minimum R. This is the equilibrium form. However, before reaching such a state, the crystal will exhibit different *Tracht* and *Habitus* determined by the relative ratio of R values (Fig. 4.7). These are the growth forms. Therefore, in order to understand

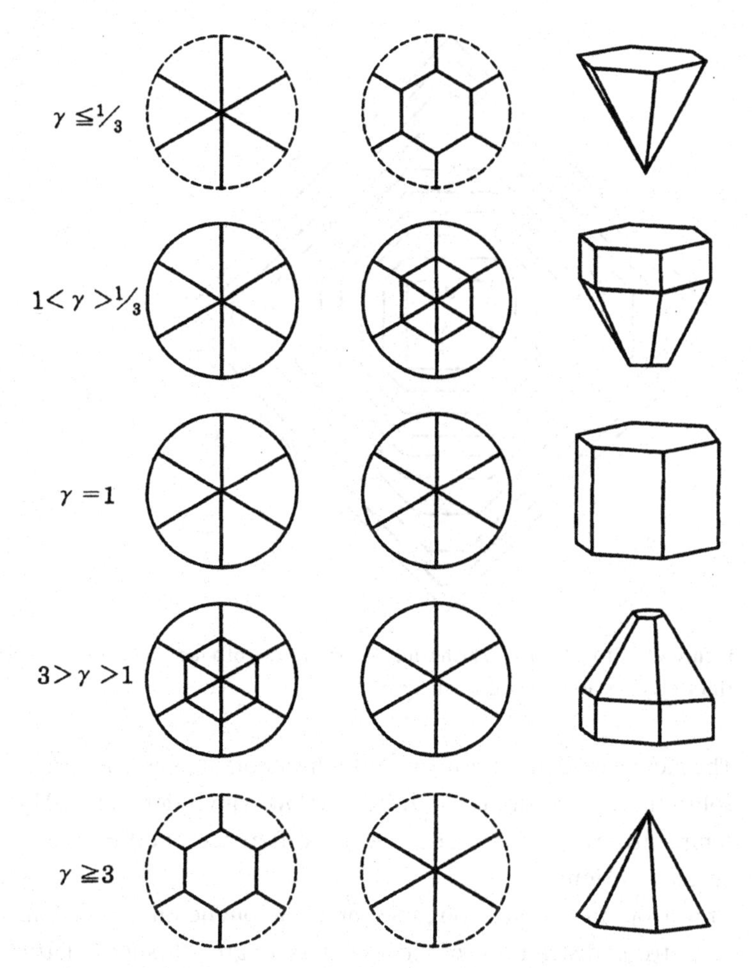

Figure 4.6. Wolff's equilibrium form [10]. AB composition of wurtzite type. The equilibrium form changes depending on the ionicity in the A–B bonding.

the origin of the variation of *Tracht* and *Habitus* in growth forms, it is necessary to analyze which factors involved in ambient phases produce which sorts of effects upon *R*.

The structural and equilibrium forms of crystals are predicted assuming that the crystal is perfect and that the ambient phase is isotropic. Growth forms, however, describe real crystals containing lattice defects growing in a real ambient phase. We should therefore consider the following factors, which may affect the growth forms.

(1) The structure of the ambient phases, i.e. the differences in the melt phase, the solution phase, and the vapor phase, and the difference of the solute–solvent interaction energies in solution growth.

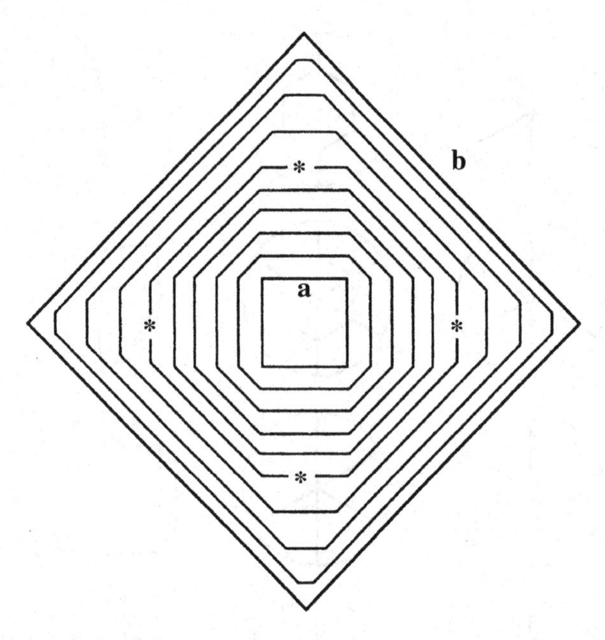

Figure 4.7. Growth forms. The normal growth rates R of **a** faces increase at *, whereas those of the **b** faces remain unchanged.

(2) The factors inducing anisotropy into the ambient phase: flows in solution, such as laminar or turbulent flow, convection induced by temperature difference, concentration difference, or difference in surface tension.
(3) Anisotropy in interface roughness and in a roughening transition.
(4) Anisotropic distribution of active centers for growth, such as lattice defects, which contribute to growth.

It should be noted that there may be other possible factors affecting the normal growth rate of crystal faces.

4.4.2 *Anisotropy involved in the ambient phase*

Let us start by examining the origin of malformed *Habitus* that deviates greatly from the structural or equilibrium forms.

As a representative example, let us investigate the *Habitus* variation of an ionic crystal growing from an aqueous solution in a beaker. In ref. [11], which was published in the seventeenth century, we see a sketch showing alum crystals grown on the bottom of a beaker taking triangular or hexagonal platy form, in contrast to the typical octahedral form of an alum crystal growing on a string in the solution, in spite of the fact that all these crystals are bounded by {111} faces only. It is clear that the reason is in the anisotropy involved in the supply of solute component. In

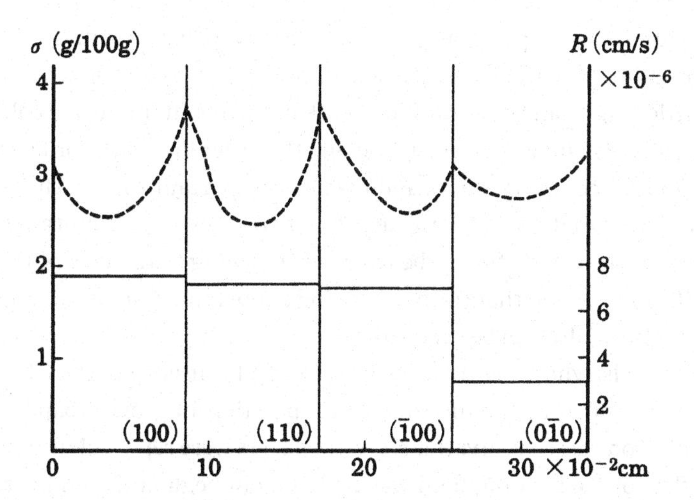

Figure 4.8. Growth rates R (solid lines, right axis) and distribution of surface supersaturation σ (dotted lines, left axis) of four {100} faces of a NaCl cubic crystal growing between two glass plates (Bunn's experiment) [12].

an isotropic environment, crystallographically equivalent {111} faces develop equally, but in an anisotropic environment the normal growth rate *R* of crystallographically equivalent faces may become anisotropic.

Figure 4.8 is taken from experimental results due to Bunn [12], who measured the growth rate and supersaturation distribution on {100} faces growing in a supersaturated aqueous NaCl solution sandwiched between two glass plates. Since the crystal is sandwiched in this way, the crystal shows a rectangular, not cubic, form. The figure clearly demonstrates that the growth rates *R* of the four crystallographically equivalent {100} faces are not the same, and that the growth rates *R* of a face with smaller Berg effect (with a smaller concentration difference between the center and edges of a face) is as small as two-fifths of that of other crystallographically equivalent faces. It has been confirmed with alum and $Ba(NO_3)_2$ crystals that the degree of the Berg effect varies due to the flow rate of the solution over a crystal face, and that the difference in the driving force between the edges and the center of a face becomes smaller as the flow rate increases [13]; see also ref. [1], Chapter 3.

From the results of analysis of the structural form of quartz crystal and the growth rate measurements on synthetic quartz, it has been well established that the difference in growth rates of $r\{10\bar{1}1\}$ and $z\{01\bar{1}1\}$ is small, $R_{10\bar{1}1} \leq R_{01\bar{1}1}$. However, it is common in natural quartz that *r* grows slightly larger than *z*, whereas $r \gg z$ in synthetic quartz. Although there may be a slight difference depending on growth conditions in the development of *r* and *z*, examples showing the reverse relation, $z > r$, have been reported only exceptionally. It has recently been confirmed that if the position of a quartz crystal with $r>z$ is reversed while growing in an autoclave, the crystal takes a form with $r<z$, associated with the reversed partitioning of an Fe ion and the depth of the resultant purple color [14]. On inspection of the structural form, we see that the difference in the order of morphological importance is very small, and r is more important than z. This experiment demonstrates that the growth rate R of two faces whose morphological importance differs by very little can be reversed according to the flow of the solution over the faces, and also that the partitioning of the impurities is influenced to such an extent that it also can be reversed.

In natural crystals, whose growth processes cannot be directly observed, the difference in R is recorded as the difference in separation in growth banding (see Chapter 6). Based on these observations, several papers were reported in which the direction of flow of ore-forming fluid was evaluated in pegmatite and hydrothermal veins. In many cases in which the natural mineral crystals exhibited extensively malformed *Habitus* from that predicted by the structural form, the malformation could be considered to be due to the remarkable anisotropy involved in the environmental conditions, such as the directional flow of the solution, which is similar to the situation of growth of NaCl from solution in between two glass plates, as discussed above.

4.4.3 *Whiskers*

Crystals showing an extremely elongated form in one direction are called whiskers. Whiskers may appear in any type of crystal, irrespective of whether the structural form is predicted to be prismatic or cubic. Special attention was first paid to whiskers when it was demonstrated that Sn whiskers grown through the coating of electric wire have a plastic strength close to that of a perfect crystal. However, mineral crystals exhibiting whisker morphology were observed long before this, examples being jamesonite ($Pb_4FeSb_6S_{14}$) and millerite (NiS). Hair silver, the name given to the elongated form of native Ag, is another example of a natural whisker. Various forms, such as coils (see Fig. 2.2(a)) and ropes, are also seen. Since whiskers have high plastic strength, they are often used as composite materials.

These remarkably elongated forms, which are not expected from the structure, have their origins in the strong anisotropy involved in either environmental conditions or in growth sites. Many models have been proposed as possible growth mechanisms of whiskers, but we refer to the following as a well established example.

When a solid particle of Au is placed on a Si substrate and SiI_2 vapor is supplied in the heated system, whiskers of Si with Au droplets at the tips are formed. We know these are whiskers of Si, because only Si (S) grows in eutectic liquid droplets of Si–Au (L) formed by Au and Si supplied in the vapor phase (V). The Au–Si eutectic liquid droplet provides the unique growth site. This mechanism is called the VLS (vapor–liquid–solid) mechanism [15]. The reason why only Si, and not Au, grows is due to thermodynamics, which states that the phase able to grow in a eutectic liquid is determined uniquely by the composition of the liquid phase. Si crystals take whisker forms as an inevitable result of the limited growth site within the liquid droplet. Various materials and systems other than Si have been known to form whiskers by the VLS mechanism. In all cases, a solidified liquid droplet is present at the tip of a whisker. The growth of the whiskers occurs at the roots.

If supersaturated NaCl (or KCl) aqueous solution is kept in a wineskin or cellophane bag in the shade, numerous NaCl (or KCl) whiskers grow on the surface of the wineskin after a few days. The explanation is that a supersaturated solution is transported by capillary action to the surface of the wineskin, which rapidly attains a highly supersaturated state, and crystallization starts. As a result, hollow tube whiskers are formed, and the supersaturated solution is transported to the tip of the whisker until the point at which the tube is closed, resulting in the formation of whiskers. There is only one growth site, and so this whisker grows at the tip [16].

The growth of Sn whiskers through the coating of electric wire occurs at the roots. Various models have been suggested to explain the mechanism of root growth of whiskers, VLS being one example. The following models also describe this growth:

(1) whiskers grow like toothpaste squeezed from a tube, because only one screw dislocation in a small crystallite acts as an active center for growth, and no growth takes place at grain boundaries, or
(2) one-directional growth results at a twin re-entrant corner (see Section 7.2) [17].

If an edge of a solid Ag_2Te or Ag_2Se crystal is pointed and electric current is allowed to pass through an electrode on the solid, whiskers of Ag start to grow from the pointed corner, since the Ag is supersaturated [18]. When the electric current is reversed, the Ag whisker is absorbed into the solid as the Ag in the solid becomes undersaturated. This is an example of root growth.

If screw dislocations outcropping on only one out of three crystallographically equivalent faces behave as active growth centers, whereas those on the other two faces are inactive for some reason, elongated whiskers will be formed in one direction. If screw dislocations on two faces are active as growth centers, a thin platy *Habitus* will appear. Depending on the growth conditions, needles and thin platy KCl crystals can coexist with ordinary cubic crystals. However, this has still not clearly answered why some screw dislocations are active growth centers whereas others are inactive.

In a crystal bounded by F and S faces, a kinetic roughening transition occurs on the morphologically less important S face prior to on the F face. Kinetic roughening may occur by adsorption of impurity elements. In the case of oxide crystals, a de-oxidation process may be involved to retain electronic neutrality by adsorption of impurity ions with different charges. This is equivalent to the transformation of an S face into a K face by breaking the PBCs. On F faces containing more than two PBCs, impurity ions behave differently from those on S faces; a de-oxidation process is not involved, and so F faces behave as smooth interfaces, whereas S faces transform into K faces. Therefore, a prismatic crystal grows in one direction and takes a needle form. If we take the Berg effect into consideration, two-dimensional nucleation occurs more easily along the edges between K faces and F faces, which are rougher than at the center of the face, leading the growth in the prismatic direction. On the other hand, the growth of F faces is controlled by the layer growth mechanism, and the crystal protrudes along the edges of the K faces. This leads to the formation of hollowed whiskers. It has been shown that with increasing impurity content, the terminal faces (i.e the faces appearing on the tips of prismatic or needle crystals) of prismatic SnO_2 and TiO_2 crystals change from flat {111} to rounded faces, and further skeletal faces occur due to preferential nucleation along the edges, and eventually hollowed needle crystals appear [19]. Figure 4.9 shows a series of scanning electron microscope (SEM) photographs showing this change. If a similar roughening transition occurs only on the side faces of platy crystals, and if the basal plane behaves like a smooth interface, a thin platy crystal will be formed. Since roughening transitions occur under higher temperature and higher driving force, in addition to the effects of impurities it is anticipated that a malformed needle or thin platy *Habitus* will appear more frequently at higher temperatures and supersaturation conditions.

Crystals with a hollow tube prismatic form and their deviations (such as scrolls or ice cream cones) have been observed among clay minerals having a sheet structure and in fullerene, encouraging research into the origin of such unusual forms. Imogolite, a clay mineral, $(OH)_6Al_4O_6Si_2(OH)_2$, occurs as a hollowed tube-like prismatic crystal; kaolin, $Al_2Si_2O_5(OH)_4$, takes the form of scrolled crystals; and chrysotile, which is formed by a hydrothermal treatment of serpentine, $Mg_3Si_2O_5(OH)_4$, looks like ice cream cones (Fig. 2.2 (b)) [20], [21]. Fullerene (C_{60} with a soccer ball form) exhibits a hollowed fibrous form called carbon nanotubes [22]. These nanotubes are of an appropriate diameter to accommodate metal atoms, and so efforts have been made to utilize them as potential materials for one-dimensional conductors or sensors.

It is not easy to provide an explanation of the origin of these unusual forms simply based on preferential growth at the tip of a prismatic crystal. Various models have been proposed, for example:

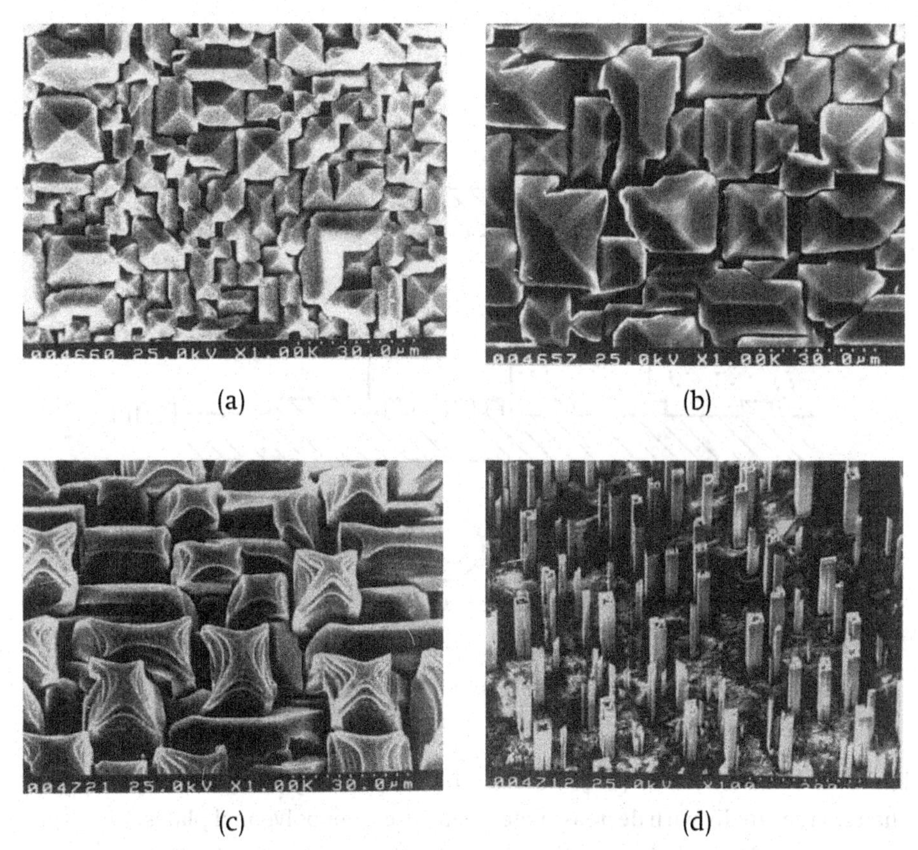

Figure 4.9. Scanning electron microscope photographs showing the roughening transition of {111} faces of a TiO_2 crystal and the formation of hollowed needle crystals as impurities are added [19]. Growth occurs by liquid phase epitaxy on a (001) substrate. Fe_2O_3 is added as an impurity in the following amounts: (a) 0%, (b) 1.3 mol%, (c) 3.1 mol%. (d) Low-magnification photograph of (c).

(1) growth layers advance in a cylindrical form owing to the structural strain involved in the sheet structure forming a scroll form, but, as soon as dislocation is induced, the strain is released and the scroll changes to platy;

(2) ice-cream cone shapes appear because two-dimensional nucleation takes place on the inner surface of a scroll.

The validation of these proposed models is a subject for future discussion.

4.4.4 Habitus *change due to temperature*

As can be seen in the Nakaya diagram in Fig. 3.23, snow crystals grown in a reduced vapor supply (smaller driving force) appear as polyhedral crystals

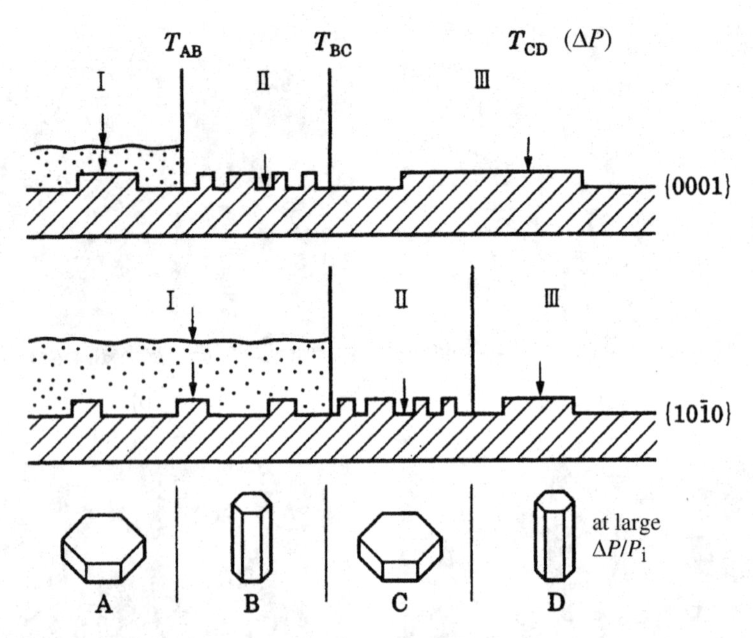

Figure 4.10. Kuroda's model explaining the repeated *Habitus* change of snow crystals [23]. Shaded areas are crystals; dotted areas are QLLs (quasi-liquid layers).

bounded by {0001} and {10$\bar{1}$0} faces. The *Habitus* of polyhedral snow crystals changes repeatedly with decreasing temperature from polygonal plates, to polygonal prisms, polygonal plates, and polygonal prisms. The reason for these repeated changes was unknown until Kuroda [23] solved the mystery by assuming the presence of a quasi-liquid layer (QLL) with different thicknesses on respective faces. His model is shown in Fig. 4.10.

The thickness of the QLLs is different on different faces, and becomes thinner as the temperature is decreased. Thus, the temperature at which a QLL disappears depends on the face. When the surface is covered by QLLs, crystal growth is regarded as a solution or melt phase growth, whereas on a naked surface it is due to vapor growth. Thus, the repeated *Habitus* change may be explained. The presence of QLLs was confirmed by ellipsometry.

4.4.5 Tracht *change*

MAJOR FACTORS

Tracht change differs from change in *Habitus* in that it describes a change in the form of a crystal grown in an isotropic environmental phase through the combination of different faces and the relative sizes of respective faces. Therefore, *Trachts* are forms determined by the relative ratio of the normal growth rate R of the different crystal faces present on the surface of a growing crystal. The identification of the factors involved in the environmental phase and how they influence R form the key points in the analysis.

Since surface structures and growth mechanisms are different at rough and smooth interfaces, the effect of environmental factors will also be different. When a crystal bounded by both smooth and rough interfaces grows, the rough interface disappears as growth proceeds, unless some factors operate to suppress the growth rate of the rough interface, and the crystal takes on a *Tracht* bounded by a smooth interface only. If factors suppress the growth rate R of a rough interface, the crystal will temporarily take on a *Tracht* bounded by rough and smooth interfaces. In extreme cases, crystals may take on a *Tracht* bounded by rough interfaces only. The most likely factor that suppresses the growth rate R of a rough interface is impurity precipitation covering the whole surface. The growth is intercepted, leading to the development of the rough interface. But this face will disappear as soon as growth continues without the effect of impurities.

The *Tracht* of polyhedral crystals is, in general, the growth form of a crystal bounded by smooth interfaces and represents the expected morphology when the crystal grows under a driving force below $\Delta\mu/kT^*$, namely under the condition where the principal method of growth is the spiral growth mechanism, as already explained in Chapter 3. Therefore, the normal growth rates R of faces appearing on the crystal are determined by the height of the growth spiral layers, h, their advancing rates, v, and the step separation, λ_0. If we could find out which factors that are involved in environmental conditions affect h, v, and λ_0, we could explain the origin of the *Tracht* variation. This analysis will be given in Chapter 5; for now, we will just point out the following major factors.

(1) Difference in the environmental phases. Since the interface roughness will be different for the same crystal species depending on whether the crystal was grown from the melt, solution, or vapor phases, different growth forms are expected for different environmental phases. This implies that the *Tracht* of the same crystal species will depend on the structure of the environmental phases, the degree of condensation, and the solute–solvent interaction.
(2) In solution or vapor phase growth, the strength of the solute–solvent interaction energies, namely the species and types of solvent component and transport agent.
(3) The growth temperature and driving force, which affect interface roughness.
(4) The surface reconstruction.
(5) Impurities and the solvent component, which affect step free energy γ.

METHODOLOGY

Many investigations have been carried out since the seventeenth century that analyze the *Tracht* variations of growth forms. These investigations fall into two principal categories: (i) experiments designed to observe *Tracht* variations in relation to growth conditions, and (ii) the analysis of *Tracht* variation of mineral crystals with respect to the geological environment in which they occur.

It had been observed that drastic changes in *Tracht* occurred in the presence of impurities, and thus experiments were originally designed that focused on impurity adsorption. Wells, however, considered other factors, and performed systematic investigations that demonstrated that solvents, growth temperatures, and supersaturation are the main factors involved in changing the *Tracht* of ionic crystals growing in the solution phase [24]. However, Wells concluded that if the observations were treated statistically, the growth forms reflected either the structural or the equilibrium form. A vast amount of data relating to the crystal morphology of natural minerals were collected, based on which the order of morphological importance was correlated to lattice types, and the relationship between *Tracht* and the growth environment and conditions was deduced based on observations of museum samples. A further type of investigation was made in which *Tracht* variations were observed *in situ* in relation to where they occurred; these results were used as guides for use in ore prospecting and in analyzing the genesis of rocks (a scientific discipline referred to as typomorphism) [25]. We will summarize the results achieved during the twentieth century in the following sections.

IMPURITIES

It was during the late eighteenth century that Romé de l'Isle [26] observed, for the first time, the *Tracht* of a crystal that had changed drastically after the addition of a small amount of impurity. He saw how NaCl crystals exhibit an octahedral form when a small amount of urea is added to the aqueous solution, in contrast to the simple cubic form commonly observed in pure solution. Motivated by this observation, many investigations were carried out, from the addition of inorganic impurities to the addition of dyes having a high adsorption power, to observe the effect caused on the *Tracht* by the impurities [27]. It has been confirmed that *Tracht* changes can occur over a wide range of impurity addition, from parts per hundred to parts per million order. The following conclusions were reached.

(1) Impurity ions adsorb selectively on a particular face and cover the whole surface, leading to retardation of the growth rate R of the face and a change of *Tracht*.
(2) There are two cases for which selective adsorption of impurities occurs: (i) over the entire surface by epitaxy, and (ii) along the steps of growth layers on the face.

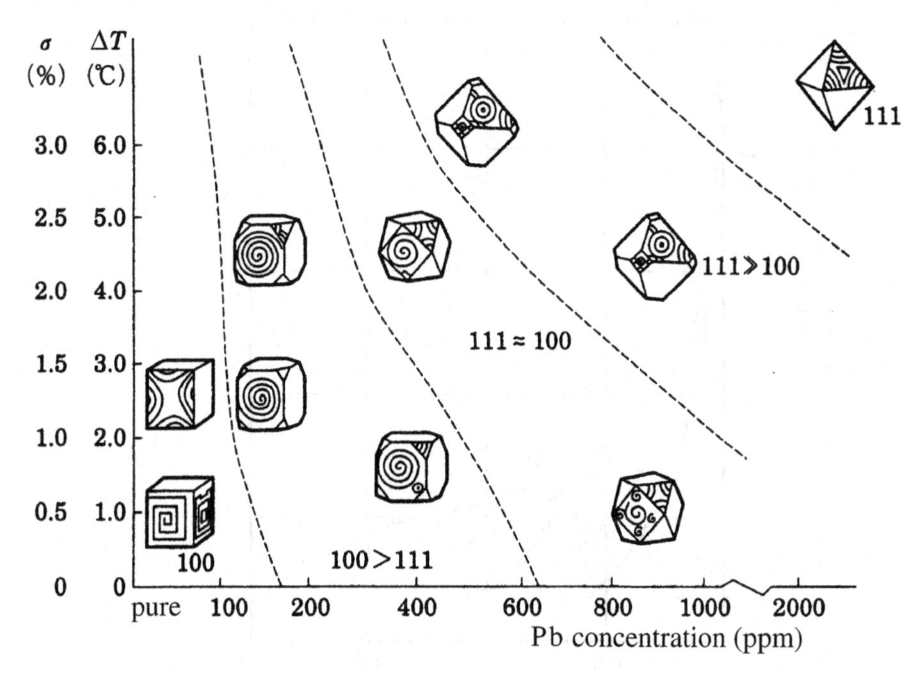

Figure 4.11. Morphodrom of KCl crystals [28]. The vertical axis is the supersaturation; the horizontal axis is the concentration of Pb ions.

(3) Impurities may act as either a promoting or a retarding force to growth rates.

When dyes are used as the impurity component, it is seen that R is diminished by the adsorption of the impurities, since only the growth sectors of faces appearing by impurity adsorption are colored. Crystal faces that appear due to impurity adsorption are, in general, the K face (by PBC analysis), indicating that the impurities adsorb at kink sites on a rough interface, resulting in a diminished growth rate. This may, in some cases, change the growth mechanism of the face. Figure 4.11 shows a morphodrom of KCl crystals grown in aqueous solution when Pb ions are added. Changes in *Tracht* and surface microtopographs of crystal faces are illustrated in relation to the driving force (supersaturation) and the concentration of Pb ions. As concentration of Pb ions is increased, the *Tracht* changes from cubic to octahedral, and the surface microtopograph of the {111} face, which is a K face by PBC analysis, changes from a hopper form to a face showing spiral growth layers that correspond to an F face. At the same time, spiral steps on the {100} face change their form from square with <001> edges to square with <110> directions.

Figure 4.12 is a morphodrom of NaCl crystals when $Fe(CN)_6$ is added as an impurity, and it shows the effect of impurity changes depending on the driving force. *Tracht* can change, however, due to driving force alone [29]. In these examples,

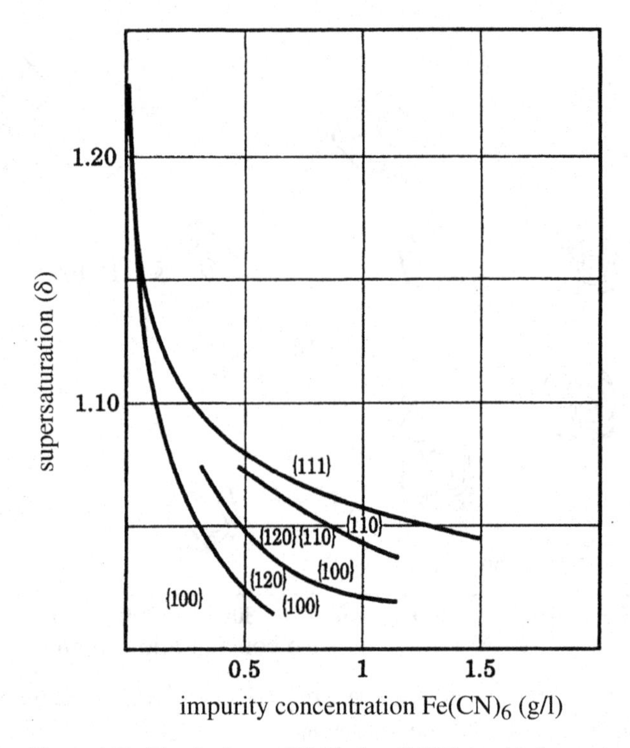

Figure 4.12. Morphodrom of NaCl when $Fe(CN)_6$ is added as an impurity [29].

impurities are adsorbed all over the surface, and act to reduce *R*. On an F face, however, impurities adsorb along steps and the advancing rate of the steps is retarded. In such cases, even a small amount of impurity may cause a significant effect.

As an important example of the impurity effect, it is necessary to give a brief explanation of the tapering phenomenon of crystals. Tapering is a phenomenon that describes how a prismatic crystal narrows in size toward the tip, and it has often been observed on piezoelectric crystals growing from the aqueous solution, such as ADP ($NH_4H_2PO_4$) or KDP (KH_2PO_4). Tapered crystals of similar type are also observed among prismatic crystals of natural quartz. It is understood that impurity ions such as Fe^{3+} are selectively adsorbed along a certain direction of growth layers on the prism faces of these tapered crystals, retarding the advancing rate of the growth steps, which results in a pile up of steps, so producing a high-index tapered face. The dog-tooth-like *Habitus* of calcite ($CaCO_3$) crystals bounded by scalenohedral $\{hk\bar{\imath}l\}$ faces is possibly formed through tapering due to impurity adsorption. When calcite crystals are synthesized from pure solution, the crystals are always rhombohedral, bounded by $\{10\bar{1}1\}$ faces, whereas a dog-tooth *Habitus* appears when an organic compound is added as an impurity. On $\{hk\bar{\imath}l\}$ faces which determine the dog-tooth *Habitus*, only striations parallel to the edges with $\{10\bar{1}1\}$ faces are observable; step patterns due to the spreading of growth layers on the face have never been observed.

The structural form of brookite (TiO_2) is expected to be bounded by {210} and {111}, both being F faces by PBC analysis, but the actual growth form observed is platy *Habitus* bounded by largely developed {100}, which is an S face. The misfit ratio between the PBC on the $(01\bar{1}0)$ face of quartz and that on (100) of brookite is the smallest among any misfit ratios between the two crystal species. From this, it was found that the platy *Habitus* of brookite arose because quartz adsorbed in an epitaxial relation on {100} of brookite, thus diminishing the growth rate *R* of (100) [30].

In addition to the cases discussed above, in which the addition of impurities decreases the growth rate, there are also cases in which the addition of impurities accelerates the growth rate. As already described in Section 4.4.3, where SnO_2 or TiO_2 crystals are synthesized in the high-temperature solution phase, if trivalent ions, such as Fe^{3+}, having a similar ionic radius to those of Sn^{4+} and Ti^{4+} are added, an extremely elongated prismatic or whisker morphology results. This is because {111} (corresponding to an S face by PBC analysis) transforms into a K face by the addition of the impurity ions. This is because the transformation from an S face to a K face occurs by breaking bonds in PBCs to maintain the electronic neutrality through the substitution by impurity ions with a different charge [19]. On the other hand, bonds in PBCs on {110}, an F face, will not be broken by this substitution, and the face behaves as an F face. However, the adsorption of impurity ions affects the advancing rate of growth steps on {110}, which diminishes. Thus, the same impurity ions can have opposing effects upon the growth rate.

Since the most drastic change in *Habitus* and *Tracht* is effected by the addition of impurities, the relationship between impurity addition and change in *Tracht* may be applied to determine the absolute structure of a crystal, such as the determination of the positive or negative axis of a polar crystal, or the right-handed or left-handed structure of an enantiomorphic crystal. Unless applying special techniques, it is not easy to determine the polarity of a crystal by the X-ray diffraction method. Generally, the axial direction is determined using either piezoelectricity or pyroelectricity. A method called tailor-made habit control has been designed to determine the absolute structure of a crystal, by observing the *Habitus* change in the presence of appropriate impurities which fit either direction of a polar axis [31].

AMBIENT PHASES AND SOLVENT COMPONENTS

It is generally observed that crystals from the same species exhibit widely different *Tracht* and *Habitus* depending on whether they grow from the vapor or solution phases.

Hematite (Fe_2O_3) crystals grow around volcanic fumaroles that formed at the most recent stage of volcanic activity. This natural process corresponds to the growth of hematite by the chemical vapor transport (CVT) method, and the crystals of this form from all over the world show a thin platy *Habitus* with a largely developed {0001} face, bounded by narrow $\{10\bar{1}1\}$ and $\{10\bar{1}0\}$ faces. In contrast to this *Habitus*, hematite crystals occurring in vein type or contact metasomatic ore deposits, which grew from a hydrothermal solution, characteristically show thick platy or nail-head *Habitus* bounded by well developed $\{10\bar{1}1\}$ and $\{10\bar{1}0\}$ faces. Corundum (Al_2O_3) crystals having the same crystal structure as hematite exhibit similar tendencies. Thin platy crystals grow on the wall of a vessel from vaporized phase by CVT, when crystals are synthesized by flux (high-temperature solution) method, using fluoride having high vapor pressure as flux. In contrast, corundum crystals grown in solution take on a thick platy *Habitus*. Natural corundum crystals formed by contact metasomatism or regional metamorphism characteristically show thick platy to prismatic *Habitus*, rather than a thin platy one. The big difference in *Habitus* observed between crystals grown from vapor and solution phases is related to the difference observed in the step separation of spiral growth layers between the two phases, which will be described in Section 5.6. Step separation observed in crystals grown from the vapor phase is up to 100 times wider than that observed in solution grown crystals.

High-temperature quartz crystals occur as phenocrysts in acidic igneous rocks with a high SiO_2 content, and show hexagonal bipyramidal *Habitus* bounded by only a $\{10\bar{1}1\}$ face, with no $\{10\bar{1}0\}$; this *Habitus* has been accepted as the typical *Habitus* of high-temperature quartz. However, when quartz crystals are synthesized above the transition temperature 573 °C in a hydrothermal solution, the crystals exhibit a hexagonal prismatic *Habitus* with $\{10\bar{1}0\}$ associated faces. The hexagonal bipyramidal *Habitus* simply represents a *Habitus* of high-temperature quartz grown in an ambient phase of acidic silicate solution.

As will be described in Chapter 9, natural diamond crystals exhibit the same *Habitus* as predicted from the structure form, on which the {111} faces always behave as smooth interfaces, whereas the {100} faces always behave as rough interfaces. Diamond crystals synthesized under high-temperature, high-pressure conditions from the solution phase, with metals or alloys as the solvent component, take cubo-octahedral *Habitus*. Although CVD diamonds grown from the vapor phase also take cubo-octahedral *Habitus*, the order of morphological importance of {111} and {100} is reversed as compared with that of diamond crystals grown under the diamond stable region, i.e. high-temperature, high-pressure conditions, in both natural and synthetic crystals (see Chapter 9).

The above examples describe the *Habitus* changes due to the difference between the vapor and solution phases, and between solvent components in solution growth. Many examples are described in refs. [24] and [27], which consider different *Habitus* of the same crystal species grown from aqueous and organic solutions. Since controlling the morphology of tiny crystals is an important aspect in utilizing the numerous tiny crystals, how to control the *Habitus* of these crystals is still an important aspect in industry even today.

Even if crystals grow from the same aqueous solution, there are differences in *Habitus*. $NaClO_3$ crystals, for example, grow easily as polyhedral crystals, whereas NH_4Cl crystals always grow as dendrites, and NaCl crystals appear as hopper crystals. If Pb or Mn ions are added, cubic crystals of NaCl bounded by flat {100} faces may be obtained quite easily, but if NaCl is grown in pure solution all crystals take a hopper form, unless great care is taken to keep the supersaturation very low. These differences occur because the solute–solvent interaction energies, and, as a result, the values of $\Delta\mu/kT^*$ and $\Delta\mu/kT^{**}$, are different for different crystals.

The reasons why we have *Habitus* variation for vapor and solution phases, and for different solvents when the crystals grow from the solution phase, can therefore be understood in terms of the α_G factor of steps because of the different solute–solvent interaction energies.

TEMPERATURE, PRESSURE, AND DRIVING FORCE

As discussed in Chapter 3, as the temperature or driving force increases a smooth interface transforms into a rough interface. When a crystal is bounded by many faces with different orders of morphological importance, it can be seen that a face with a lower order of morphological importance will show a roughening transition at a lower temperature and driving force than does a face with higher order of morphological importance. As a result, it is expected that changes in *Habitus* and *Tracht* accompanying changes of temperature and driving force will be observed. Pressure will also affect the *Tracht* and *Habitus*, owing to its relation with the driving force, and the degree of impurity concentration will also show a temperature dependence. Changes of this type are seen not only in changes from polyhedral to hopper to dendritic shown in Fig. 3.21, but also changes in *Habitus* and *Tracht* of polyhedral crystals.

TRACHT *CHANGES ASSOCIATED WITH GROWTH*

Figure 4.13 shows statistical results on the frequency of appearance of different *Tracht* of a few hundred pyrite crystals occurring in approximately a handful of clay, depending on crystal size. The pyrite crystals were formed together with clay minerals by the reaction of hydrothermal solution with rocks near the sea bottom, and so can be regarded as being crystallized under nearly identical conditions of temperature, pressure, and driving force. Figure 4.13 indicates that smaller crystals show a higher frequency of appearance of simple cubic form bounded by

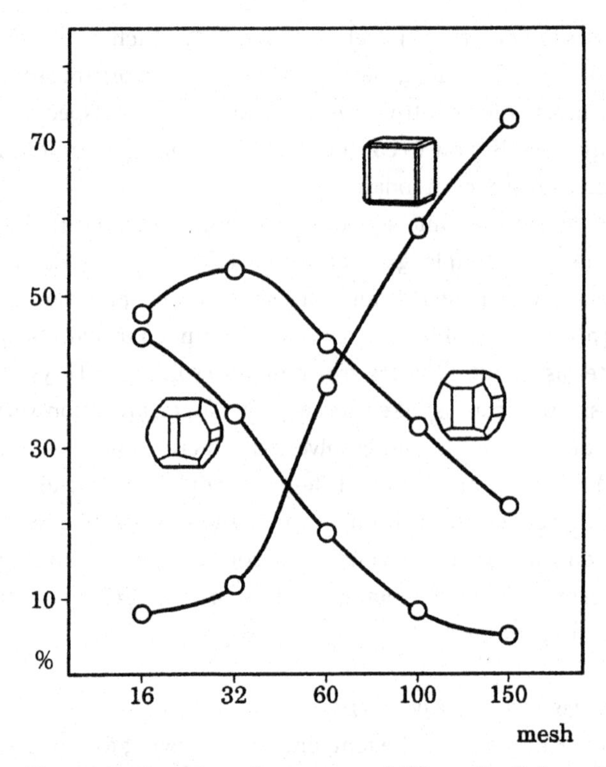

Figure 4.13. Frequency of appearance of different *Tracht* depending on crystal size. Statistical measurements were made on a few hundred pyrite crystals occurring in approximately a handful of clay. The mesh axis denotes the number of openings in a unit area of the mesh filter. See ref. [2], Chapter 2.

only {100} faces, and, as the grain size increases, that the frequency of appearance of pentagonal dodecahedral form mainly bounded by {210} faces increases. This implies that the *Tracht* changes from cubic to pentagonal dodecahedral as the crystals grow larger. This change in *Tracht* associated with growth is represented as striations on pentagonal dodecahedral faces. Figure 4.14 indicates that pentagonal dodecahedral faces appear by the piling up of steps of rectangular growth layers on {100} faces as the grain sizes increase. Pentagonal dodecahedral faces are characterized by the development of striations parallel to the edge with a {100} face, and also consists of not only {210} but also pentagonal dodecahedral faces such as {430}, {540}, etc. The most well developed {210} face is characterized by striations only, and does not show growth layers parallel to the face, as can be seen in Fig. 11.2. In contrast, there are cases in which {210} faces exhibit striations perpendicular to the edge; for example, in pyrite crystals from different modes of formation such as those from Elba. In this case, striations appear due to freely developed elongated growth layers on the {210} face. The same {210} faces grow differently depending

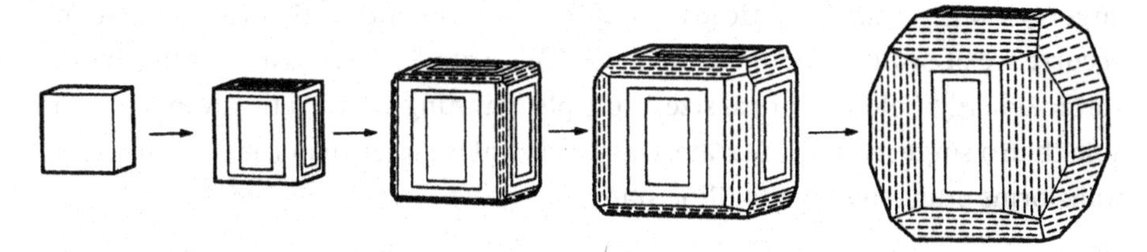

Figure 4.14. Changes observed on *Tracht* and surface microtopographs of {100} and {210} faces of pyrite crystals as growth proceeds (see ref. [1], Chapter 2).

on growth conditions. Similar changes are noticed on {10$\bar{1}$0} faces between natural and synthetic quartz crystals.

Similar observations of *Tracht* variation as growth proceeds were also made on garnet crystals in regional metamorphic rocks (a {110} *Tracht* at the earlier stage changes to {211} as the crystals grow in size) [32]. Smaller crystals show a dodecahedral *Tracht* bounded by {110}, and as sizes increase by growth the *Tracht* changes to tetragonal trioctahedron bounded by {211}. In contrast to the fact that growth layers of rhombic form are universally observed on {110} faces, {211} shows only striations parallel to the edges with {110}.

4.4.6 *Ultra-fine particles*

The discussions summarized in Sections 4.4.2–4.4.5 relate to the morphologies of crystals of size down to a few millimeters. Will extremely tiny crystals of sizes ranging from a few to 100 nm take polyhedral forms bounded by flat faces, or will they show forms bounded by curved faces due to the size effect? It was theoretically predicted that, as the size of solid materials decreases, a change is expected to occur in the physical properties. Also, considering the presence of ultra-fine particles in the cosmos, the morphology exhibited by ultra-fine particles attracts a special concern.

Investigations into ultra-fine particles may be regarded as a Japanese specialty. In Japan, there is a fifty-year-long history of investigation on the forms and structures of ultra-fine particles of metals and chemical compounds using gas evaporation techniques, electron microscopy, and selected area electron diffraction; the results are summarized in ref. [33]. In this reference, the *Habitus* and *Tracht* exhibited by single crystals, twinned, and multiply twinned particles are summarized in relation to the growth conditions on hcp metals (Mg, Zn, Cd, Be), bcc metals (Fe, V, Nb, Ta, Cr, Mo, W), group IVB elements (C, Si, Ge, Sn), and Sb and Bi. Surprisingly, ultra-fine particles of nanometer size show a polyhedral *Habitus* bounded by flat faces in most cases, and their *Habitus* and *Tracht* change depending on growth conditions, just as in the case of larger crystals. However, there are still many unsolved problems in understanding how these *Habitus* and *Tracht* changes occur. As compared to larger crystals, a characteristic feature shown by ultra-fine particles is the frequent occurrence of multiply twinned particles. This is understood as a size effect upon surface energy: below a critical size, multiple twinning is energetically more favorable than a single crystalline state. Once a crystal is larger than the critical size, it transforms into a single crystalline state.

4.4.7 *Factors controlling growth forms*

Examples of changes in growth form were introduced in Section 4.4.5 where we discussed the factors involved in causing the changes.

Since the growth forms of crystals are determined by various inter-relating factors, it is inevitable that to form a clear picture we would need to analyze the origin of each example case by case. However, by studying real examples, we see that the factors determining the forms of polyhedral crystals may be generalized as: (1) the relative normal growth rates of interfaces; (2) the interface roughness; (3) the fact that polyhedral crystals are forms exhibited by a crystal grown by the layer-by-layer growth mechanism or the spiral growth mechanism; (4) the degree of smoothness, which depends on crystallographic direction. Therefore, factors which affect these points will affect the growth forms of polyhedral crystals, and to understand them it is necessary to analyze the step patterns observed on crystal faces. In Chapter 5, we will analyze the problem by consideration of surface micro-topography.

References

1 A. Bravais, Les systemes formes par des pointes distribues regulierement sur un plan ou dans l'espace, *J. Ec. Polytech.*, **XIX**, 1850, 1–128
2 J. D. H. Donnay and D. Harker, A new law of crystal morphology extending the law of Bravais, *Am. Min.*, **22**, 1937, 446–7
3 P. Hartman, Modern PBC, in *Morphology of Crystals*, Part A, ed. I. Sunagawa, Dordrecht, D. Reidel, 1987, pp. 269–319
4 P. Bennema and J. P. van der Eerden, Crystal graphs, connected nets, roughening transition and the morphology of crystals, in *Morphology of Crystals*, Part A, ed. I. Sunagawa, Dordrecht, D. Reidel, 1987, pp. 1–75
5 X. Y. Liu and P. Bennema, Prediction of the growth morphology of crystals, *J. Crystal Growth*, **166**, 1996, 117–23
6 J. W. Gibbs, On the equilibrium of heterogeneous substances, in *The Scientific Papers of J.W. Gibbs, 1*, London, Longman Green & Co., 1906
7 P. Curie, On the formation of crystals and on the capillary constants of their different faces, *J. Chem. Edcn.*, **47**, 1970, 636–7 (translation of *Bull. Soc. Franc. Min. Cryst.*, **8**, 1885, 145–50)
8 G. Wulff, Zur Frage der Geschwindigkeit des Wachstums und die Auflosung der Kristallfachen, *Z. Krist.*, **34**, 1901, 449–530

9 R. Kern, The equilibrium form of a crystal, in *Morphology of Crystals*, Part A, ed. I. Sunagawa, Dordrecht, D. Reidel, 1987, pp. 77–206

10 G. A. Wolff and J. D. Broder, The role of ionicity, bonding and adsorption in crystal morphology, in *Adsorption et Croissance Cristalline*, CNRS, 1965, pp. 172–94

11 R. Hooke, *Micrographia*, London, Royal Society, 1665

12 C. W. Bunn, Crystal growth from solution, II, Concentration gradients and the rates of growth of crystals, *Disc. Faraday Soc.*, no. 5, 1949, 132–44

13 K. Onuma, K. Tsukamoto, and I. Sunagawa, Measurement of surface supersaturation around a growing K-alum crystal in aqueous solution, *J. Crystal Growth*, **98**, 1989, 377–83

14 V. S. Balitsky, H. Iwasaki, and I. Sunagawa, Growth morphologies and their computer simulations in quartz crystals synthesized under various growth conditions, collected abstract of ICCG 13/ICVGE 11, 2001, 345

15 R. S. Wagner and W. C. Ellis, Vapor-liquid-solid mechanism of single crystal growth, *Appl. Phys. Lett.*, **4**, 1964, 89–90

16 Y. Aoki, Growth of KCl whiskers on KCl crystals including the mother liquids, *J. Crystal Growth*, **15**, 1972, 163–6

17 R. S. Wagner, On the growth of germanium dendrites, *Acta Metall.*, **8**, 1960, 57–60

18 T. Ohachi and I. Taniguchi, Growth control of silver whisker on α-Ag_2Se and α-Ag_2Te, *Jpn. J. Appl. Phys.*, **8**, 1969, 1062

19 F. Kawamura, I. Yasui, and I. Sunagawa, Investigations on the growth and morphology of TiO_2 in the TiO_2-$Na_2B_4O_7$ system with and without impurities using a new LPE method, *J. Crystal Growth*, **231**, 2001, 186–93

20 K. Yada and K. Iishi, Serpentine minerals hydrothermally synthesized and their microstructures, *J. Crystal Growth*, **24/25**, 1974, 627–30

21 K. Yada and K. Iishi, Growth and microstructure of synthetic chrysotile, *Am. Min.*, **62**, 1977, 958–65

22 S. Iijima, Helical microtubules of graphitic carbon, *Nature*, **354**, 1991, 56–8

23 T. Kuroda, *Kinetik des Eiswachstums aus der Gasphase und seine Wachstumsformen*, Thesis, Braunschweig, Germany.

24 A. Wells, Crystal habit and internal structure, I, II, *Phil. Mag. Ser.* 7, **37**, 1946, 184–236

25 I. Kostov and R. I. Kostov, *Crystal Habits of Minerals*, Sofia, Pensoft, 1999

26 Romé de l'Isle, *Essai de Cristallographie*, Paris, 1772

27 H. E. Buckley, *Crystal Growth*, New York, John Wiley & Sons, 1951

28 Li Lian, K. Tsukamoto, and I. Sunagawa, Impurity adsorption and habit changes in aqueous solution grown in KCl crystals, *J. Crystal Growth*, **99**, 1990, 1156–61

29 M. Bienfait, R. Boistelle, and R. Kern, Formes de croissance des halogenures alcalius dans un solvant polaire, Les morphodoromes de NaCl en solution et l'adsorption d'ions etrangers, in *Adsorption et Croissance Cristalline*, CNRS, 1965, pp. 515–35, 577–94

30 P. Hartman, Habit variation of brookite in relation to the paragenesis, in *Adsorption et Croissance Cristalline*, CNRS, 1965, pp. 597–614

31 Z. Berkovitch-Yellin, J. van Mit, L. Addadi, M. Idelson, M. Iahav, and L. Leiserowitz, Crystal morphology engineering by "tailor-made" inhibitors, A new probe to find intermolecular interactions, *J. Am. Chem. Soc.*, **107**, 1985, 3111–22

32 A. Pabst, Large and small garnets from Fort Wrangler, Alaska, *Am. Min.*, **28**, 1943, 233–45

33 R. Uyeda, Crystallography of metal smoke particles, in *Morphology of Crystals*, Part B, ed. I. Sunagawa, Dordrecht, D. Reidel, 1987, pp. 367–508

Suggested reading

G. G. Lemmlein, *Morphology and Genesis of Crystals, Collected Papers of G. G. Lemmleim*, Moscow, Nauk, 1973 (in Russian)
E. I. Givargizov, *Highly Anisotropic Crystals*, Dordrecht, D. Reidel, 1986
I. Sunagawa (ed.), *Morphology of Crystals*, Parts A and B, Dordrecht, D. Reidel, 1987
I. Sunagawa (ed.), *Morphology of Crystals*, Part C, Dordrecht, D. Reidel, 1994
I. Kostov and R. I. Kostov, *Crystal Habits of Minerals*, Sofia, Professor Martin Drinov Academic Publishing House and Pensoft

5

Surface microtopography of crystal faces

Step patterns or etch figures which represent the final stage of growth or etching after the cessation of growth, respectively, are observed on flat crystal faces comprising a polyhedral crystal. We refer to these as surface microtopographs of crystal faces; they possess information at the atomic level relating to the mechanism of growth or dissolution and perfection of the crystal. This is because we can observe the spiral growth layers, either *ex situ* or *in situ*, with step heights of nanometer order. Since crystal faces are unique places where growth or dissolution of crystals occur, the analysis is directly connected to the concept of growth forms.

5.1 The three types of crystal faces

Crystal faces bounding a polyhedral crystal are broadly classified into three types according to their surface microtopographs: (i) those appearing as mirror-flat faces; (ii) those characterized by striations; and (iii) those showing rugged or rounded forms (see Fig. 5.1). If a crystal face which is large enough to control the *Habitus*, with a mirror-flat surface, is observed by methods capable of detecting differences in levels at the nanometer scale, step patterns, resembling the contour lines on a topographic map, can be seen. These characteristic features on crystal faces are called surface microtopographs or are referred to as the surface morphology.

Crystal faces that show only striations are called vicinal faces with high indexes. Most typically, they appear as side faces of polygonal growth hillocks developing on low-index flat crystal faces, or as high-index faces appearing between a set of neighboring flat crystal faces. These correspond to S faces in Hartman–Perdok (HP) theory. Vicinal faces grow in a vertical direction to the face

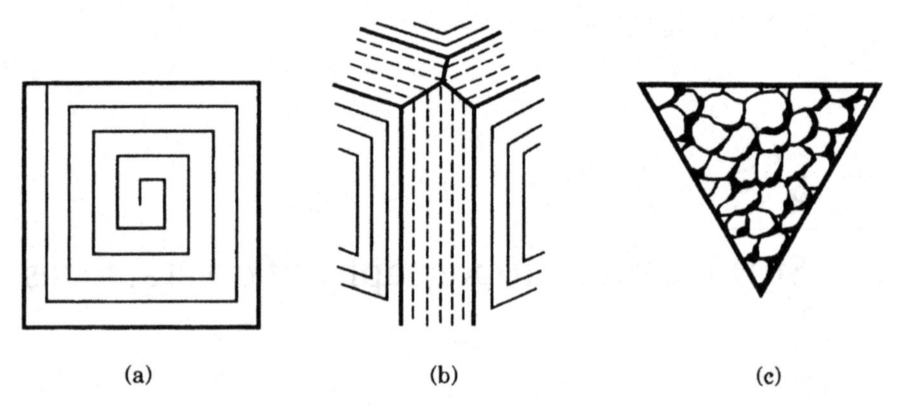

Figure 5.1. Surface microtopographs seen on three types of crystal faces (Kossel crystal). (a) F face; (b) S face; (c) K face.

quickly enough to disappear from the surface, and they do not develop large enough to control the *Habitus*; also, crystal faces on which step patterns occur due to layer growth are not expected to be seen. The substrate surface called an off-facet face in epitaxial growth corresponds to this type of face.

The $\{hk\bar{i}l\}$ faces, which determine the dog-tooth (scalenohedral) *Habitus* of calcite crystals, and the prismatic $\{hk\bar{i}l\}$ face of tourmaline crystals show striations only, never step patterns. These faces are S faces, by PBC analysis, and they appear due to a pile up of steps developing on the neighboring F faces. Yet they develop as large as those that determine the *Habitus*.

There are also crystal faces that show either step patterns or striations only, depending on the growth conditions, in spite of the fact that the faces correspond to an F face in PBC analysis. For natural quartz crystals or synthetic quartz grown in NaCl solution, the $\{10\bar{1}0\}$ faces are characterized by striations running parallel to the edge between $\{10\bar{1}1\}$ and $\{10\bar{1}0\}$; however, on synthetic quartz grown from NaOH or KOH solution, the $\{10\bar{1}0\}$ faces are characterized by polygonal growth hillocks, and no striations are observed. The $\{210\}$ face of pyrite crystals is characterized by striations either running parallel to the edge with the neighboring $\{100\}$ face, or running perpendicular to this direction, depending on their localities. In the former case, no growth layers are observable on the $\{210\}$ face, whereas, in the latter case, striations are due to step patterns of elongated growth layers. The reason why the same crystal face behaves differently depending on growth conditions, and changes its characteristic from an F face to an S face, remains unanswered (see Chapters 10 and 11).

Crystal faces with curved or wavy surfaces, not exhibiting either striations or step patterns, are rarely encountered. In most cases, these faces appear by dissolution. Rough interfaces grow by the adhesive-type growth mechanism, their normal

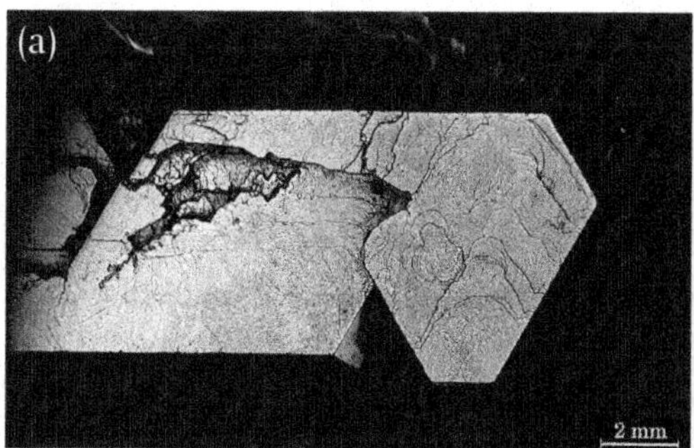

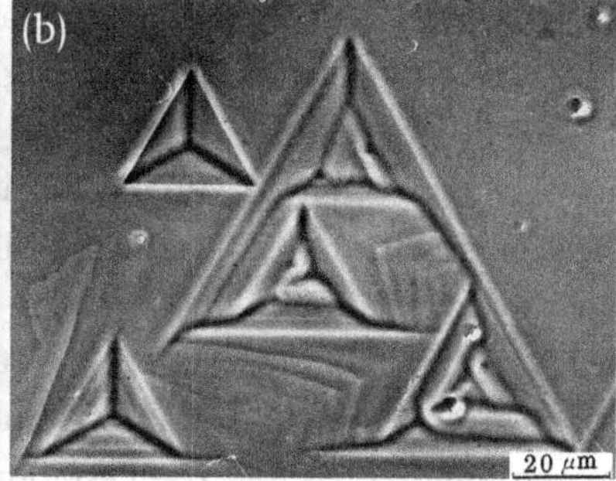

Figure 5.2. (a) Reflection-type photomicrograph showing step patterns due to growth observed on low-index (0001) face of hematite. (b) Phase contrast photomicrograph showing etch figures on (111) face of diamond.

growth rates are high, and therefore the faces soon disappear from the crystal. If such faces become large, there must be a reason why these faces have survived.

In contrast to striated or curved faces, on low-index mirror-flat crystal faces, step patterns are observed if an appropriate method of observation is applied, which indicates that the face grew by two-dimensional layer spreading or spiral-step-like spreading of growth layers parallel to the face. These steps range from elemental steps with unit cell height to thick macro-steps formed by the bunching of elemental steps; circular, polygonal, and irregular steps are also observed. Step patterns range from those with distinct spiral steps with a wide step separation at the center to those appearing as conical or polygonal pyramids (growth hillocks) due to narrow step separations. Etch figures, such as etch pits and etch hillocks, may also be seen on this type of face (Fig. 5.2).

There is a great deal of information relating to the growth or dissolution of crystals contained in surface microtopographs. They contain more direct information relating to the growth of crystals than does the bulk morphology of crystals, and they provide information that allows us to form the basis of analysis of the origin of variations in the *Habitus* and *Tracht* of polyhedral crystals. The reason for this is clear: the surfaces of crystals are unique in that they are the places where growth or dissolution may take place. In this chapter, we will summarize the information obtained from surface microtopographs of crystal faces, and the methods required to decode this information will be discussed.

5.2 Methods of observation

The earliest interest in surface microtopographs observable on crystal faces developed in the 1920s; these observations were made using reflection-type microscopes on etch figures seen in natural mineral crystals [1], [2]. At that time, the main point of interest was the relationship between the symmetries of the etch figures and the crystal faces. Through these observations, etch pits (depressions formed by etching) and etch hillocks (protrusions caused by etching) were distinguished, and both were collectively called etch figures. However, a clear distinction between growth hillocks and etch hillocks was not established. Arguments as to whether trigons (triangular pits with orientation opposite to the triangle of a (111) face; see Chapter 9) commonly observed on the {111} faces of natural diamonds are due to growth or dissolution continued for many years (see Chapter 9 and Fig. 5.2 (b)). Although it has now been established that trigons are etch pits that arise due to the strain field associated with dislocations or point defects, there was at one time a strong assertion that they were of growth origin. Once dislocation theory was established, the observation of etch pits was utilized as a powerful methodology to investigate dislocation movement.

A renewed interest in the surface microtopographs of crystal faces developed after the interferometric observations by Volmer, who confirmed for the first time the two-dimensional spreading of interference fringes on a crystal face growing from the vapor phase (see ref. [10], Chapter 3). This observation introduced not only the concept of surface diffusion, but also formed a starting point for the layer growth theory later proposed by Kossel and Stranski (see refs. [8] and [9], Chapter 3). The spiral growth theory by Frank provided further stimulation in this subject (see ref. [7], Chapter 3). It was around this time that phase contrast microscopy, which can detect extremely small step heights (of nanometer order), and multiple-beam interferometry, by which differences in levels of this order can be measured, were invented. This enabled the first verification of the theory, achieved by the observation of a horse's hoof step pattern on $\{10\bar{1}0\}$ faces of natural beryl, $Be_3Al_2Si_6O_{18}$, whose step height was measured to be of unit cell order by multiple-beam interferometry (see ref. [18], Chapter 3).

Thus, spiral step patterns served as excellent subjects, and many observations were reported using these new techniques [3], [4]. It was also around this time that the movement of spiral growth layers spreading on the (0001) face of CdI_2 growing in aqueous solution was first observed *in situ*. By using these optical techniques, spiral growth layers with monomolecular height (0.23 nm) were observed and measured on natural hematite crystals [5].

Later, differential interference microscopy was developed, enabling the detection of difference in levels as sensitively as phase contrast microscopy, and, because this technique was easier to use, it came to be used in preference to the former techniques [6]. Differential interference microscopy is superior to phase contrast microscopy in the observation of vicinal or curved surfaces, which are impossible to observe under a phase contrast microscope because the contrast is too high.

Optical microscopy, such as phase contrast or differential interference contrast,

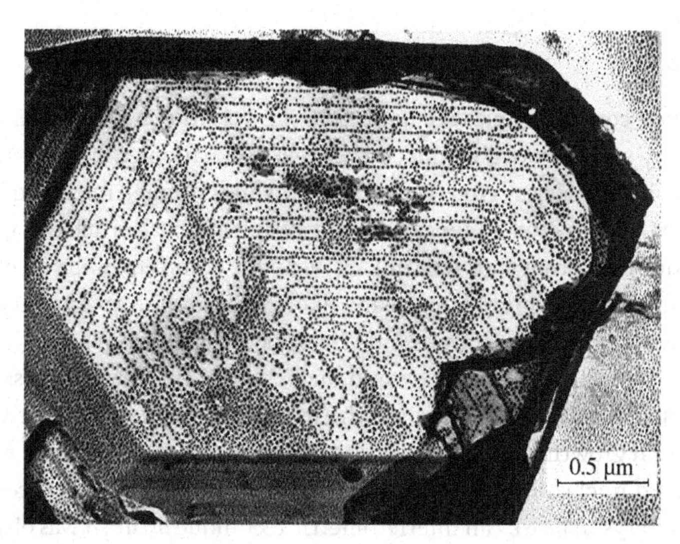

Figure 5.3. Spiral steps observed on the mineral kaolin using transmission electron microscopy with the decoration method [7].

increases the resolution in the vertical direction, but the horizontal resolution is limited to the wavelength of visible light. To observe surface microtopographs of tiny crystals of micrometer order, such as clay minerals, a decoration method [7] was developed using transmission or scanning electron microscopy. This involved the evaporation of gold in a vacuum onto dispersed samples; the gold selectively nucleates along the steps, and the surface microtopographs of tiny crystals become observable. By these means, it was demonstrated that tiny muscovite or kaolin minerals also grow by the spiral growth mechanism. An example is shown in Fig. 5.3.

Powerful methods that have been developed more recently, and are currently used to observe surface microtopographs of crystal faces, include scanning tunnel microscopy (STM), atomic force microscopy (AFM), and phase shifting microscopy (PSM). Both STM and AFM use microscopes that (i) are able to detect and measure the differences in levels of nanometer order; (ii) can increase two-dimensional magnification, and (iii) will increase the detection of the horizontal limit beyond that achievable with phase contrast or differential interference contrast microscopy. The presence of two-dimensional nuclei on terraced surfaces between steps, which were not observable under optical microscopes, has been successfully detected by these methods [8], [9]. *In situ* observation of the movement of steps of nanometer order in height is also made possible by these techniques. However, it is possible to observe step movement *in situ*, and to measure the surface driving force using optical microscopy. The latter measurement is not possible by STM and AFM.

5.3 Spiral steps

An F face containing more than two PBCs grows either by two-dimensional nucleation or by spiral growth mediated by screw dislocations. Since the growth source is not fixed in layer growth through two-dimensional nucleation, and growth centers will fluctuate, appearing and disappearing during the growth process, it is not likely that a growth hillock with one summit will be formed by this mechanism. It is conjectured that island-like layers may be formed on terraced surfaces of layers of growth spirals, but the appearance of pyramidal growth hillocks with single summits will be exceptional. There is, however, the possibility of repeated nucleation on the same site, which may possibly result in the formation of pyramidal growth hillocks, but this model is unrealistic unless the mechanism of repeated nucleation is verified. When observing the growth process of a dislocation-free crystal face by an interferometric technique *in situ*, it was observed that growth centers appear, disappear, and reappear at different sites [10]. Although island-like steps on terraced surfaces may sometimes be observed by AFM, step patterns resembling contour lines on a geographic map (which are exclusively spiral growth steps originating from screw dislocations) are commonly observed on low-index F faces, which develop to determine the *Habitus*.

Within spiral growth steps, there exist the following types: elemental spirals with step heights equal to the unit cell or of monomolecular size, originating from isolated single screw dislocations; composite spirals spreading from screw dislocations; macro-steps originating from screw dislocations with large Burgers vector; and macro-steps formed by the bunching of elemental steps during their advancement.

Features characterizing these steps include

(1) whether an elemental spiral is circular or polygonal, and, if it is polygonal, its symmetry;
(2) the separation between the neighboring steps;
(3) the morphological changes that occur when elemental steps bunch together to form macro-steps.

These three features correspond to the two-dimensional morphology of a crystal, and are directly related to the problems of the three-dimensional morphology of polyhedral crystals, *Habitus* and *Tracht*. This is because the normal growth rate R which determines *Habitus* and *Tracht* is related in the following way to the height of a step, h, the advancing rate of the step, v, and the step separation, λ_0:

$$R = hv/\lambda_0$$

(see Fig. 5.4). The height of an elemental spiral step is determined by the Burgers vector of the screw dislocation. Since this corresponds to the size of a unit cell or

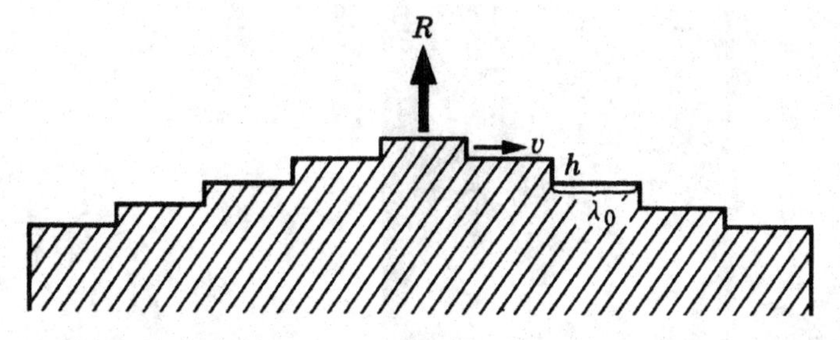

Figure 5.4. Relationship between normal growth rate R of a crystal face and the step height h, the advancing rate v, and the step separation λ_0 of a growth spiral.

monomolecule, the height depends on the crystal face. In general, h becomes smaller as the morphological importance increases (the Bravais empirical rule; see Section 4.2).

Growth spirals may also sometimes be formed from screw dislocations with a large Burgers vector, but since screw dislocations are energetically disadvantageous, they tend to dissociate into elementary steps. In crystals having zigzag stacking in the structure, spiral steps originating from a dislocation with a Burgers vector of unit cell height dissociate into two or three spiral steps with opposite orientations, forming interlaced spiral patterns, which will be described in Section 5.5.

5.4 Circular and polygonal spirals

The steps of spiral growth layers are one-dimensional interfaces. If the step is rough, the advancing rate is isotropic, forming a circular step pattern. The spiral pattern will be an Archimedean-type spiral. If a step is smooth, the spiral will be polygonal. Since the roughness of a step is controlled by strong bonds in the plane, i.e. PBCs, in addition to the growth parameters, the polygonal form will follow the symmetry elements involved in the crystal face. So, a square form with a four-fold symmetry axis will be expected on a {100} face of a crystal belonging to the cubic system m3m; a triangular form with a three-fold symmetry axis will be expected on the {111} face of a crystal belonging to the cubic system; a hexagonal form will appear on the {0001} face of a hexagonal system; a triangular form with a three-fold symmetry axis will be expected on the {0001} face of a trigonal system; and a polygonal spiral containing a symmetry plane only on the {001} face of a monoclinic system or on the $\{10\bar{1}1\}$ faces of hexagonal and trigonal systems. A few examples of polygonal growth spirals are shown in Fig. 5.5.

Since the one-dimensional roughness of the steps determines whether a spiral takes circular or polygonal form, these morphologies may be treated similarly to the roughening transition of an interface, as described in Chapter 3. It is possible to predict interface roughness either by Jackson's α factor or by Bennema–Gilmer's generalized α_G factor (see Section 3.8). The coefficients which determine the α

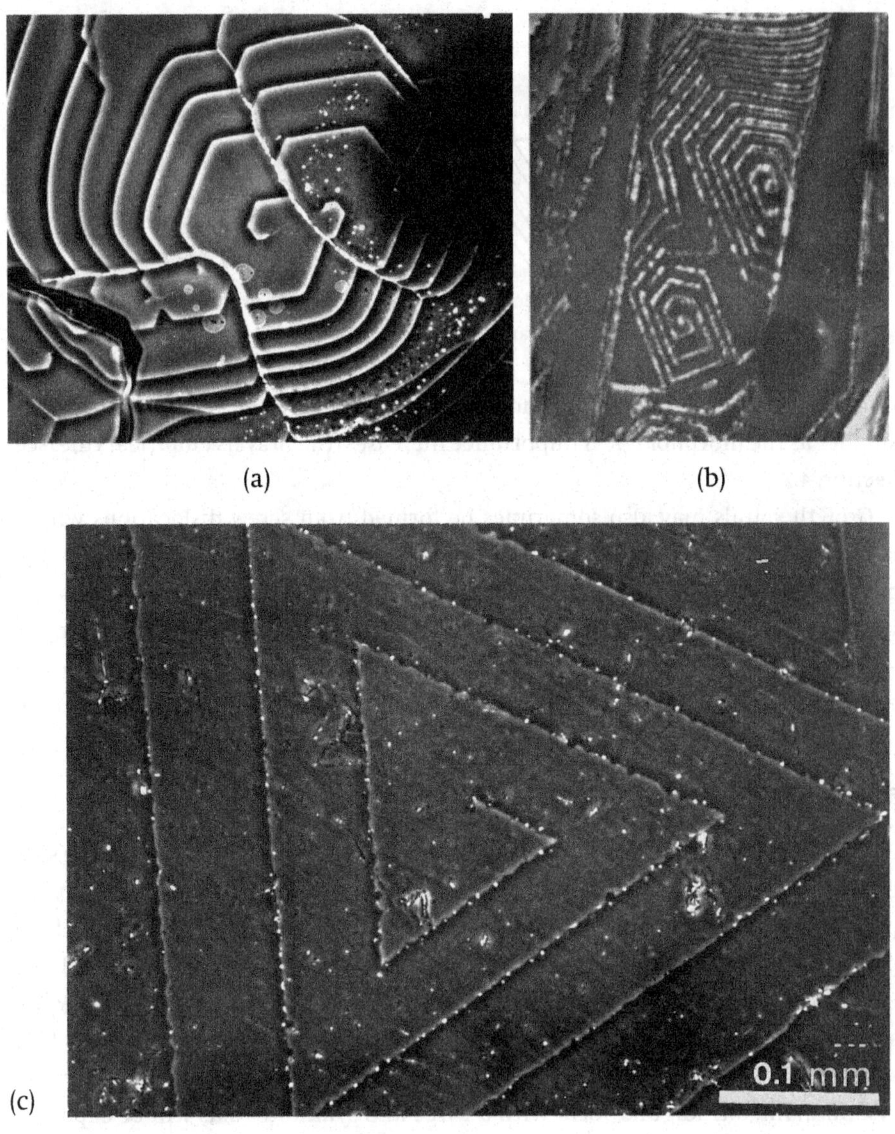

Figure 5.5. Examples of polygonal spirals. (a) SiC, 6H polytype, (0001). (b) Phlogopite, 1M polytype, (001). (c) SiC, 15R polytype, (0001).

factor are the orientational factor ξ, the latent heat L, the melting point T_M (in Jackson's formula), the bonding energy in the solid and liquid, φ_{ss} and φ_{ff}, respectively, the solute–solvent interaction energy φ_{sf}, and the growth temperature T_G (in Bennema–Gilmer's formula). The morphology of the elemental spirals will be modified depending on the factors that influence these coefficients. It is anticipated that polygonal spirals will be seen on crystal faces with a higher order of morphological importance, and that more rounded spirals will be observed as the order of morphological importance lowers. On hexagonal prismatic crystals of

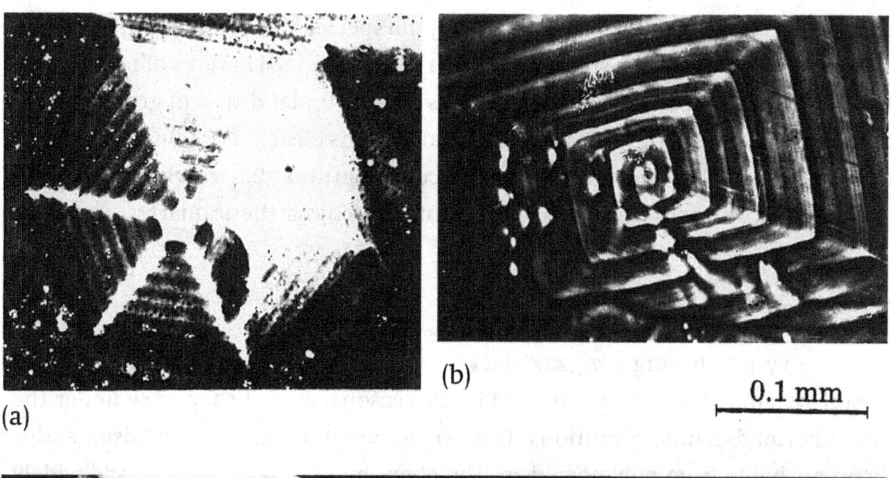

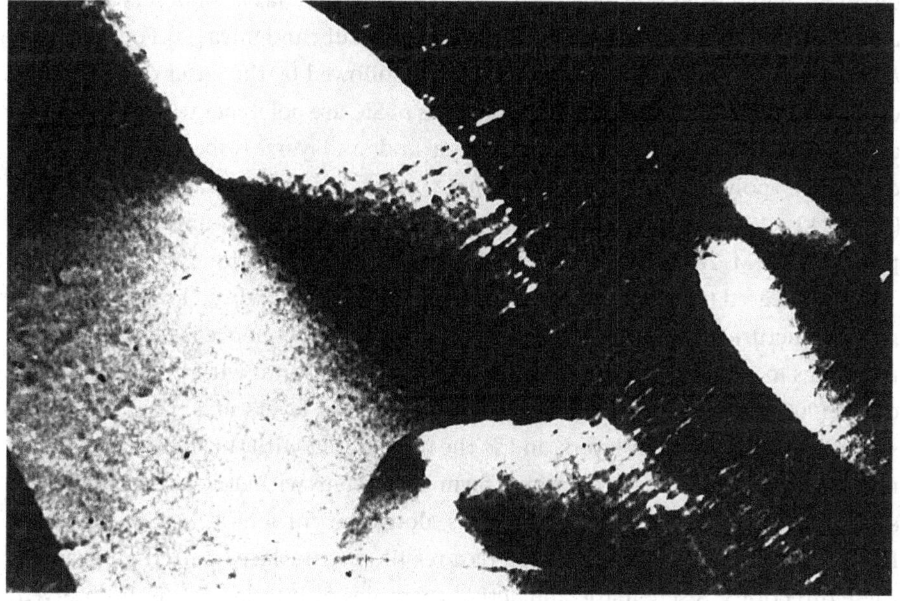

Figure 5.6. Differential interference contrast photomicrographs showing the morphology of growth spirals observed on (a) (0001), (b) $(21\bar{3}1)$, and (c) $(10\bar{1}0)$ faces of synthetic emerald.

hydrothermally synthesized emeralds, the {0001} face shows spirals with regular hexagonal form, rectangular spirals with rounded corners on $\{10\bar{1}0\}$, and a spindle-like form on $\{21\bar{3}1\}$ (see Fig. 5.6) [6].

On quartz crystals synthesized in a hydrothermal solution of NaOH, KOH series, $\{10\bar{1}0\}$ faces show polygonal spiral steps, whereas rounded step patterns are observed on $\{10\bar{1}1\}$ and $\{01\bar{1}1\}$.

The effect of factors which affect the α factor, such as growth temperature or solute–solvent interaction, may be observed by comparing the surface microtopographs of the same face on the same crystal species grown under different conditions. If we compare the spiral steps observed on the $\{10\bar{1}1\}$ faces of natural and synthetic quartz, or among natural crystals formed under different growth conditions, we see these differences in growth conditions clearly. Polygonized step patterns are generally observed on $\{10\bar{1}1\}$ faces of natural quartz, whereas circular patterns are commonly observed on the same faces of synthetic quartz.

5.5 Interlaced patterns

Crystals having a zigzag stacking in their structure can take different crystal systems or structures due to kinetic reasons, even if they grow under the same thermodynamic conditions. This phenomenon is called polytypism, and is distinguishable from polymorphism. The phenomenon has been observed widely among crystals with a layer structure, such as SiC, CdI_2, and mica [11]. Polytypes are represented by the number of stacking layers followed by the letter denoting the crystal system. For example, 4H, 6H polytypes of SiC are polytypes with unit cells of a hexagonal (H) system, consisting of four and six layers, respectively; and 15R denotes a polytype consisting of three layers of a 32 stacking sequence ($(3+2)\times 3=15$ layers altogether), forming a rhombohedral (R) unit cell. In mica, polytypes of 1M, $2M_1$, $2M_2$, 3T, and higher orders, are known.

In this layered type of structure, for instance the upper two and lower two layers in the structure of the 4H polytype of SiC, the stacking sequence is reversed. Since it belongs to a hexagonal system, we expect a growth spiral with hexagonal form on the (0001) face, but the steps of the upper and lower layers advance as two oppositely oriented triangular layers, and so the upper layer, with the higher advancing rate, will unite with the lower layer to form single steps with unit cell height in the six edge directions. In the six directions along the corners of hexagonal form, however, two layers will never unite. As a result, an interlaced pattern appears in these directions. A schematic and actual examples of interlaced spirals observed on magnetoplumbite and SiC 6H polytype are shown in Fig. 5.7. In the mica polytypes 2M, 2O, and 3T, the elemental form of a single layer is five-sided, i.e. it is a truncated hexagon, and stacking occurs by rotation of the elemental layer by 60°, 180°, and 120°, and the resultant interlaced patterns become more complicated than those of SiC. Since the patterns of the steps of elemental growth spiral reflect very well the characteristics of the structure, it is possible to identify polytypes based on observations of these steps. Since the advancement of the steps is isotropic for circular spirals, the interlaced patterns will not appear unless the isotropy is violated.

In crystals containing no zigzag stacking, the orientations of the polygonal elemental spiral steps may be used as a criterion to identify whether the crystal is twinned or contains stacking faults.

(a)

(b)

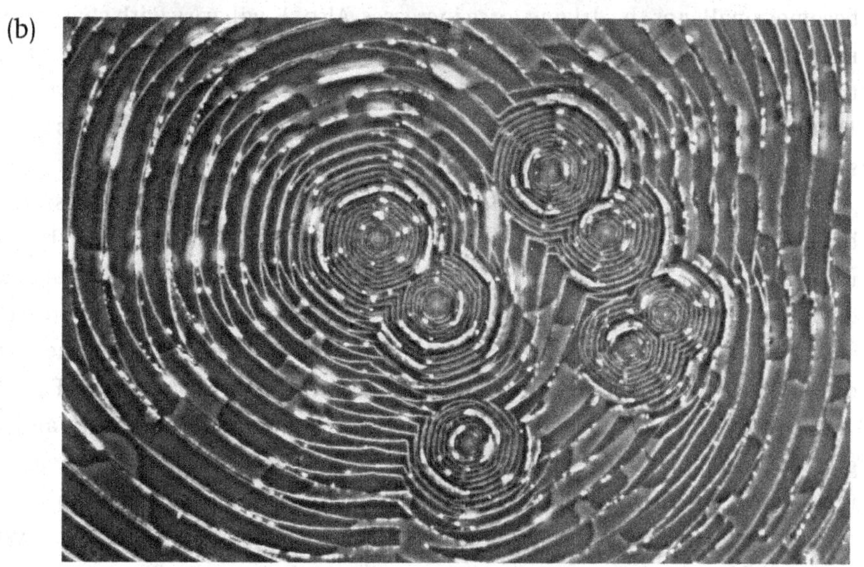

(c)

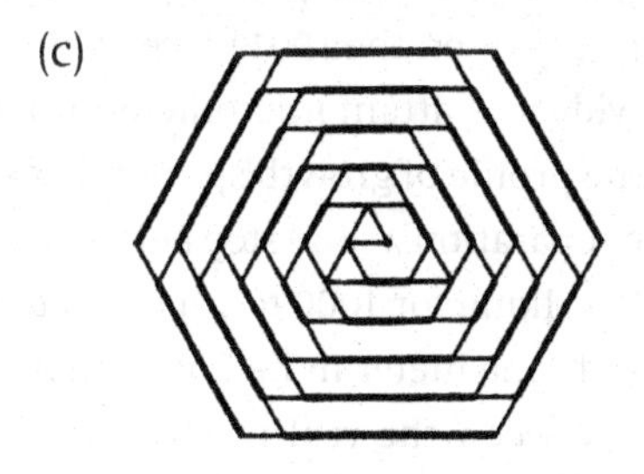

Figure 5.7. Phase contrast photomicrographs depicting interlaced spiral steps observed on (0001) faces of (a) magnetoplumbite and (b) SiC 6H polytype. (c) Schematic figure. The very narrow step separations observed at the center of the spiral in (b) are due to a sharp increment of supersaturation at the final stage due to the discontinued electrical supply.

5.6 Step separation

Elemental growth spiral layers originating from an isolated dislocation can advance, keeping the step separation constant, unless factors which affect the advancing rate of the spiral steps, such as a local fluctuation in driving force or impurity adsorption, takes place. The step separation of a spiral, λ_0, is related to the critical radius of two-dimensional nuclei, r_c, in the following manner (see ref. [11], Chapter 3):

$$\lambda_0 = 19r_c,$$

and r_c has the following relation to step free energy and the driving force:

$$\Delta G\,(r_c) = 16\pi\gamma^3(v)^2/(3\Delta\mu^2),$$

where v is the molecular volume of the bulk nucleated phase. Therefore, λ_0 becomes smaller as the driving force increases. Also, λ_0 will vary with changes in γ reflecting the difference in temperatures, impurities, and solute–solvent interaction, and this results in the variation in the normal growth rates of crystal faces. Therefore, the factors affecting the *Tracht* and *Habitus* of polyhedral crystals are the same as those that affect the roughness of the steps and λ_0.

When the step separation is wide enough, typical spiral step patterns observable by optical microscopy may appear, but if the separation becomes narrower than the resolution power of the optical microscope, the spirals appear in the forms of polygonal pyramids or conical growth hillocks. Even if spiral patterns are not directly observable, we may assume that these growth hillocks are formed by the spiral growth mechanism. Examples representing the two cases are compared in Fig. 5.8.

Step separations vary depending on crystal faces, environmental phases, and the growth conditions. The step separation λ_0 on a crystal face of higher morphological importance is wider than that on a face of lower morphological importance. This may be compared with the difference observed for different faces of a crystal grown under the same growth conditions, polygonal or circular (Fig. 5.6). The λ_0 of growth spiral layers on the {0001} face of SiC or CdI_2, or on the {001} face of mica, which have strong anisotropy in bonding, are wide and attain micrometer order. Since the step heights are of nanometer order, the profile of growth spiral hillocks on these faces attain a ratio of order $10^{3\sim4}$ in step separation λ_0 and step height h. To get a feel for the scale of this, we could imagine walking for 1000 to 10 000 meters on an atomically flat surface before meeting a cliff one meter in height. It should be remembered that there is a big difference between the real profile and the images deduced from a schematic figure of spiral growth or optical photomicrographs of growth spirals.

The values of λ_0/h are different for different ambient phases and growth conditions. The ratio is of the order $10^{3\sim4}$ for crystals grown from the vapor phase,

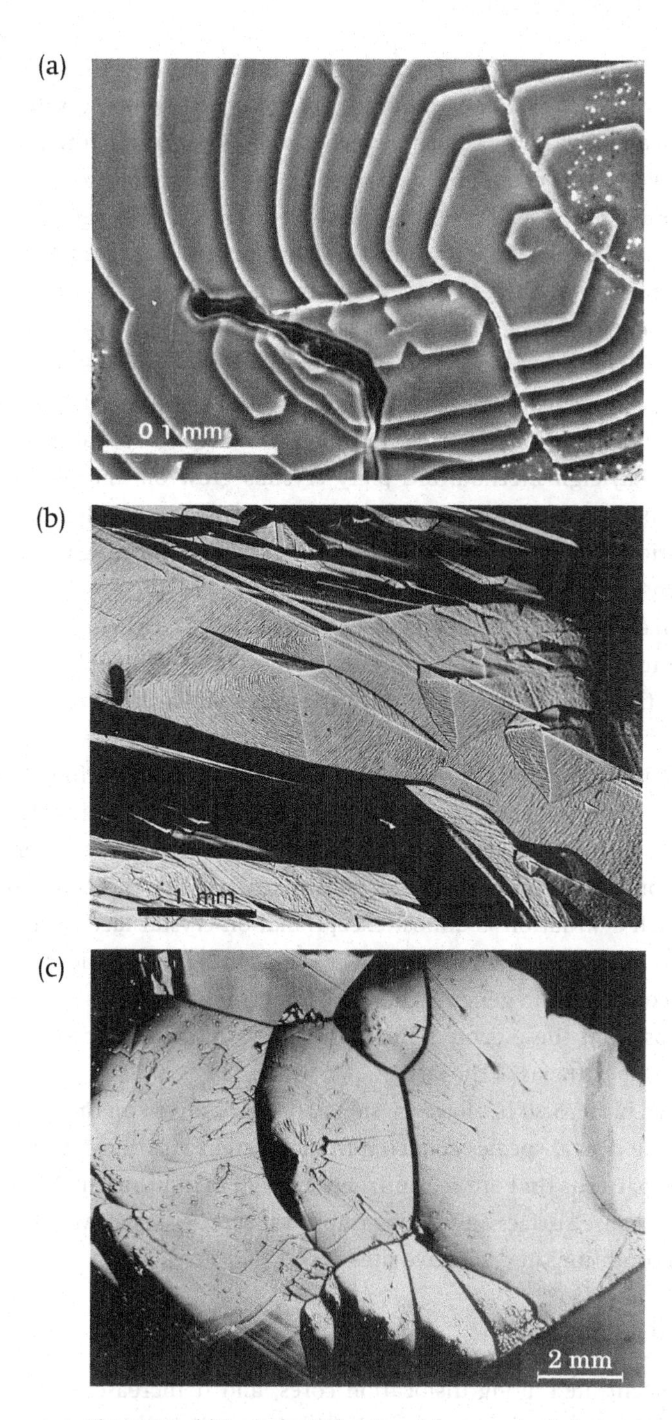

Figure 5.8. (a) Typical spiral pattern (phase contrast photomicrograph of (0001) face of SiC grown from the vapor phase), and spiral growth hillocks which appear as (b) polygonal and (c) conical pyramids due to narrow step separation. Part (b) is a differential interference photomicrograph, $(10\bar{1}0)$, and part (c) is a reflection photomicrograph, $(10\bar{1}1)$, of hydrothermally synthesized quartz.

whereas it is typically $10^{2\sim3}$ on crystals grown from the solution phase (because the step separation is narrower) [12]. When the ratios are compared among crystals grown from the same ambient phase, it can be seen that λ_0 becomes narrower as supersaturation increases. It is often observed that SiC crystals grown from the vapor phase show a much narrower λ_0 at the spiral center than anywhere else (see Fig. 5.7(b)). This can be understood as being due to a sharp increase in supersaturation at the final stage of growth. There are cases of spirals originating from a screw dislocation that show extreme eccentricity. This is explained as being due to the supersaturation gradient over the surface owing to flow of vapor. Figure 5.9 shows an example of spiral patterns on a (0001) face of a SiC crystal synthesized by the Rayleigh method [13]. The direction of narrowing of λ_0 is the direction towards the crucible wall. In the Rayleigh method, SiC platy crystals grow on and attach obliquely to the wall of the crucible, which is made of sintered SiC powder. As a result, a supersaturation gradient is formed on the (0001) surface, forming eccentric spiral step patterns.

It is extremely unlikely that the whole surface of a crystal face is covered by step patterns with constant λ_0. In general, λ_0 becomes narrower towards the edge. This is due to the fact that the driving force is higher at the edge than at the center, i.e. the Berg effect.

It is often observed in tiny crystals of micrometer order, such as clay minerals, that the entire surface of a crystal face is covered by elementary spiral layers originating from one screw dislocation (Fig. 5.3). Figure 5.10(a) shows such an exceptional case observed on a (0001) face of a SiC crystal synthesized by the Acheson method. However, such a situation is almost exceptional on crystal faces larger than millimeter size, and is encountered only on crystals synthesized under very precisely controlled conditions. In general, there are many growth centers on one crystal face, and steps from these centers bunch together to form macro-steps, which constitute the step patterns of the face.

An example is shown in Fig. 5.10(b). However, step patterns observed on the same crystal face of the same crystal species collected from different locations usually exhibit characteristic patterns that are recognizable enough to indicate the locality of origin. This is because surface microtopographs reflect very sensitively the difference in growth conditions at respective sites.

5.7 Formation of hollow cores

Strain is concentrated along dislocation cores, and it increases as the Burgers vector of dislocations increases. It was Frank who predicted that above a critical value a dislocation is energetically more stable if a free surface is created along the dislocation core. This critical value is expressed by

$$b > 8\pi^2\theta\gamma/\mu,$$

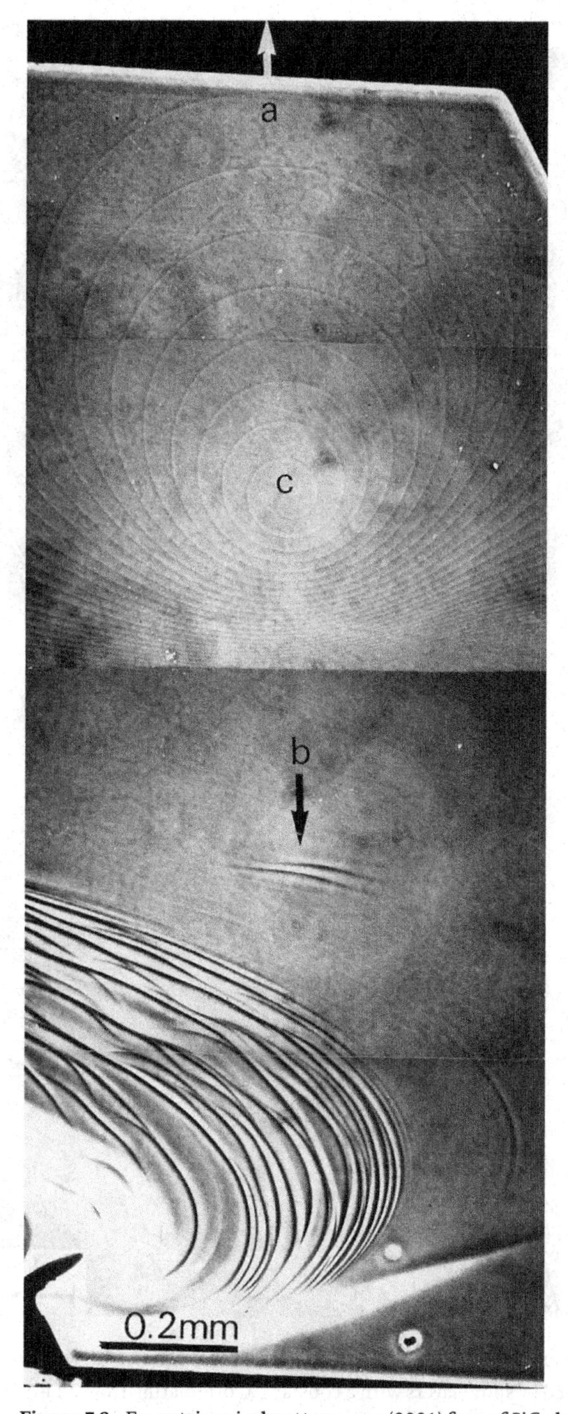

Figure 5.9. Eccentric spiral pattern on a (0001) face of SiC, due to a supersaturation gradient over the surface [13]. The spiral center is indicated by c, and the arrows a, b show the directions of eccentricity (where the white arrow refers to the center of the crucible, and the black arrow refers to the wall).

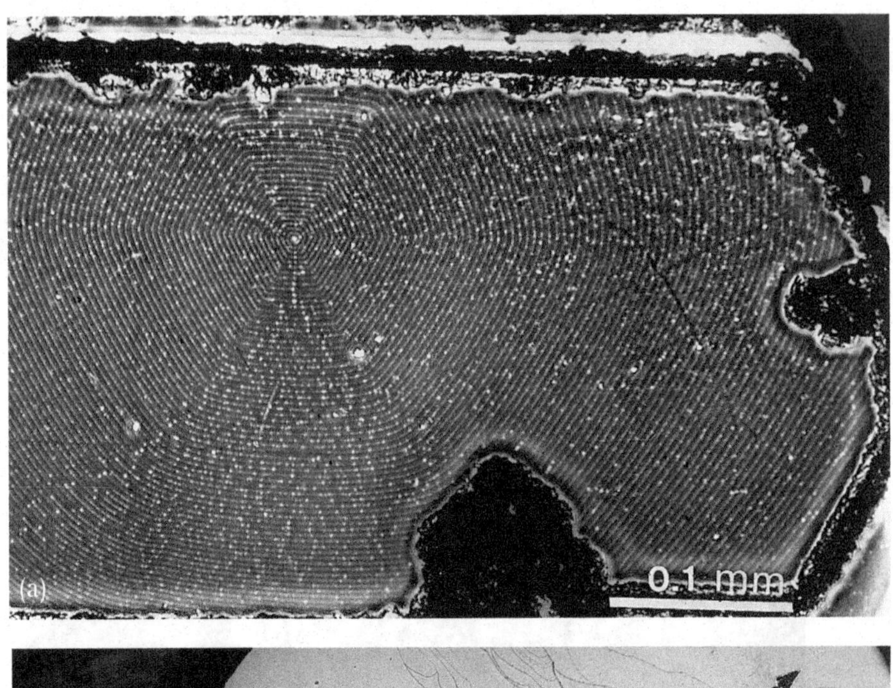

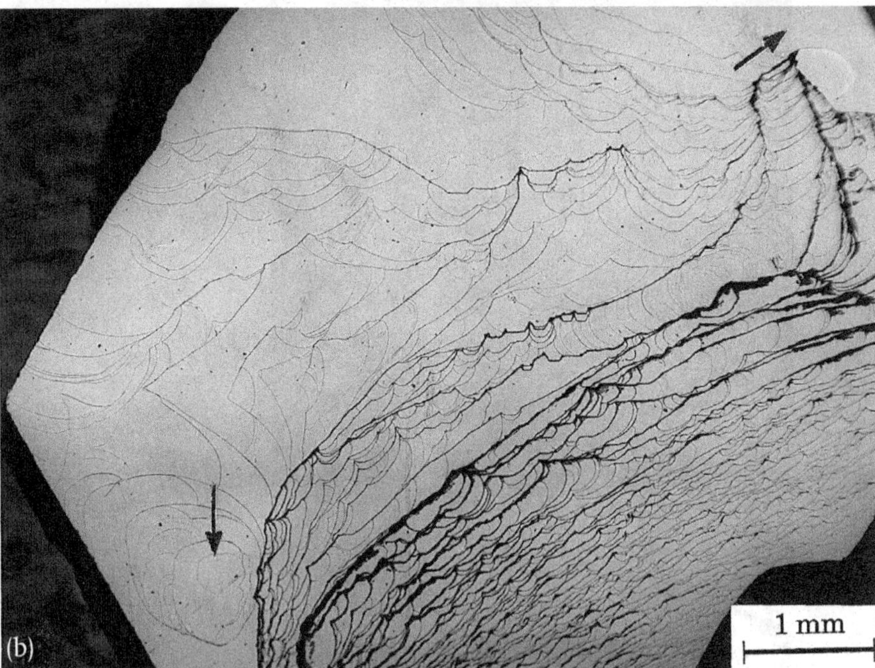

Figure 5.10. (a) An exceptional case showing a wide (0001) surface of SiC with an area of 0.7×0.4 mm covered by spiral layers with a constant λ_0 originating from an isolated single screw dislocation. The step height is 1.5 nm. (b) Reflection photomicrograph at low magnification showing a commonly observed step pattern in hematite, (0001). The arrows indicate growth centers, where elemental spiral steps are observed. All observed macro-steps originate from these points, from where the elemental spiral steps advance.

where $\boldsymbol{b}$ is the Burgers vector, μ is the rigidity, γ is the surface free energy, and θ is a factor close to unity. In the case of crystals with large μ, such as SiC, a hollow core is expected for a dislocation with $\boldsymbol{b} > 2$ nm, and a hollow core with a diameter of ~100 nm is expected for $\boldsymbol{b} > 10$ nm. Indeed, hollow cores were observed at the centers of growth spirals [3].

That a hollow core is formed by the creation of a free surface along a dislocation core implies that the curvature of the spiral step is reversed due to the strain field along the dislocation core. The effect of a strain field upon the advancement of a step was theoretically treated by Cabrera and Levine [14], [15].

Only a few crystals exhibit hollow cores at the centers of growth spiral layers. However, on the (0001) faces of SiC, which has a large μ value, hollow cores due to growth have often been observed. According to the summary by Sunagawa and Bennema [16], various degrees of the effect of the strain associated with dislocation cores have been observed depending on the sizes of $\boldsymbol{b}$ and the concentration of dislocations.

In the case of growth spirals originating from dislocations with large $\boldsymbol{b}$, hollow cores with diameters of micrometer order are observed at the spiral center; however, when a number of dislocations with small $\boldsymbol{b}$ concentrate in a narrow area, a basin-like depression appears at the central area of the composite spirals, since the curvature of advancement of the spiral steps is reversed near the center. A straight step may appear near the spiral center as an intermediate state in the reversal of step curvature. Several examples are shown in Fig. 5.11.

Hollow cores associated with dislocations may be formed in both growth and etching processes.

5.8 Composite spirals

Elemental spiral steps originating from dislocations form various composite step patterns through cooperation or repulsion depending on the sign of dislocations and the distance between neighboring dislocations.

When spiral growth layers advance from dislocations with the same sign (Burgers vector) and with distance smaller than one-half of the radius of the critical two-dimensional nuclei ($\lambda_0/2 = 19r_c/2$), a composite step pattern consisting of the same number of spiral steps as dislocations will be formed (Fig. 5.12). At the center, spiral steps will advance by mutually interchanging the centers. If the separation between a pair of dislocations among many neighboring dislocations is narrower than the others, the advancement of the spiral steps from other dislocations is influenced by these steps, and bunching of steps will occur rhythmically. Figure 5.13 is a positive phase contrast photomicrograph of composite spirals on a (0001) face of a hematite crystal. In positive phase contrast, the bright contrast appears exclusively on the higher side of a step, and so it is possible to determine

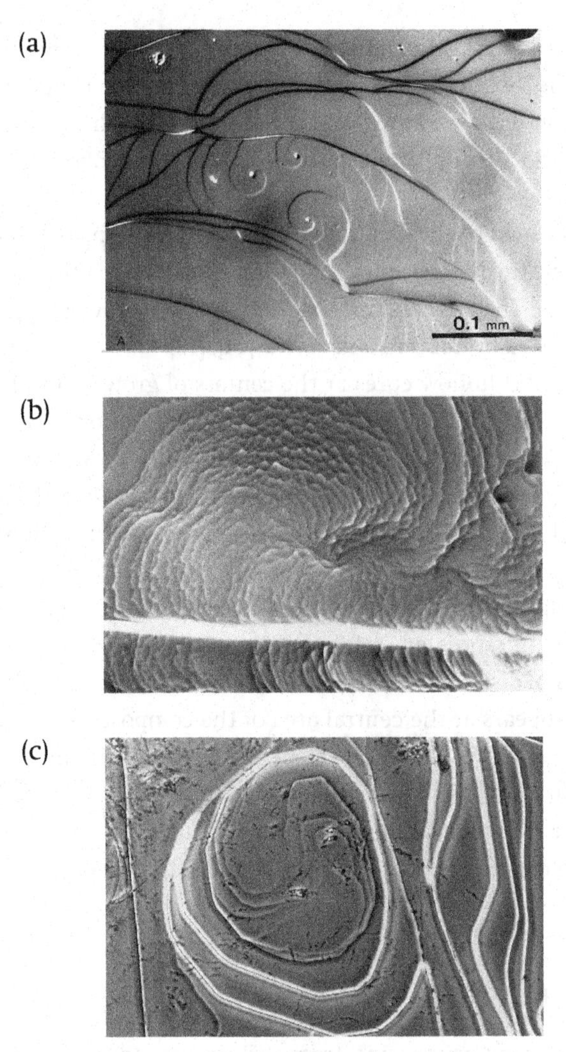

Figure 5.11. Various step patterns appear because the advancing rate and the curvature of the spiral layers are affected by the strain field at dislocation cores. (a) Central hollow core (SiC). (b) Basin-like depression formed at the central area of composite spirals (hematite). (c) Straight step near the center (hematite).

unambiguously which side is higher. If we check the photograph from this viewpoint and start from a step and go successively to the higher side, we see that, as we rotate around the center, that the lowest step eventually corresponds to the highest step. This is comparable to Escher's famous staircase. To achieve Escher's staircase, it is necessary to have a terrace inclined from the horizontal surface, and this is just what happens at the center of a spiral pattern, as we can clearly see in the optical photomicrograph of Fig. 5.13.

It is easily understood that the rotation of a spiral step created by a dislocation is reversed depending on the sign of the dislocation. If two dislocations with the

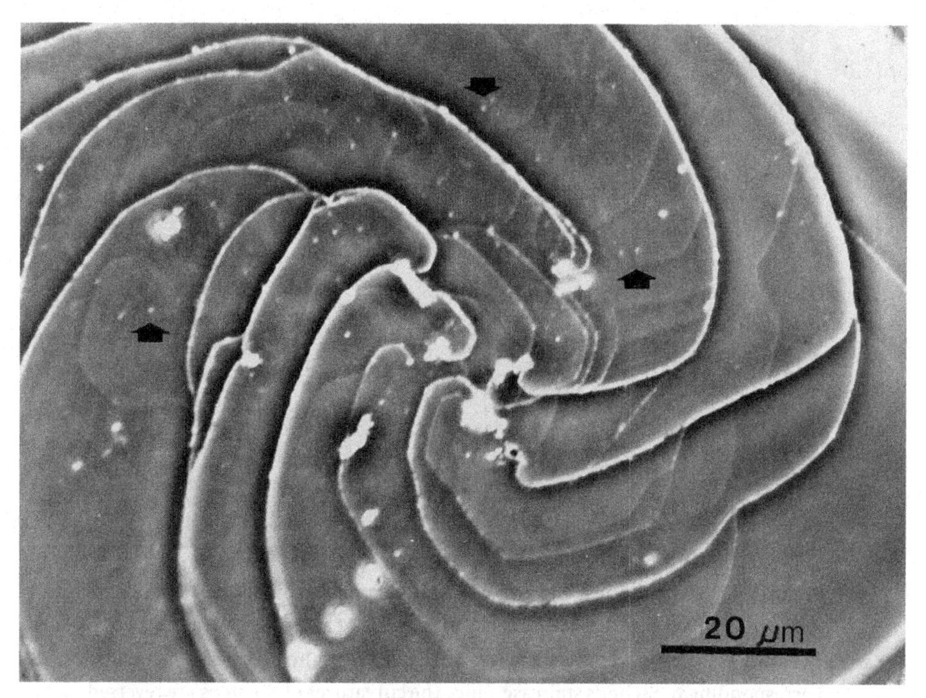

Figure 5.12. Phase contrast photomicrograph of a composite spiral in SiC, (0001). White patches indicate impurities selectively adhering at the outcrops of dislocations. Each white patch corresponds to one dislocation. Arrows indicate an elemental spiral.

same Burgers vector, but opposite signs, occur close together, spiral steps from the two dislocations will coalesce to form a common terrace surface and will advance as a closed loop. If the separation between the two dislocations is wider than $19r_c$, each forms a spiral pattern with a few turns, after which the steps will advance as a closed loop. From screw dislocations situated closer together than $19r_c$, a closed loop appears from the beginning. Therefore, it is not advisable to come to the rash conclusion, based simply on the observation that no spiral pattern is observable at the center of step pattern, that the growth is not by the spiral growth mechanism.

If a new dislocation is introduced while a spiral step is advancing, the new step from the new dislocation will be affected by the advancement of the earlier spiral step, and can have only one-half or a couple of turns. This results in the coexistence of spirals with a small number of turns with those covering a wide area; Fig. 5.14 is such an example.

5.9 Bunching

When we observe the process of advancement of elemental spiral steps *in situ*, it is often noticed that two steps bunch together to form a step with the height of two layers as they advance. The advancing rate of the bunched layer is retarded

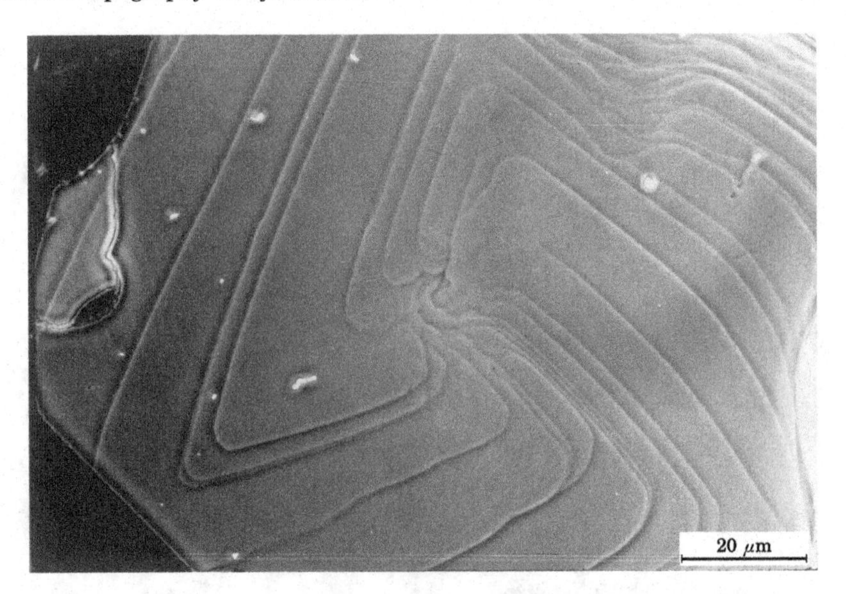

Figure 5.13. Positive phase contrast photomicrograph of composite spiral in hematite, (0001). A bright contrast appears on the higher side of a step. By tracing the route from the lowest step to higher steps, the lowest step becomes the highest step after one turn, corresponding to Escher's staircase. Since the curvatures of the steps are reversed at the center of a group of dislocations, a depression appears due to the associated strain field (refer to Section 5.7). See also Fig. 5.11.

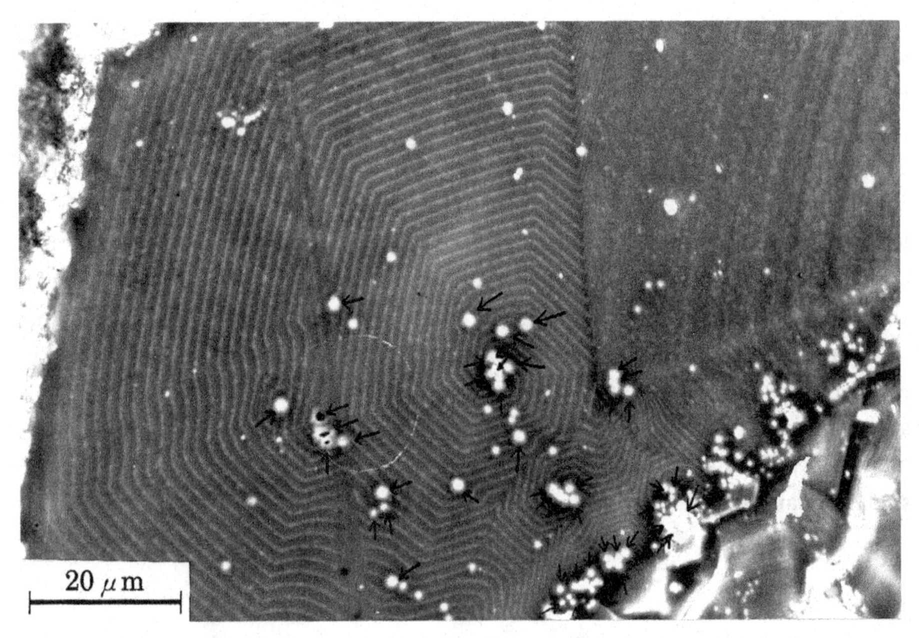

Figure 5.14. Phase contrast photomicrograph of dominant and dominated spirals (arrows) in SiC, (0001).

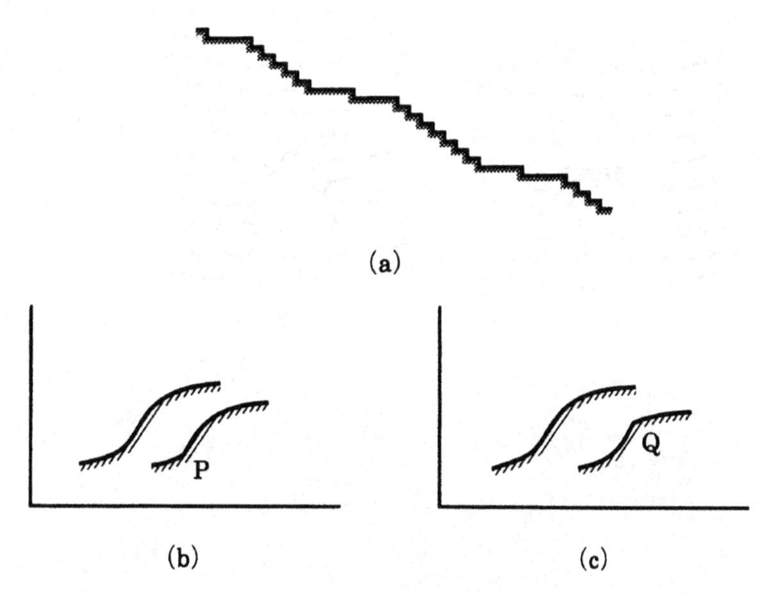

Figure 5.15. The profile of a kinematic shock wave [17]: (a) growth; (b), (c) dissolution. In (b) and (c), dissolution steps advance from left to right, and shock waves appear at P and Q.

compared with that of a single step, and the form is perturbed. As the bunching progresses and the height increases, the forms become more perturbed. Step bunching may be regarded as a similar phenomenon to a traffic jam in which vehicles initially advance at a constant rate, then bunch. The wave profile of the macrosteps appearing due to a traffic-jam-like effect was solved by Frank [17]. Such a wave is called a kinematic shock wave, and the profile is as shown in Fig. 5.15.

There can be many reasons why step bunching occurs: fluctuation of the supersaturation on the surface; asymmetry in the surface diffusion between the upper and lower terraces of a step; impurity adsorption; or the presence of foreign or differently oriented particles may affect the rate of advancement of a step, and act as reasons for step bunching to occur. Supersaturation affects directly the rate of step advance. Asymmetry in surface diffusion is called the Schwoebel effect [18], which results in fluctuations in the step advance rate due to the asymmetry between the incorporation of the adsorbed molecules into steps coming from the upper terrace and those on the lower terrace. The effect of impurity adsorption at a step upon the step advancement rate was analyzed theoretically by Cabrera and Vermilyea [14], [15] in terms of the pinning effect. Impurity components that are difficult to incorporate into the crystal cause drastic effects upon the advancing rate of a step at parts per million order. Similar effects may be seen in the cases of foreign particles and differently oriented particles precipitated on a growing surface, around which the advancing rate of a step is retarded and the step takes on a curved form,

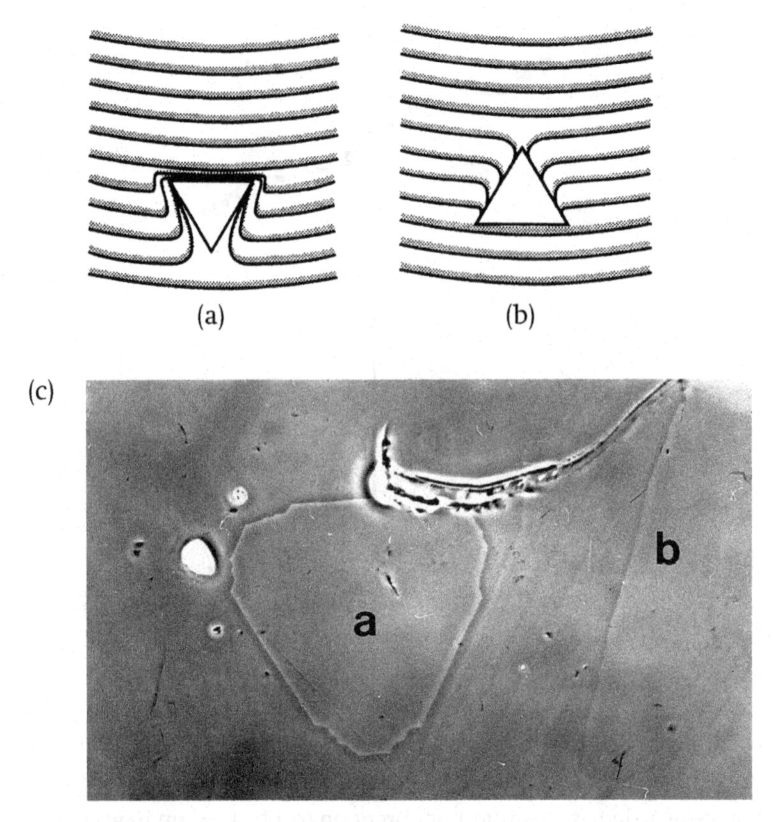

Figure 5.16. Difference in step patterns for the case in which (a) foreign particles or the same crystal particles with different orientation precipitate on a growing surface, and (b) in which the same crystal particles precipitate in the same orientation. The shaded side indicates the higher side of the steps. (c) Actual example of cases shown in (a) and (b), observed on a (0001) face of hematite.

engulfing the particles. When a particle of the same crystal precipitates with the same orientation on the growing surface, the step advancement rate increases as it approaches the particle, and a step pattern appears as if the step is sucked in by the particle. Two cases are shown in Fig. 5.16.

In this way, elemental steps bunch together, forming irregular macro-steps in the process of advancement (Fig. 5.17). In real crystal growth processes, it is usual that growth centers are present on one crystal face, and so interference and cooperation occur among the steps, resulting in the formation of step patterns resembling the contour lines on a geographic map (see Fig. 5.10).

5.10 Etching

We may assume that the surface microtopographs that appear due to growth are symmetrical to those that appear by dissolution, since the processes of

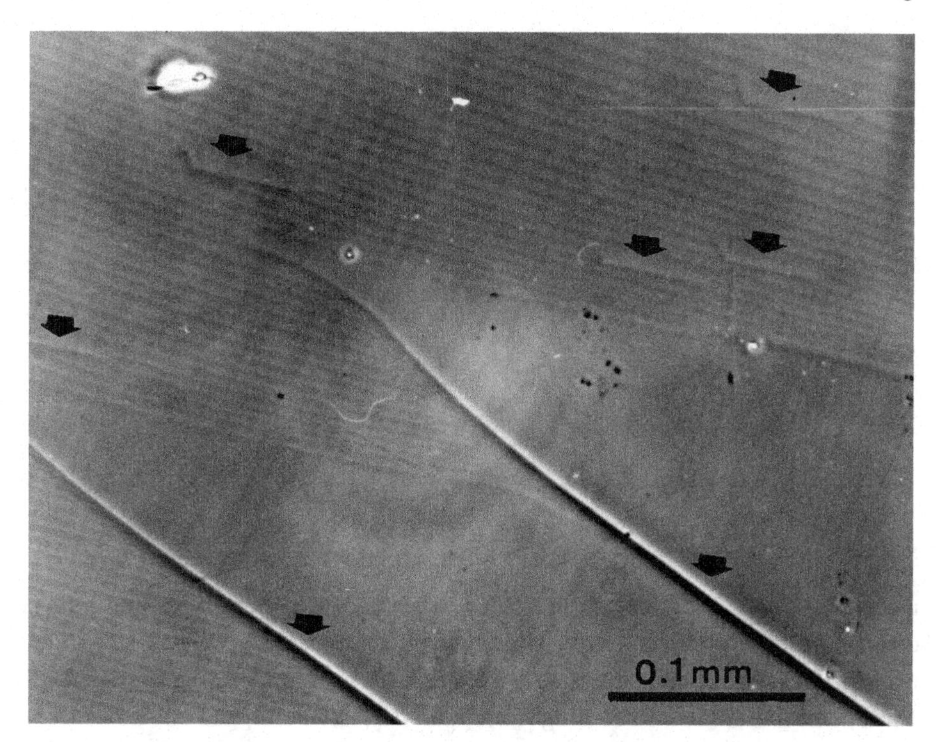

Figure 5.17. Phase contrast photomicrograph showing bunching of elemental spiral steps (arrows) in hematite, (0001) face.

growth and etching are assumed to be symmetrical. However, the role played by dislocations, impurities, and point defects in the respective processes is not the same. This is because the role of the strain field is different in growth and dissolution. For example, point defects cannot be active growth centers, but they do act as active centers in etching. Screw and edge dislocations play the same role in etching, but not in growth.

With regard to the attachment and detachment energies, the corners of a crystal or a rough interface that is constructed by kinks alone are sites where the process proceeds most quickly, whereas the low-index crystal faces, corresponding to smooth interfaces, represent the direction with the minimum rate of normal growth and dissolution. As a result, if a single crystalline sphere is dissolved in an isotropic environmental phase, a dissolution form bounded by both flat and curved crystal faces appears. This is called the dissolution form, which is not the same as the growth form.

If a crystal face with step patterns appearing by spiral growth is etched in a weak dissolution process, after growth the steps resulting from the growth process will retard two-dimensionally, essentially forming a step pattern by etching. Irrespective of whether the crystal is of circular or polygonal form, the growth steps are smooth, whereas steps formed by two-dimensional etching usually show minute

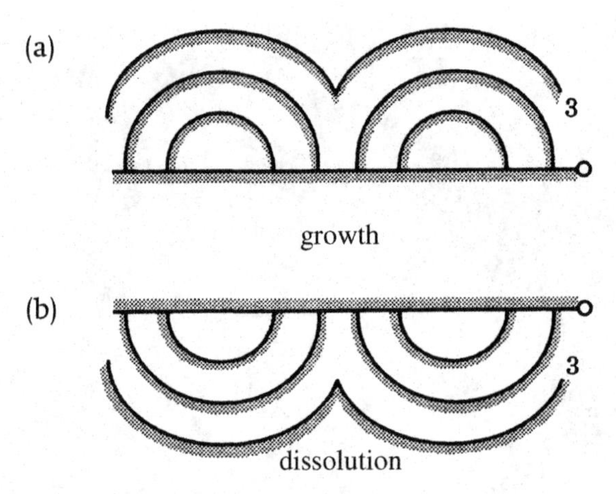

Figure 5.18. Difference between (a) growth steps and (b) steps formed by two-dimensional etching.

unevenness. This is because strain fields associated with impurities and point defects affect the advancement of etching steps. As illustrated in Fig. 5.18, steps due to etching show a chopping wave form, different from protruded growth steps [19]. Through cooperation with etching from point defects outcropped on the terrace surface, a surface with one level lower than the surface of the growth step will appear between the position of the original step and the etch step. It was confirmed on a {0001} face of a natural hematite crystal that the depth of the etched surface corresponds to a few unit cells. This was analyzed and was found to correspond to the outermost layer, which had the highest concentration of point defects [19].

Since the strain fields present on crystal surfaces are preferential sites for etching to proceed, strain fields are sites where etch pits are formed at the sites of strain fields. There are two types of etch pit: flat-bottomed (F-type) and point-bottomed (P-type). Their morphology varies from circular to polygonal. Etch pits associated with point defects are F-type, whereas those associated with linear defects are P-type; both can coexist on the same surface (see Fig. 5.2). When etch pits are polygonal, the symmetry follows that of the face. Polygonal etch pits, irrespective of whether they are of P- or F-type, formed on a face of a single crystal by etching under certain conditions, take the same orientation, but, if the crystal is twinned, the orientations of the etch pits are reversed on the common surface of the two individuals. On the other hand, if the etching conditions are changed, polygonal etch pits with opposite orientations may appear, and sometimes etch pits in intermediate form, circular, and reversed forms may appear. The morphology of etch pits varies, like that of growth forms, depending on the conditions.

Since there is a strain field associated with a dislocation, etching originating from the outcrop at first forms a P-type etch pit. As etching proceeds further, a tunnel is formed along the dislocation, starting from the bottom of the P-type etch pit, with a reversed trumpet form. Depending on the type of dislocation and the etching conditions, an etch tunnel with a hollow core will be formed along a dislocation line.

By the utilization of etching that selectively proceeds at strain fields associated with dislocations, etch pits have been utilized as a good method of measuring dislocation density and of investigating the movement of dislocations [20]. Etching agents and conditions were developed to form P-type etch pits by selectively etching dislocations. Sirtle etchant, $HF : H_2O = 1:1 + CrO_3$ is an example of such an agent, and it is used to investigate Si crystals.

Since dislocations are linear strain fields, if a crystal is treated in an appropriate atmosphere, impurity ions selectively precipitate along the dislocation. These can be detected by infra-red microscopes, and so the method was used to prove the presence of dislocations during the early period of dislocation studies. If a dislocation is decorated by metallic elements, the dislocations act as a resistance against etching, and only the portion apart from dislocations is etched, and decorated dislocations remain as protrusions. The resulting protrusions are etch hillocks.

5.11 Surface microtopography: *Tracht* and *Habitus*

The *Tracht* and *Habitus* of polyhedral crystals bounded by flat faces are determined by the relative ratio of normal growth rates of faces appearing, and these morphologies are expected to appear under a driving force condition below $\Delta\mu/kT^*$, where the spiral growth mechanism principally operates. Therefore, factors which affect the *Tracht* and *Habitus* of polyhedral crystals will also affect the forms and advancement rates of spiral growth steps on smooth interfaces. As summarized in this chapter, these factors are determined by whether steps are rough or smooth, by the edge free energy of the steps, and the driving force. So, anything that has an impact on these factors will also affect the *Tracht* and *Habitus* of polyhedral crystals. These factors will be summarized in the following.

Since the edge free energies, γ, are different for the vapor and solution phases, and particularly for solute–solvent interaction energies, the same crystal species will exhibit different *Tracht* and *Habitus* in different ambient phases and different solvents. If impurities are present in the system, this affects γ and the advancing rates of steps. There are two opposite cases in impurity effects, and, depending on the interface state, some will promote growth, whereas others will suppress growth.

The driving force, $\Delta\mu/kT$, affects the advancement rate of steps, as well as the state of the interface. Growth temperature (and pressure) may modify the morphology of growth spiral layers and may affect *Tracht* and *Habitus* through the modification of the interface state, but the effect will appear through $\Delta\mu/kT$.

References

1 S. Ichikawa, Studies of etch figures of Japanese minerals, I, II, *Am. J. Sci., Ser.* 5, **6**, 1923, 53–72, 137–56

2 A. P. Honess, *The Nature, Origin and Interpretation of the Etch Figures of Crystals*, New York, John Wiley & Sons, 1927

3 A. R. Verma, *Crystal Growth and Dislocations*, London, Butterworths, 1953

4 W. Dekeyser and S. Amelinckx, *Les Dislocations et la Croissance des Cristaux*, Paris, Masson et Cie, 1955

5 I. Sunagawa, Step height of spirals on natural hematite crystals, *Am. Min.*, **46**, 1961, 1216–26

6 I. Sunagawa, Surface microtopography of crystal faces, in *Morphology of Crystals*, Part A, ed. I. Sunagawa, Dordrecht, D. Reidel, 1987, pp. 321–65

7 I. Sunagawa and Y. Koshino, Growth spirals on kaolin group minerals, *Am. Min.*, **60**, 1975, 407–12

8 J. J. De Yoreo, C. A. Orme, and T. A. Land, Using atomic force microscopy to investigate solution crystal growth, in *Advances in Crystal Growth Researches*, eds. K. Sato, Y. Furukawa, and K. Nakajima, Amsterdam, Elsevier, pp. 361–80

9 K. Onuma, T. Kameyama, and K. Tsukamoto, In situ study of surface phenomena by real time phase shift interferometry, *J. Crystal Growth*, **137**, 1994, 610–22

10 K. Maiwa, K. Tsukamoto, and I. Sunagawa, Direct observation of 2-D nucleation on a solution grown barium nitrate crystal, *Proc. 4th Topical Meeting on Crystal Growth Mechanism*, Tokyo, Japan Society for the Promotion of Science, 1991, pp. 67–70

11 A. R. Verma and P. Krishna, *Polymorphism and Polytypism in Crystals*, New York, John Wiley & Sons, 1966

12 I. Sunagawa, Vapor growth and epitaxy of minerals and synthetic crystals, *J. Crystal Growth*, **45**, 1978, 3–12

13 I. Sunagawa, J. Narita, P. Bennema, and B. van der Hoek, Observation and interpretation of eccentric growth spirals, *J. Crystal Growth*, **42**, 1977, 121–6

14 N. Cabrera and M. M. Levine, On the dislocation theory of evaporation of crystals, *Phil. Mag.*, **VII**, (1), 1956, pp. 450–8

15 N. Cabrera and D. A. Vermilyea, The growth of crystals from solution, in *Growth and Perfection of Crystals*, eds. R. H. Doremus, B. W. Roberts, and V. Turnbull, New York, John Wiley & Sons, 1958

16 I. Sunagawa and P. Bennema, Observations of the influence of stress fields on the shape of growth and dissolution spirals, *J. Crystal Growth*, **53**, 1981, 490–504

17 F. C. Frank, On the kinetic theory of crystal growth and dissolution process, in *Growth and Perfection of Crystals,* eds. R. H. Doremus, B. W. Roberts, and V. Turnbull, New York, John Wiley & Sons, 1958

18 R. E. Schwoebel and E. J. Shipsey, Step motion on crystal surfaces, *J. Appl. Phys.*, **37**, 1966, 3682–6

19 I. Sunagawa, Mechanism of natural etching of hematite crystals, *Am. Min.*, **47**, 1962, 1332–45

20 K. Sangwal, *Etching of Crystals – Theory, Experiment and Application*, Defects in solid, 15, eds. S. Amelinckx and Nihoul, Amsterdam, North-Holland, 1987

Suggested reading

A. R. Verma, *Crystal Growth and Dislocations*, London, Butterworths, 1953

W. Dekeyser and S. Amelinckx, *Les Dislocations et la Croissance des Cristaux*, Paris, Masson et Cie, 1955

W. R. Wilcox (ed.), *Preparation and Properties of Solid Materials*, vol. 7, New York, Marcel Dekker Inc., 1982

K. Sangwal and R. Rodriguez-Clemente, *Surface Morphology of Crystalline Solids*, Zurich, Trans. Tech., 1991

I. Nakada, *Crystal Growth Seen at Molecular Level*, Tokyo, AGNE Technical Center (in Japanese)

6

Perfection and homogeneity of single crystals

Real single crystals are not perfect or homogeneous; they contain various imperfections and inhomogeneities, such as growth sectors, vicinal sectors, growth banding, spatially distributed dislocations, and inclusions. Because these imperfections are induced either during or after growth, they provide us with informative records of growth processes. Crystals formed at higher growth rates may show lineage structure or split growth. Since we cannot observe the growth or post-growth processes directly in natural or synthetic crystals, these imperfections serve as an important information source, and they may be used to decode the growth process. In this chapter, we summarize and analyze how these imperfections and inhomogeneities are induced in single crystals.

6.1 Imperfections and inhomogeneities seen in single crystals

It would be foolish to assume that in real systems a constant growth rate is maintained throughout the whole crystal growth process, starting from nucleation through to the final stage. This is just as true in synthetic crystals grown under precisely controlled conditions as it is in natural crystallization, in which conditions are not controlled. The growth rates may vary for various reasons, resulting in fluctuations in impurity concentration and point defect density. Such fluctuations are recorded in single crystals in the form of "growth banding," which arises due to fluctuating concentrations of point defects, impurities, and inclusions. Since growth rates are dependent on crystallographic direction, perfection and homogeneity will be different among "growth sectors" formed by the growth of respective faces. Growth sectors may be closed within a crystal, or may appear repeatedly. Since the forms of growth sectors are basically reversed pyramidal, they are referred to as growth pyramids in the Russian literature. In this book, we call them growth sectors, so as to avoid confusion with growth hillocks formed by the spiral growth mechanism.

There are, in most cases, many spiral growth hillocks on one crystal face grown by the spiral growth mechanism. A polygonal growth hillock is bounded by a number of vicinal faces. Since step advancement rates are different in different directions, small growth sectors are formed corresponding to individual vicinal faces. Such growth sectors are called vicinal sectors or intra-sectorial sectors. Since impurity concentrations can depend on the difference in growth rates, optical properties such as the optic axis, extinction angles, color, and intensity of luminescence are different among different growth sectors and intra-sectorial sectors.

In many cases, macroscopic defects larger than those mentioned above are induced in the growth process. Minute crystallites of the same, or a foreign, species growing simultaneously with the host crystal will precipitate on the growing surface and will be trapped in the crystal as solid inclusions. A solution of the ambient phase is also often trapped as a liquid phase inclusion. Such inclusion occurs in the advancing process of macro-steps. Liquid inclusions transform, through the decrease in temperature and pressure, into vapor–liquid two-phase, or vapor–liquid–solid multi-phase inclusions, which often generate dislocations. Dislocations may be generated from the surface of the nucleus or seed crystal, or from defects such as a twin composition plane, and they tend to run nearly perpendicularly to the growing surface for energetic reasons. As a result, dislocation bundles originating from the center of the crystal or from inclusions and running perpendicularly to the habit faces are often observed in single crystals grown from the solution phase.

When growth rates are high, it is often observed that an apparent single crystal is composed of small blocks with slightly different mutual orientations starting from the center and reaching out to the surface of the crystal. This is referred to as a "lineage structure," implying handing down to posterity [1], [2]; it is also referred to as "split growth."

Although growth banding, growth sectors, intra-sectorial sectors, lineage structures, and split growth may sometimes be observed by the naked eye or by polarizing microscopy, often these imperfections and inhomogeneities can only be detected using methods such as X-ray topography, which can detect physical imperfections such as lattice defects; electron probe micro-analysis (EPMA) and X-ray fluorescence (XRF), with which it is possible to obtain data of chemical compositions of very small area, with a detection limit of parts per million order; or cathodoluminescence or laser tomography, which can detect the distribution of luminescent centers. Before methods such as these were developed, it was assumed that garnet crystals formed in metamorphic rocks were homogeneous, since they had reached the equilibrium state. Since chemical fluctuations have been detected in these crystals, the importance of kinetics has become recognized generally.

Representative growth banding, growth sectors, and the distribution of dislocations observable in single crystals are shown in Fig. 6.1. A cross-section of the prismatic crystal of tourmaline, $Na(Mg, Fe)_3Al_6(BO)_3Si_6O_{18}(OH, F)$, seen by transmitted light is shown in Fig. 6.1(a). The banding detectable by changes in color is the growth banding corresponding to compositional fluctuation. Figure 6.1(b) shows growth banding in quartz, which is discernible by the naked eye, corresponding to fluctuations in the density of two-phase inclusions. Figure 6.1(c) shows banding due to fluctuations of etch pit density, revealed by etching in KNO_3 of a (100) section of a diamond crystal, which appears to be perfect to the naked eye. The banding is growth banding, and it clearly shows changes in the morphology of the crystal during the growth and post-growth processes [3]. Figure 6.1(d) is a cathodoluminescence tomograph of a diamond, in which the growth history of the crystal can be decoded from fluctuations in the brightness of the luminescence. Figure 6.1(e) is an X-ray topograph of a section perpendicular to the *c*-axis of a quartz crystal, and it records the spatial distribution of growth banding, growth sectors, dislocations, and inclusions. Figure 6.1(f) is a sketch showing the lineage structure seen in a galena (PbS) crystal. All these figures are examples of observations on mineral crystals formed by natural crystallization in completely uncontrolled conditions. However, if highly sensitive detection methods are applied, imperfections and inhomogeneities of the types mentioned above may be detectable even in synthetic crystals grown under well controlled conditions. For example, the lineage structure is important in controlling the perfection of Si single crystals grown from the melt phase.

The inhomogeneities and imperfections seen in single crystals provide us with excellent records that indicate how the crystal developed through its growth history, how the growth rates or element partitioning fluctuated, and how the morphology varied. These records serve as important information sources relating to natural and synthetic crystals, whose growth process and post-growth changes are impossible to see directly. To decode the information properly, it is necessary to understand how such inhomogeneities and imperfections are induced into a growing crystal. We shall present the fundamental concepts of these processes in the following sections.

6.2 Formation of growth banding and growth sectors

Growth banding, growth sectors, intra-sectorial sectors, etc., are formed through the coupling of fluctuations in diffusion rates and growth rates, and the incorporation of impurities and point defects. Therefore, understanding the relationship between growth rate and element partitioning will provide us with a firm

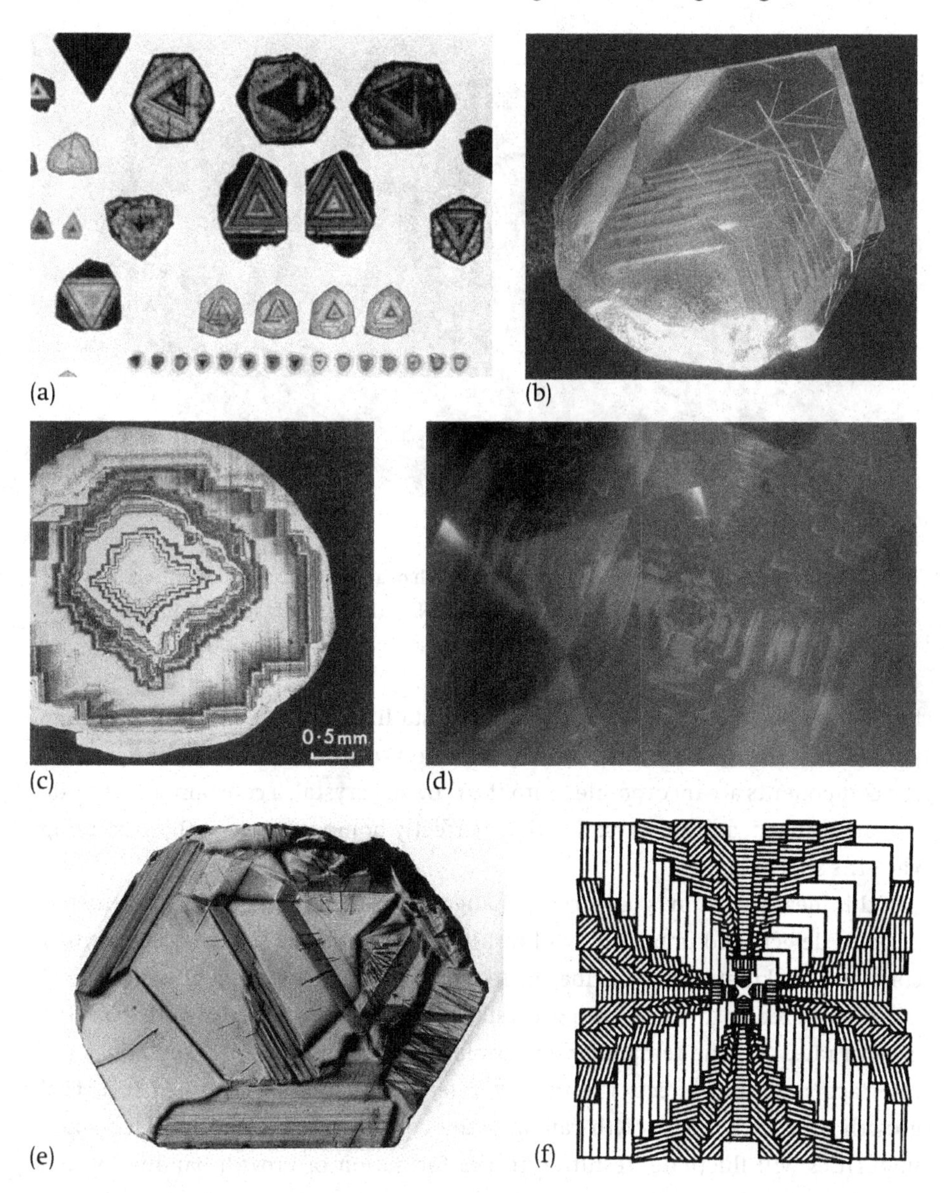

Figure 6.1. Representative examples of growth banding, growth sectors, and dislocation distribution seen in single crystals. (a) Color banding seen in cross-sections of prismatic tourmaline crystals (transmitted light). (b) Growth banding observed in a single crystal of quartz, generally known as ghost quartz. (c) Growth banding in a diamond crystal revealed by etching (ref. [2], Chapter 9). (d) Cathodoluminescence tomograph of a diamond-cut stone. The growth process can be decoded by fluctuations in cathodoluminescence intensity. (e) X-ray topograph of a section perpendicular to the *c*-axis of a quartz crystal from a pegmatite. Note the distribution of growth sectors, growth banding, inclusions, and the distribution of dislocations. Note also the direction of solution flow, the fluctuation in the growth rate, and the incorporation of inclusions. (Photograph by T. Yasuda.) (f) Lineage structure seen in galena crystals.

Figure 6.2. Cathodoluminescence tomograph of an emerald crystal synthesized by the high-temperature solution growth method. Note the regularly spaced faint growth banding appearing in between two distinct bands. (By courtesy of T. Miyata.)

foundation on which to build our ideas. If crystallizing (solute) and impurity components are transported constantly by diffusion from the ambient phase and all the components are incorporated into the growing crystal, a constant growth rate may be realized, and the crystal will be perfectly homogeneous, with no banding patterns appearing.

If this ideal situation is perturbed, changes in growth rates and in the densities of point defects and impurities will result. The partition of impurity elements is controlled by kinetics, as described in Section 3.14. Impurity elements with $K_{eff} < 1$ are distributed into the growing crystal, in the proportion controlled by the growth rate, and the remainder accumulate on the interface. This modifies the interface state and affects the growth rate; since these are coupled in actual growth processes, the growth rate and the concentration of point defects and impurities will fluctuate, resulting in the formation of growth banding. When growth parameters, such as temperature, change abruptly, we expect the formation of distinct bands due to an abrupt change in growth rate, but the above analysis also predicts that there is a possibility that bands will form, even in the absence of abrupt changes in conditions. It is often observed that there are thin, but regularly spaced, growth bands in between two distinct growth bands. Figure 6.2 shows a cathodoluminescence tomograph of an emerald crystal grown by the high-temperature solution (flux) method, on which regularly spaced, faint growth bands are seen in between distinct growth bands that correspond to abrupt temperature changes [4].

Growth bands are formed parallel to the interface. Straight banding is expected in the growth sector of a smooth interface, and curved, or hammock, banding is expected in the growth sector of a rough interface. If a roughening transition occurs on a smooth interface due to changes in condition, the corresponding banding will appear. When growth is intermittent and a curved interface appears by etching, which is followed by regrowth, a pattern consisting of high-index facets or alternating facets of a smooth interface will appear in the transitional process from a curved to a smooth interface. The micro-facets observable in Fig. 6.1(d) are examples of this. If macro-steps appear during the growth of a smooth interface, impurities concentrate there, and banding inclined to the interface will appear along the loci of concentrated impurities. This is called Type 2 banding, and is distinguished from banding that is parallel to the interface.

As described in Chapter 3, there are three types of growth mechanism that occur depending on whether the interface is rough or smooth, and the growth rate versus driving force relations are different depending on the respective growth mechanisms. Even if a crystal is growing under the same bulk driving force condition, the normal growth rates are different for different crystallographic directions. This is the reason for the formation of growth sectors. Therefore, the partitioning of impurity elements with $K_{eff} < 1$ will also be different in different growth sectors, and, as a result, impurity concentrations are different both in different growth sectors and intra-sectorial sectors. Impurity elements are not homogeneously distributed in a single crystal; they are distributed differently in respective growth sectors. The mode of growth banding will also be different in different sectors due to the same reason.

It might be expected that the concentration of impurities will be higher in the faces with higher growth rates, and thus that the impurity concentration will be higher in those growth sectors formed by a rough interface having the highest normal growth rate than in those formed by a smooth interface. However, the step advancing rate v on a smooth interface is much higher than the normal growth rate R of a rough interface. As a result, impurity elements with $K_{eff} < 1$ concentrate in the growth sectors of smooth interfaces rather than in those of rough interfaces. The partitioning of impurity elements with $K_{eff} < 1$ between growth sectors of rough and smooth interfaces was first noticed in the partitioning of Bi in Si single crystals grown from the melt phase, in which Bi concentrates in the growth sector formed by the growth of a smooth (111) interface. By extension, this difference in element partitioning between rough and smooth interfaces may lead us to presume that different partitioning occurs among different degrees of roughness of a smooth interface, i.e. depending on the difference in the order of morphological importance of the crystal faces. It is anticipated that the concentration of impurity elements with $K_{eff} < 1$ will be higher in the growth sector with the

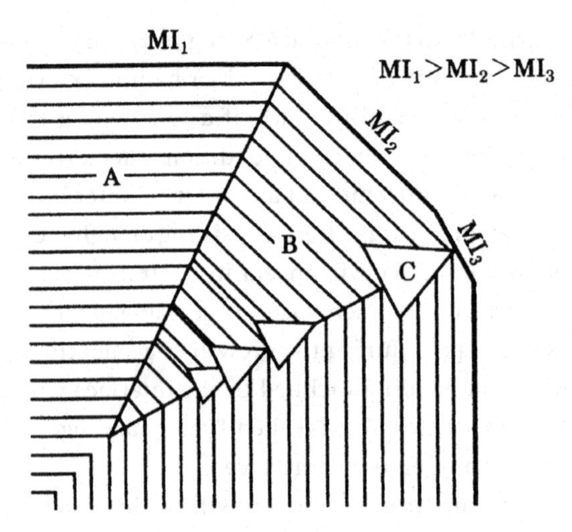

Figure 6.3. Expected difference in the features of growth banding in respective growth sectors depending on the order of morphological importance ($MI_1 > MI_2 > MI_3$). A, B, and C indicate growth sectors corresponding to respective faces and the features to be observed in growth banding. The distribution of dislocations is not indicated.

highest order of morphological importance. It is due to this that in diamond crystals nitrogen with $K_{eff} < 1$ concentrates more in the {111} growth sectors than in the {110} or {100} growth sectors. Impurities with $K_{eff} > 1$ behave in the opposite manner, and concentrate on the rough interfaces. In this case, both solute and impurity components are incorporated into the crystal at kinks through an adhesive process.

Growth sectors (including intra-sectorial sectors) form because growth mechanism, growth rate, and the associated partitioning of impurities and point defects are different. The boundary between the neighboring growth sectors corresponds to the place where strain concentrates, and so can be visualized by contrast on, for example, an X-ray topograph (refer to Fig. 6.1(e)). These boundaries often trap inclusions selectively, or act as sites that reflect or bend dislocations in one growth sector. The accumulation of strain occurs not only along the boundary between growth sectors of different faces, but also between growth sectors of crystallographically equivalent faces. This is because crystallographically equivalent faces do not necessarily have the same growth rates.

Growth banding seen in different growth sectors will have different features depending on the order of morphological importance. This is due to environmental conditions.

Figure 6.3 illustrates schematically features seen in the growth banding patterns in respective growth sectors, in relation to the order of morphological importance of the crystal faces. A rough interface disappears as growth proceeds, but the

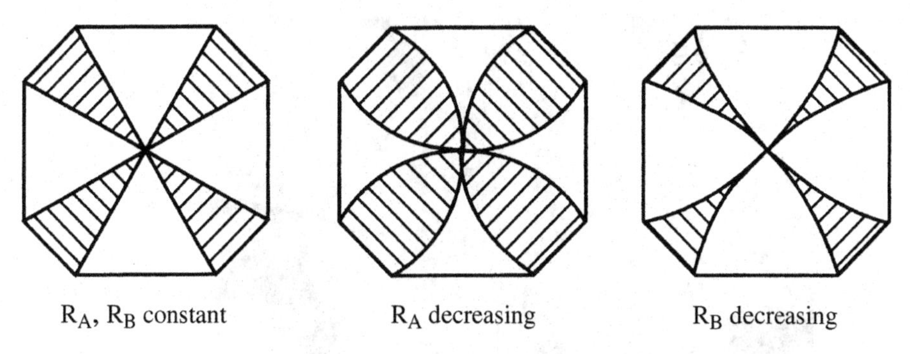

Figure 6.4. Forms of growth sector boundaries depending on the relation between R_A and R_B.

growth will be interrupted by impurity adsorption and grows again with a high growth rate. As a result, growth banding appears intermittently, and the growth sector will appear, but will soon disappear (because of the high normal growth rate). In the growth sector formed by the morphologically most important crystal face, growth at near steady state is realized and growth banding is indistinct; this is because fluctuations in environmental conditions affect the growth rate far less than in growth sectors of rough or morphologically less important faces.

If conditions fluctuate during the growth process, the degree to which the growth rates are modified will change on different crystal faces. As a result, the growth sector boundaries will change from straight, to barrel, to reverse trumpet type in crystals bounded by two faces A and B, depending on whether $R_A = R_B$, $R_A > R_B$, or $R_A < R_B$, respectively.

6.3 Origin and spatial distribution of dislocations

Dislocations are linear distortions of the lattice, and, unlike point defects, they are energetically undesirable. However, real crystals contain dislocations in the order $10^{6\sim8}/cm^2$, most of which are induced into the crystals during growth, and they also contribute to promoting the growth. As long as something occurs to generate a mismatch of the lattice planes, a dislocation will be generated, even though it is energetically unfavorable. In Fig. 6.5, X-ray topographs of two crystals, one grown from a seed with etched surface and another grown from an as-grown seed are compared (ref. [11], Chapter 7).

In Fig. 6.5(b), it is seen that dislocations in the seed are inherited, and that only a small number of dislocations generate from the seed surface, whereas in Fig. 6.5(a) most of the dislocations are newly generated on the seed surface. These newly generated dislocations start from initial liquid inclusions trapped at etch pits. The dislocations originate due to the failure of matching of lattice planes when inclusions

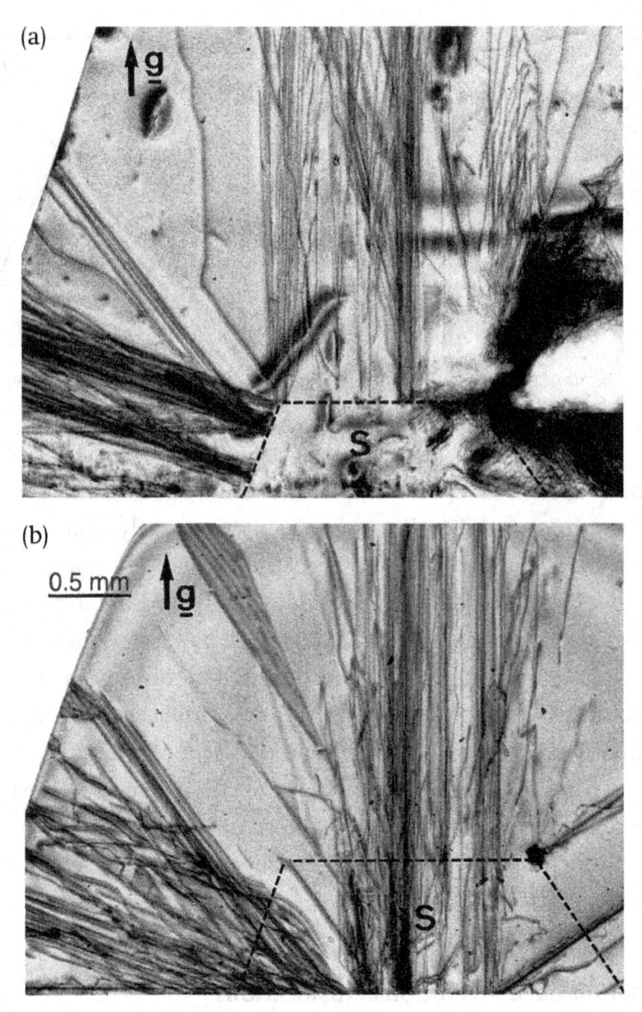

Figure 6.5. X-ray topographs demonstrating the differences in inheritance and generation of dislocations when (a) an etched crystal and (b) an as-grown crystal are used as seed crystals (ref. [11], Chapter 7). The crystal here is $Ba(NO_3)_2$, grown from aqueous solution. g = Burgers vector; S = seed crystal.

are enclosed. Dislocations generated from inclusions are seen in the X-ray topograph of quartz shown in Fig. 6.1(c).

Phenomena similar to inclusions (in respect to the origin of dislocation generation) are encountered at various stages of real crystal growth processes. For example, (1) when neighboring side branches meet in dendritic growth, (2) where two adjacently growing crystals come into contact, (3) at grain boundaries of slightly misoriented components in lineage structure or split growth, and (4) where one macro-step meets another macro-step while advancing on a smooth surface from different growth centers. Depending on the situation, dislocations

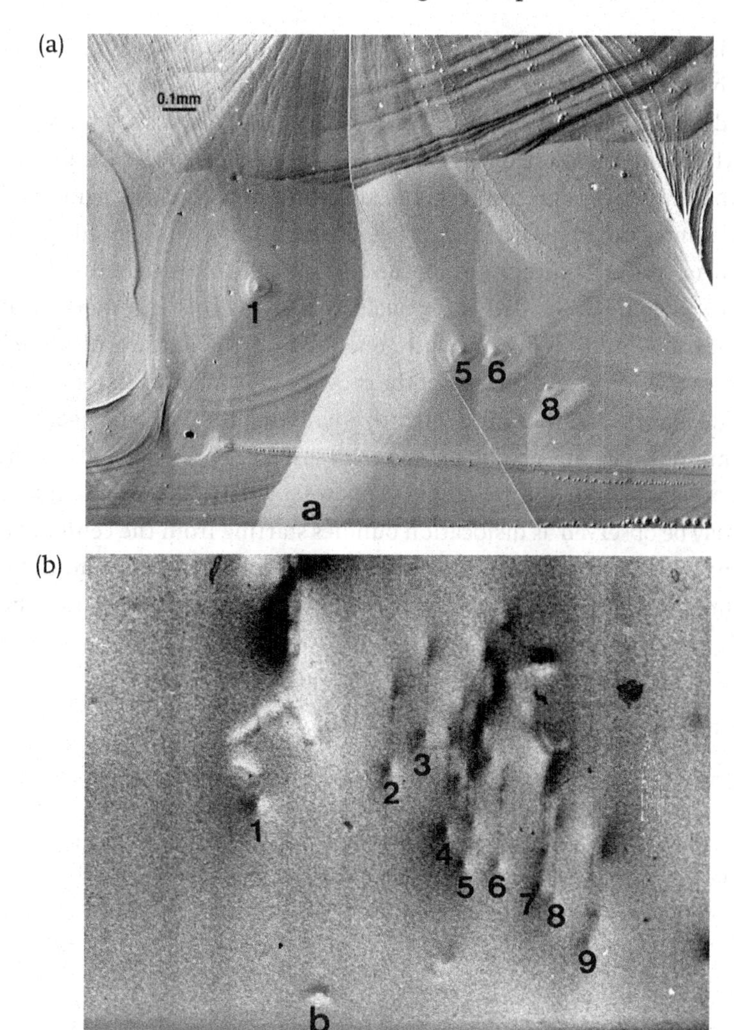

Figure 6.6. (a) Spiral growth hillocks observed on a (111) face of a $Ba(NO_3)_2$ crystal grown from aqueous solution. (b) Corresponding X-ray topograph. Only numbered dislocations act as active growth centers; the others do not contribute to the growth (ref. [11], Chapter 7).

with large Burgers vectors will be generated due to large lattice mismatch, and macro-steps with large step heights will advance from the outcrop on a smooth surface. However, such dislocations are energetically unfavorable, and tend to dissociate into elemental dislocations with small Burgers vectors and elemental spiral steps.

Not all the dislocations outcropped on a crystal surface are active growth centers. In Fig. 6.6, the distribution of growth hillocks on a (111) face of an as-grown $Ba(NO_3)_2$ crystal is compared with the distribution of dislocations revealed by X-ray topography. It is clearly seen in this figure that, among many dislocations, only some act as active growth centers.

The spatial distribution of dislocations in single crystals may differ greatly depending on the mode of growth of the crystal. In crystals grown from the melt, a condensed ambient phase using seed crystals, dislocations are mainly generated from the surface and from inside the seed. These dislocations are inherited by the growing crystal unless an appropriate method is applied to prevent this from occurring. Thus, a technique called Dash necking is used. This is a method of removing dislocations inherited from and generated on the surface of a seed crystal; it involves thinning the diameter of the growing crystal, until the crystal is of an appropriate length, and then subsequently increasing the diameter again.

The most commonly encountered distribution of dislocations in crystals grown from solution or vapor phase (dilute ambient phases) by natural nucleation and without seed may be observed as dislocation bundles starting from the center of a crystal and running nearly perpendicular to the habit faces. In addition to these dislocations, smaller dislocation bundles originating from inclusions may be observed. See Figs. 6.1(e) and 6.5 for examples.

References

1 M. J. Buerger, The significance of block structure in crystals, *Am. Min.*, **17**, 1922, 177–91
2 M. J. Buerger, The lineage structure of crystals, *Z. Krist*, **89**, 1934, 195–220
3 M. Seal, Structure in diamond as revealed by etching, *Am. Min.*, **50**, 1965, 105–23
4 T. Miyata, K. Tsubokawa, and M. Kitamura, Observation of synthetic emeralds by scanning cathodoluminescence method, *J. Gemmol. Soc., Japan*, **18**, 1993, 3–14

7

Regular intergrowth of crystals

Twinning, epitaxy, syntaxy, etc., describe the joining together or "intergrowth" of two individual crystals along a regular crystallographic orientation. Regular types of intergrowth (intergrowth relations) may also result when a crystal is decomposed or exsolved into two phases with different compositions, starting from one phase at the time of growth and forming a lamellar texture through changes of temperature and pressure. In all these intergrowths a new interface is introduced between the two individuals, and so they are in a higher energetic state than a single crystal by the amount of energy corresponding to the newly introduced interface. In some twinning relations, a re-entrant corner (edge) is also introduced between the two individuals, where preferential growth may take place. As a result, twinned crystals often show a different morphology from that of the coexisting single crystals. The aim of this chapter is to analyze the reasons for the features that result from the regular intergrowth of crystals.

7.1 Regular intergrowth relations

Crystals often intergrow with respect to a certain crystallographic orientation. The following intergrowth relations of two or more single crystals of the same species have been identified:

(1) twin;
(2) coincidence site lattice (CSL);
(3) parallel intergrowth.

Preferential nucleation along edges or at corners should also be mentioned. The following are examples of regular intergrowth between crystals of different species:

(1) epitaxy;
(2) syntaxy or topotaxy;
(3) exsolution or precipitation;
(4) spinodal decomposition.

7.2 Twinning

7.2.1 *Types of twinning*

Twinning is an intergrowth relation between two individuals of the same crystal species by a symmetry operation which is not included in the single crystal, formed through the processes of growth, transformation, and deformation. They are referred to as growth twinning, transformation twinning, and deformation, mechanical or secondary twinning, respectively.

Two different concepts have been introduced as symmetry elements that introduce the twin relation: one allows a center of symmetry, a symmetry plane, and two-fold rotation axis only; the other allows a rotation axis higher than three-fold [1]. In this book, we follow the first concept, and consider twinning by higher than three-fold rotation as a repeated symmetry operation of a two-fold axis. Symmetry elements that produce twin relations are referred to as the center of twin, the twin plane, and the twin axis. During a twinning operation, a composition plane and a re-entrant corner are introduced between the two individual crystals. A twinning relation in which the twin plane and the composition plane are the same is called a contact twin, and the composition plane is, in general, a low-index crystal face. There are also twinning relations in which two individuals are mutually penetrated, and in this case the composition plane of the two individuals is an irregular boundary plane. These are called penetration twins. When a twinning relation is repeated, a lamellar twin, an elbow twin, or a cyclic twin is formed. Examples are shown in Fig. 7.1.

7.2.2 *Energetic considerations*

A twin, in which a composition plane is newly introduced between two individuals, is in a state such that the energy corresponding to the interface is newly added to the energy state of the single crystal, which is in the minimum energy state. To form a twin, a driving force is required to overcome this additional energy, which automatically leads to a prediction that the smaller the interface energy, the higher the probability that twinning will occur.

The interface energy comes from a mismatch of atomic positions on the interface. If there is good coherency in atomic positions and bonds between the two individual crystals, the additional energy is small, and the probability of forming a twin is high. In the situation with less coherency or continuity and larger mis-

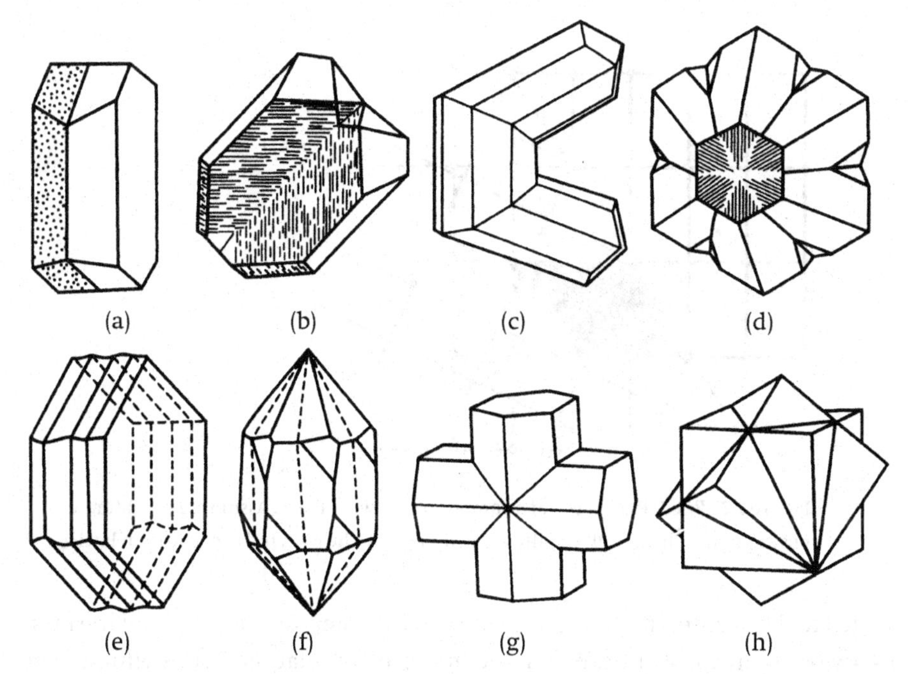

Figure 7.1. Various growth twins. (a) Contact twin (albite); (b) inclined twin (quartz); (c) elbow twin (rutile); (d) cyclic twin (chrysoberyl); (e) lamellar twin (albite); (f) penetration twin (quartz); (g) penetration twin (staurolite); (h) penetration twin (fluorite).

match on the interface, the interface energy is large, and the probability of intergrowing in a twin relation is low.

Before interface energy was understood, the concept was explained by Friedel [2] in terms of a compound or twin lattice. This may be explained as follows (see Fig. 7.2). If the lattice of one individual crystal is extended to superpose that of the other, where both are projected onto the same plane, a new lattice consisting of common lattice points results. This lattice is called a twin lattice or compound lattice, and the twin index is defined by the number of multiples of the unit cell size of a single crystal. The smaller the twin index, the higher the probability that twinning will occur.

The concept of the coincidence site lattice (CSL) due to Kronberg and Wilson [3] is an extension of Friedel's twin lattice. If two lattices are superimposed and one is rotated about an axis, a series of compound lattices consisting of coincident sites appear as the rotation angle changes. The compound lattices appearing in this way are larger than the basic lattice, and their sizes change according to rotation angle. Kronberg and Wilson explained that the rotation angle producing the smaller CSL is the one that leads to a higher probability of intergrowth. This is

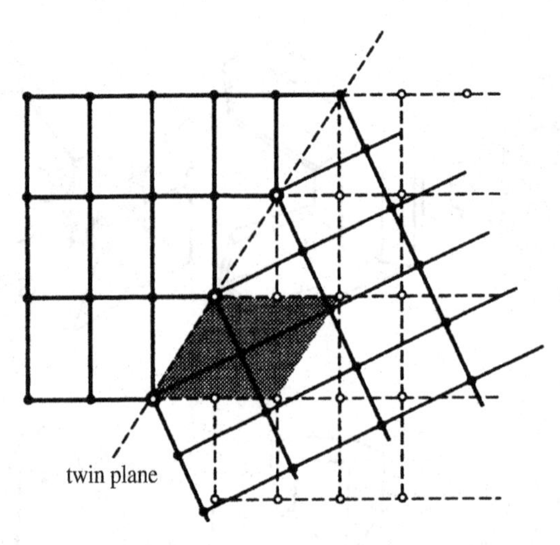

Figure 7.2. Friedel's concept of a compound or twin lattice. Assuming that atomic positions are adjusted elastically at lattice points, the twin index of the twin lattice is 2.

related to the value of interface free energy at the composition plane and the probability of producing a crystallographic intergrowth relation. Both twinning and CSL involve probability, and the probability increases if the size of the common lattices is smaller. It should be understood that the term "twin" is applied to the interactions occurring at higher probability than other intergrowth relations.

Aragonite is a high-pressure polymorph and calcite is a high-temperature polymorph of $CaCO_3$. Aragonite belongs to the orthorhombic system mmm, and calcite is part of the trigonal system, $\bar{3}$m. Aragonite occurs, in most cases, as a cyclic twin showing a pseudo-hexagonal form by repeated twinning on {110} as the composition and twin plane, whereas calcite occurs less frequently as a twin. The reason for this discrepancy has been accounted for in terms of the twin lattice or the difference in how many unit cells there are in a CSL.

Both Friedel's twin lattice and Kronberg and Wilson's CSL are related at unit cell level; both explain well, at least qualitatively, the twin relations, including the probability. At the structural level within the crystal, this corresponds to a problem in continuing the structure of two individual crystals across the composition plane; put another way, we have a problem of whether there is a good coherency or not. Figure 7.3 is taken from Bragg and Claringbull [4], which explains the twin of aragonite. There is not an unreasonably large distortion in the positions and continuation of ions of Ca and CO_3 across the boundary plane.

In close-packed structures, a twin relation is introduced by a stacking fault. In this case, there are no obstacles to the continuation of the structure, other than the stacking fault itself. However, in some twinning relations, strain due to mis-

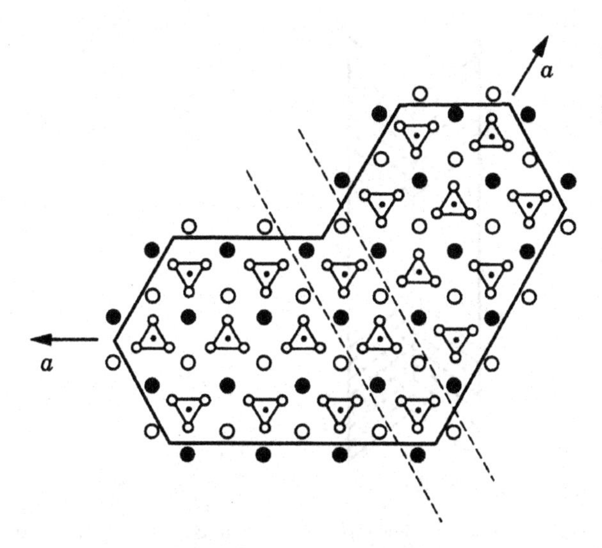

Figure 7.3. Structure of aragonite twin [4]. Triangles denote CO_3, and the solid and open circles represent Ca.

match is released by the introduction of dislocations. The Brazil twin of quartz is such an example, in which a dislocation is always present at the termination of the twin lamellae (see Chapter 10).

To form a twin that is in a higher energy state than a single crystal, by an amount equal to the interface energy, we require a higher driving force, which will overcome this excess energy. If nucleation takes place under a higher driving force condition than that required for the nucleation of a single crystal, the probability of formation of a twinned nucleus increases. From an energy standpoint, the prediction is that the probability of formation of a growth twin will be highest in the nucleation stage under a high driving force condition, i.e. at high-supersaturation condition, and the probability will diminish sharply in the growth stage when the driving force is lowered. This analysis was made by Buerger [5]. Two individual single crystals joined in a twin relation during the nucleation stage (i.e. a growth twin) should therefore develop to become equal in size; observations confirm this prediction. A high driving force not only results from a high-supersaturation condition, but also for other reasons, such as the presence of appropriate impurity components, or mechanical shock in the nucleation stage.

Although the probability of twin formation occurring during growth is far lower than that during nucleation, there is a possibility that growth twins, formed by the conjoining of two individuals during growth, will result. In this case, it is necessary that the two individual crystals have polyhedral forms bounded by flat faces. This flat face acts as the composition plane, which is equal to the twin plane. The

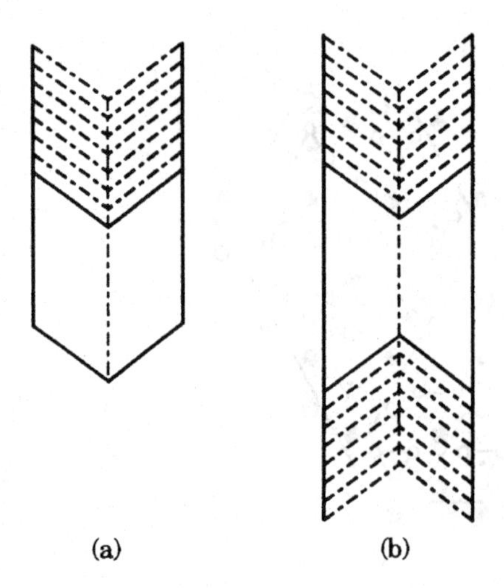

Figure 7.4. The re-entrant corner effect. (a) One re-entrant corner; (b) two re-entrant corners, one at each end.

sizes of the two individual crystals may be different. The probability of joining an individual in a twinning relation along the edges or at the corners of another larger crystal is much lower than the probability of twinning on a flat face.

7.2.3 *Re-entrant corner effect and pseudo re-entrant corner effect*

A re-entrant corner (edge, see Fig. 7.5(c)) is introduced between two individuals by twinning. Due to the presence of a re-entrant edge (corner), a re-entrant corner is an energetically more favorable site for two-dimensional nucleation than is a flat face. This results in a different morphology for a twinned crystal than for a single crystal. In twin relations with only one or two re-entrant corners (Fig. 7.4), the twinned crystal takes an elongated or ribbon form along this direction, and in twin relations with three re-entrant corners (Fig. 7.5), the twin takes a triangular form, as opposed to the octahedral form of the single crystal. If several twin planes are introduced repeatedly in this twin, re-entrant corners appear in six directions, and thus preferential growth sites do not disappear for ever, resulting in the formation of a ribbon crystal as long as nutrient is supplied. (This principle is used in the synthesis of Ge ribbon crystals [6].) The re-entrant corner effect [7] was proposed to account for the origin of a characteristic morphology of twinned crystals that is different from that of single crystals in terms of preferential growth at the re-entrant corners.

As another example of utilizing platy forms of twinned crystals, we should mention a recent remarkable increase in the photosensitivity of photographic film. Photographic emulsion film consists of minute AgBr crystallites, with octahedral or cubo-octahedral *Habitus*. The photosensitivity is determined by the area

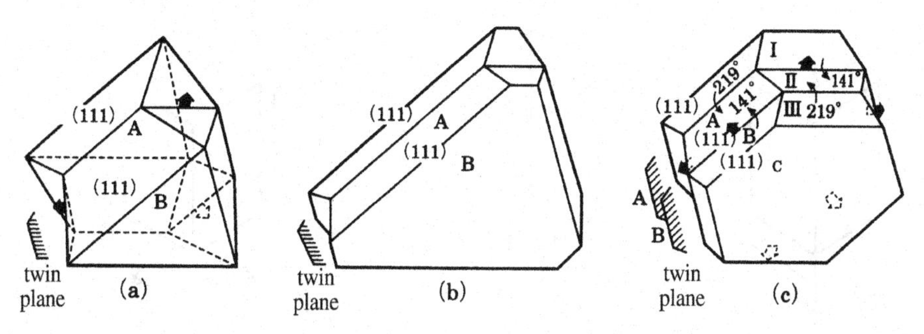

Figure 7.5. (a) Twin with re-entrant corners in three directions (shown by arrows) and (b) its expected re-entrant corner effect. (c) Morphological change is expected for repeated twinning.

covered by these crystals. However, if the probability of twinning is increased by controlling the conditions, the area covered by the crystals increases greatly, since twinned crystals take triangular or hexagonal platy *Habitus*, so increasing the photosensitivity.

Real crystals contain dislocations, which act as self-perpetuating growth sites. In some twin relations, dislocations are present in the twin plane or concentrate in the composition plane as growth proceeds. Since growth mediated by dislocations is energetically more advantageous than that mediated by re-entrant corners, either a unidirectional elongated form appears, or the twin grows in a similar form in real crystals, as shown schematically in Fig. 7.6, because both re-entrant and protruded sides grow similarly. This means that, in real crystals, the re-entrant corner effect, in its original meaning, cannot be expected to operate. However, many actual twins have sizes and morphologies different from those of coexisting single crystals. A good example may be seen in quartz twinned crystals, which adhere to the Japan law, which grow larger and show a more flattened characteristic V-shape or fan-shape than do hexagonal prismatic forms of coexisting single crystals, which will be discussed in detail in Chapter 10. This is because dislocations converge in the composition plane $\{11\bar{2}2\}$ and promote the growth of $\{10\bar{1}1\}$ faces adjacent to the composition plane above that on crystallographically equivalent $\{10\bar{1}1\}$ faces on the opposite, non re-entrant corner, side. On the opposite protruding side, with no re-entrant corner, dislocations diverge as growth proceeds from the composition plane, and so preferential growth at the twin junction does not occur. Since this type of re-entrant corner effect is due to preferential growth caused by re-entrant + concentration of dislocations, the effect is called the pseudo or apparent re-entrant corner effect [8].

In close-packed structures or fcc crystals, twinning is introduced by stacking faults. In this case, the excess energy at the twin plane (= composition plane) is small, and so neither generation nor concentration of dislocations is required. In

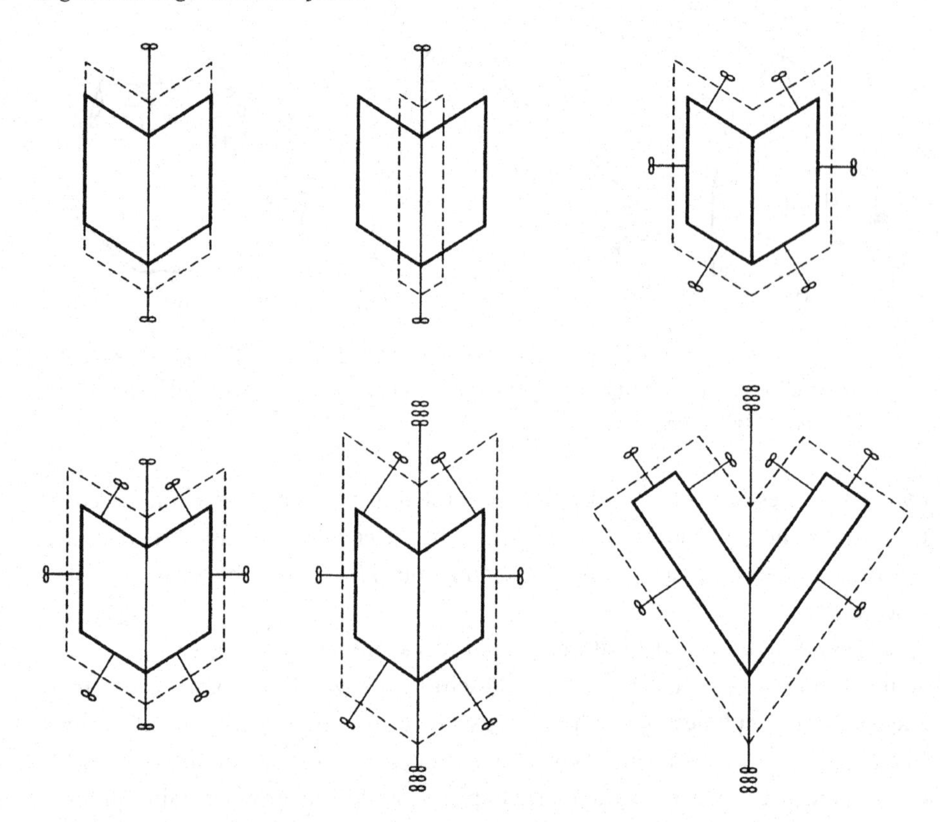

Figure 7.6. Morphological variation expected in twins of real crystals. The ∞ symbols indicate dislocations. Only similar forms or unidirectional growth are expected.

such a case, the mechanism proposed to continue preferential growth by twinning is alternately repeating the re-entrant corner effect and rough interface growth [9]. Figure 7.7 shows the situation in which a stacking fault is introduced and a twin relation is generated in an fcc structure. It is seen that through the twinning operation, re-entrant corners and rough {100} faces appear alternately at an angle of 60° to the interface edge. As a result of preferential growth at re-entrant corners, rough {100} faces appear, and, in the following step, re-entrant corners appear as a result of growth on rough {100} faces. Since this alternation continues forever and is self-perpetuating, a thin platy twinned crystal appears in this type of crystal, even if the crystals do not have dislocations.

7.2.4 *Penetration twins and contact twins*

Penetration twins have an irregular composition plane, and the twin plane and the composition plane are not the same. As a representative example, the {111} twin of fluorite, CaF_2, is shown in Fig. 7.8.

On the {100} faces of individual A, the <111> corner bounded by three {100}

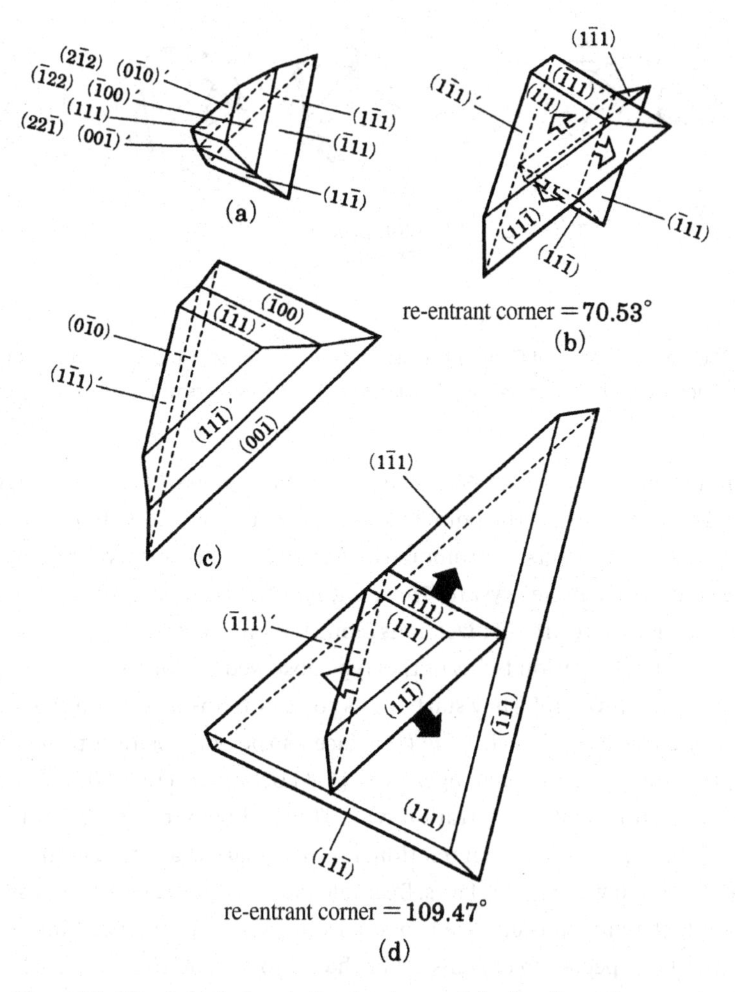

Figure 7.7. Morphological evolution due to growth by the alternating re-entrant corner effect and rough interface growth [8].

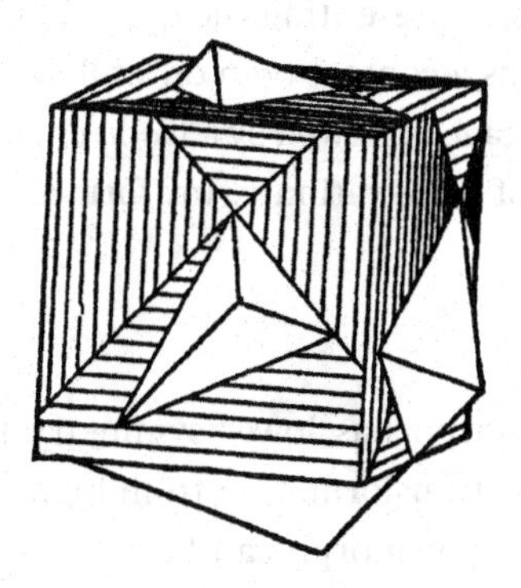

Figure 7.8. Penetration twin of fluorite. Note the origin of growth layers on the {100} face.

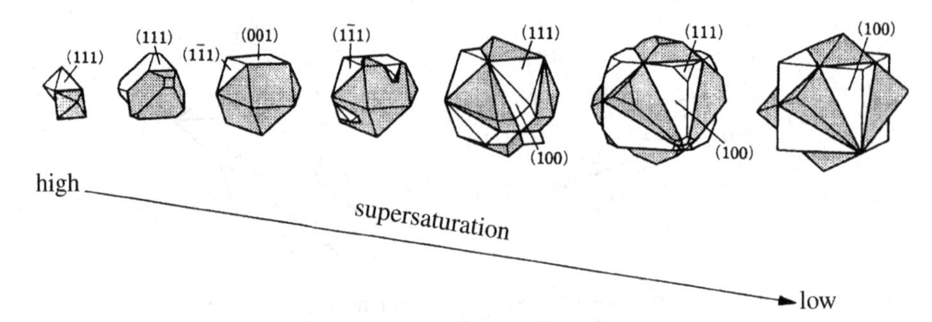

Figure 7.9. Process of formation of a penetration twin, starting from a contact twin, followed by *Tracht* change associated with growth (as supersaturation decreases).

faces of individual B protrudes. All growth layers on the {100} faces of both A and B individuals originate from the points of contact of the two individuals, indicating that the twin relation acts to promote the growth. As a result, twins grow larger than the coexisting single crystals. However, no change is observed in their morphology, because there are six directions along which to accelerate growth.

The origin of penetration twins of this type observed in fluorite and galena may be analyzed as follows [10]. They start out as contact twins by the joining together of two octahedral crystals, with the (111) face as the composition plane, and, as growth proceeds, their *Tracht* changes from {111} octahedral to {100} cubic, resulting in the appearance of a penetration twin. This is shown in Fig. 7.9. For two individuals to join together in a twin relation, it is necessary that the crystal is already bounded by well developed flat faces. Examples such as these, where twins formed first as contact twins and changed to apparent penetration twins due to *Tracht* change, may be expected in crystals other than fluorite and galena. The characteristic V-shape or fan-shape morphology of quartz crystals following the Japan law was previously explained as representing the upper half of an X-shape twin (penetration twin), but may now be accounted for as representing the upper half of a Y-shape twin (contact twin), which started out as a contact twin on well developed {10$\bar{1}$1} faces that appeared at the nucleation stage, followed by a *Tracht* change (see Chapter 10). There must be many examples of penetration twins formed by this process.

7.2.5 *Transformation twin*

In general, a low-temperature polymorph has lower symmetry than a high-temperature polymorph. So, in the phase transformation from high- to low-temperature polymorph, the low-temperature polymorph can have two orientations, which leads to twin formation. This is called the transformation twin. To minimize the interface energy between twinned individuals through transformation, transformation twins generally show lamellar cross-hatch patterns parallel to the low-index faces of a high-temperature polymorph crystal. There are, however, cases in which the boundary plane (composition plane) takes a non-crystallographic irregular form. The extreme case of the former type is called a martensitic transition, in which a twin relation may appear without any precursor phenomenon under a small driving force. In the latter type, a precursor phenomenon is observed. An example is seen in the Dauphiné twin of quartz formed by a phase transition from high-temperature to low-temperature polymorph. If a transmission-type electron microscope with a temperature gradient is used, numerous minute twin domains, followed by their conjoining to form large Dauphiné twins, may be observed.

The appearance of a twin nucleus associated with phase transitions occurs preferentially, in many cases, on free surfaces, internal surfaces of inclusions, boundary surfaces of solid inclusions, or in dislocations where strain is concentrated.

7.2.6 *Secondary twins*

A mechanical, or secondary, twin is a twin relation that appears due to the displacement of one part of a crystal in a twin relation with respect to the other part, when an external force is applied, and can be regarded as being due to glide. Glide is a phenomenon that may be described as the displacement of part of a crystal by the Burgers vector on a close-packed glide plane, without loss of adhesive force, and has been accounted for in terms of dislocation movement. Twinning due to glide is a phenomenon that occurs when part of a crystal is displaced, as a body, with respect to another part by an external force applied from a certain direction, and the two parts are in a twin relation. Glide plane, glide direction, and glide twin plane (and cleavage plane) may or may not be the same crystal plane. In calcite, $CaCO_3$, both the glide plane and the mechanical twin plane are $(01\bar{1}2)$ faces, but the cleavage face corresponding to the close-packed plane is $(10\bar{1}1)$, whereas in galena, PbS, both the cleavage and glide planes are (100), but many faces, (221), (332), (441), etc., have been observed to be mechanical twin planes. Deformation twinning has been universally observed in calcite crystals constituting marble, and also in corundum crystals in basalt which have experienced similar mechanical stress histories. Thus, we see that mechanical twinning is useful in decoding the stress history experienced by minerals and rocks.

7.3 Parallel growth and other intergrowth

After the completion of a series of growth processes forming polyhedral crystals, and a period of cessation of growth, a new stage of growth may recommence from newly supplied solution. In such a case, various intergrowth relations

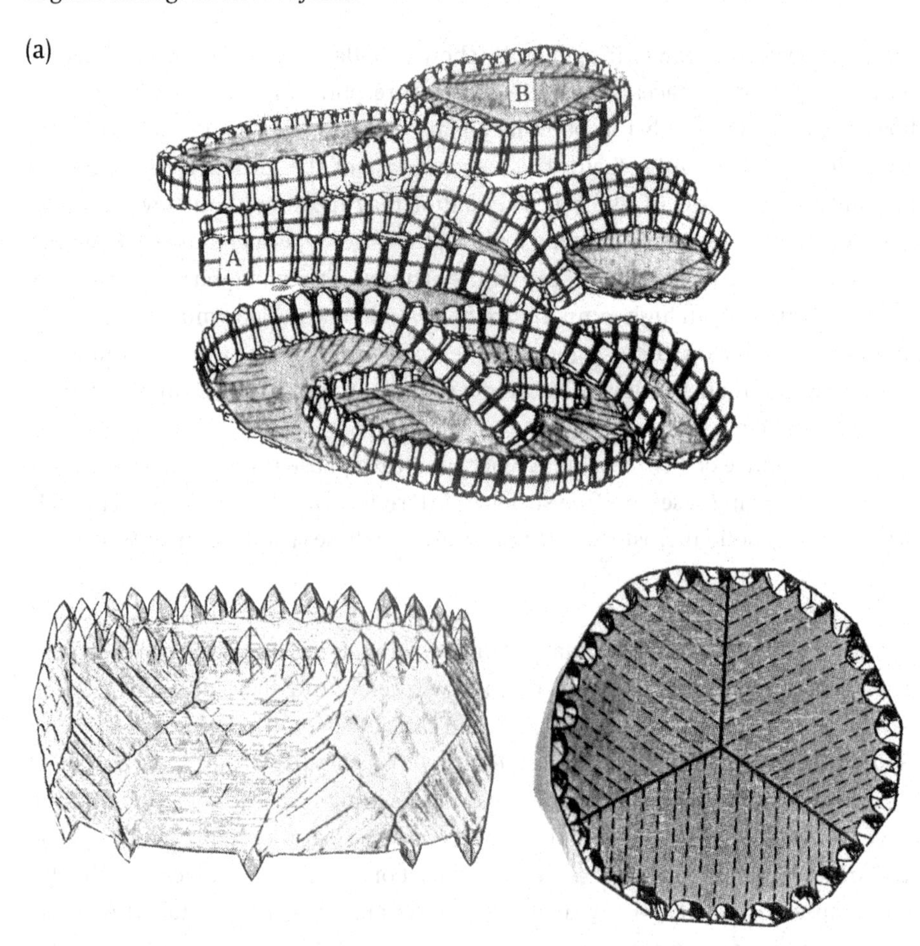

Figure 7.10. (a) Sketches and (b) photograph of parallel growth due to preferential nucleation at the corners and along the edges of existing calcite crystals (ref. [3], Chapter 11). Note the *Habitus* change between earlier and later formed crystals.

may appear depending on the state of the earlier formed crystal surfaces. What results depends on the origins of various intergrowth relations, such as crystal growth using seed, epitaxial growth, parallel growth and preferential nucleation along edges and corners, and unusual morphology such as platy *Habitus*.

New growth may continue in a similar manner to previous growth because the active sites for growth on the crystal surface remain active for new growth to proceed. This corresponds to growth using a seed. In single crystal synthesis, the surface of the seed is usually etched to activate growth, and so etch pits are formed on the seed surface. Dislocations are newly generated on the seed surface in the process of enclosing the inclusions. In general, more dislocations than those already present in the seed are newly generated on the seed surface. If as-grown,

(b)

Figure 7.10 *(cont.)*

non-etched seed crystal is used, dislocations in the seed are inherited which constitute the major part of dislocations in the newly formed crystal [11]; see also refs. [1] and [2], Chapter 3, and Fig. 6.5.

If most of the active growth centers on the surface become inactive because the surfaces are covered by other crystallites that were precipitated later or because a high driving force is achieved, the edges and corners of the older crystals act as preferential nucleation sites, and a large number of tiny crystallites preferentially nucleate at corners or along edges, all with the same orientation. As seen in Fig. 7.10, such a situation may give the impression that a large polyhedral crystal is formed by the coagulation of minute crystallites acting as growth units. Parallel growth occurs through a similar process. The reason why both the older and newly

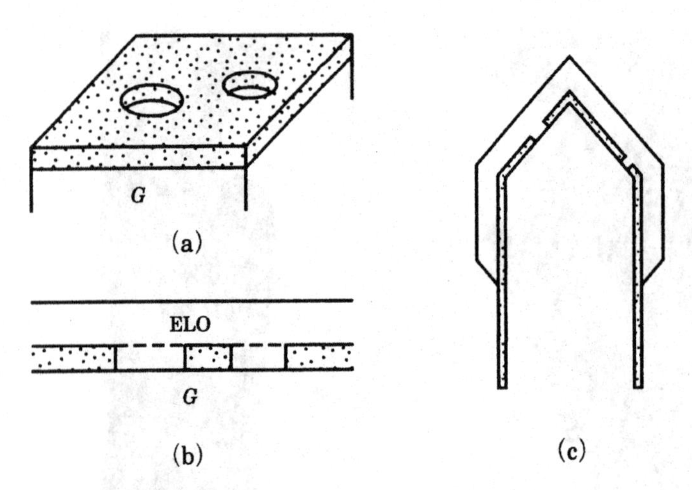

Figure 7.11. Schematic illustrations of (a), (b) epitaxial lateral overgrowth, ELO, and (c) the mechanism of formation of scepter quartz.

formed crystals have a common crystallographic orientation is that there are channels in the layer of foreign material covering the surface of the older crystal, through which the host and guest crystals are related. By extension, we may predict that the growth of a new layer proceeds over the covered layer in the same direction as the older or substrate crystal. This mechanism is called epitaxial lateral overgrowth* (ELO) and has been used as a method of controlling dislocations in liquid phase epitaxy (LPE) (see refs. [18] and [19], Chapter 10), a method that produces a dislocation-free epitaxial layer by masking the area with dislocation outcrops and allowing epitaxial growth only in the non-masked area without dislocation outcrops, followed by ELO on the masked area (Fig. 7.11). The same mechanism operates among natural quartz crystals in the formation of scepter quartz. (The details will be discussed in Chapter 10.)

7.4 Epitaxy

The term "epitaxy" is a compound word formed from the Greek επι (over) and τάξιζ (orderliness), and was first used by Royer [12] to describe the phenomenon of growth of a guest crystal with a certain crystallographic orientation over a low-index crystal face of (a different) host crystal which acts as the substrate. This phenomenon has been observed widely in minerals; examples are shown in Fig. 7.12. Royer used the cleavage surfaces of mineral crystals as host crystals, and systematically investigated the relationship between host and guest crystals by growing and observing the growth of guest ionic crystals from liquid droplets

* Recently, this has been given the name "microchannel epitaxy."

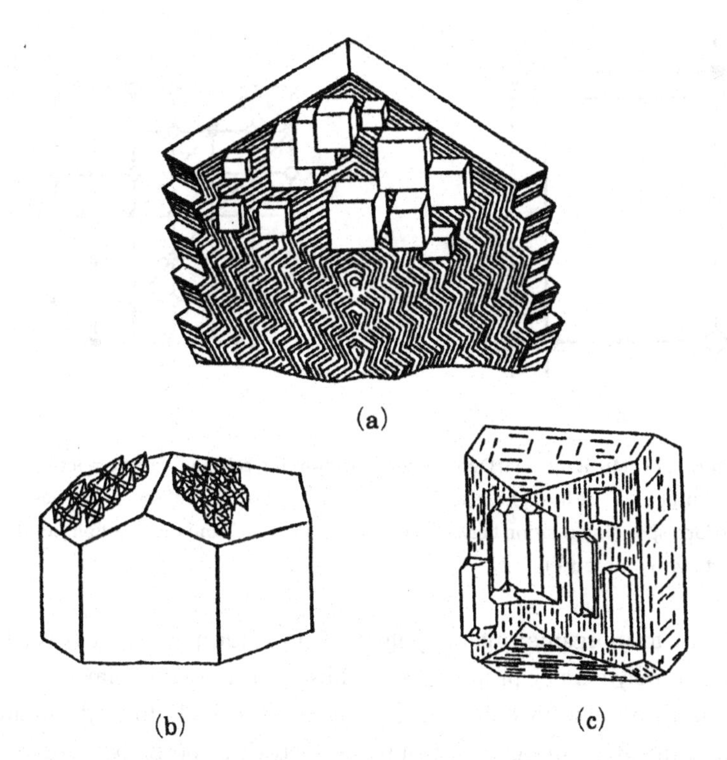

Figure 7.12. Examples of epitaxy observed in minerals. (a) Pyrite on marcasite; (b) quartz on calcite; (c) albite on orthoclase.

placed on the host surface. He systematized the observed mutual relations using the concept of a misfit ratio. Epitaxial growth has become widely used in industries that require pn junctions or that make semiconductor devices. The term epitaxial growth, which was originally applied to the relationship between different crystal species, has now been expanded to describe the relationship between p and n types of the same crystal with different impurity content. Consequently, misunderstandings have arisen about the original meaning of epitaxy. The term "hetero-epitaxy" is specifically used to express the relation between different crystal species, and "homo-epitaxy" is used for the relation between crystals of the same species.

As epitaxy has become used extensively in the semiconductor industry, so various methods of epitaxial growth have been developed. Growth methods from the vapor phase, such as metal–organic chemical vapor deposition (MOCVD), and those from the high-temperature solution phase, such as liquid phase epitaxy (LPE), have been widely investigated and used in practical applications. Also, host–guest relations have been expanded from inorganic:inorganic, such as metal:metal, metal:semiconductor, oxide:semiconductor, and oxide:van der

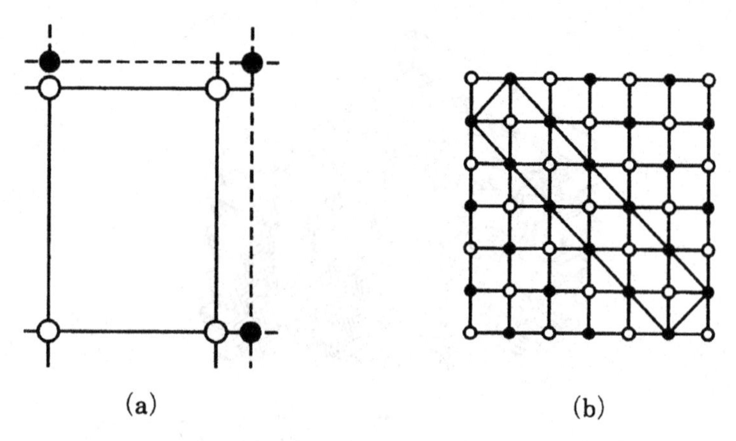

Figure 7.13. Misfit ratio. (a) One-to-one corresponding relation. Open circles indicate the unit cell of the host phase, and the solid circles show that of the guest phase. (b) Corresponding relation between the host phase, NaCl, and the guest phase, alizarin (rectangular unit cell).

Waals crystals, to inorganic:organic polymer, inorganic:protein. It is predicted that the use of polymers or proteins as the host and inorganic materials as the guest will be developed. In doing this, we hope eventually to understand the growth mechanism of inorganic crystals constituting tooth, bone, and shell formed by physiological mechanisms.

An interface is introduced by epitaxy between the host and guest crystals. As in the case of twinning and CSL relations, it is expected that the growth of the host and guest crystals in a certain crystallographic orientation is expected if the interface energy is small, but a definite relation will not be realized if the interface energy is large. Thus, Royer introduced the concept of a misfit ratio. To express the amount of misfit between the host and guest lattices, the degree of disagreement is expressed by the misfit ratio, taking the lattice size of the host crystal to be 100. If this ratio is a small percentage, an epitaxial relation may be realized, but above this the probability of the host and guest growing in an epitaxial relation diminishes sharply. However, there are exceptional cases in which an epitaxial relation is realized, even though the misfit ratio exceeds 10%.

The corresponding relation between the host and guest crystals when evaluating the misfit ratio may be a one-to-one lattice relation in the same direction ($a \times b$ to $a \times b$ axes), or in different axial directions ($a \times b$ axes versus $a \times \langle 110 \rangle$ axes), or on the basis of one unit cell versus a few unit cell sizes (see Fig. 7.13). Royer's misfit ratio is generally a two-dimensional correspondence, but Hartman [13] extended this relation to the misfit ratio in PBCs (see Section 4.2), which is a one-dimensional correspondence. Royer's epitaxial relations correspond to a relation between the F faces of the host and guest crystals containing more than two PBCs, and an epitaxial relation is not allowed between S faces or K faces. In Hartman's analysis, relations between one-dimensional PBCs become important, permitting epitaxial relations on an S face as well. In this way, epitaxial relations unexplained by Royer's misfit ratio were accounted for.

The freedom of the dangling bonds on the crystal surface increases with increasing temperature. As a result, there is a critical temperature below which an epitaxial relation cannot be realized. This temperature is called the epitaxial temperature, and it depends on interface orientation. If the misfit ratio is small, the epitaxial temperature is low; if the misfit ratio is large, the epitaxial temperature is high, and an epitaxial relation will not be achieved unless the temperature is higher than the epitaxial temperature.

Another element controlling the epitaxial relation is the state of the dangling bonds on the surface of the host crystal. When the dangling bonds are no longer active owing to adsorbed molecules, the epitaxial relation is not realized. This is why host surfaces cleaved in air have different epitaxial temperatures from those cleaved in a vacuum.

The preceding arguments are based on the direct relation between host and guest crystals. If the surface of a host crystal is covered by an adsorbed species, or if the presence of an intermediate layer, such as a transition layer, is allowed between the host and the guest through which the misfit ratio is diminished, it is possible to achieve an epitaxial relation between host and guest crystals with a large misfit ratio. Epitaxial growth via transition layers has been successful in compound semiconductors. The morphology of crystals grown by epitaxy is different from that of freely grown crystals. This is, in some way, phenomenologically similar to anisotropic development of crystallographically equivalent faces when crystals grow in an anisotropic environmental phase, but the reason is different and is attributed to the presence of interface energy. Also, crystals elongated along PBC directions will appear when the growth of the guest is controlled by the PBCs in the host.

In the above discussion, three-dimensional nucleation of a guest crystal on the surface of a host crystal is presumed to be an epitaxial growth mechanism.

The following three mechanisms are suggested for the growth of a guest crystal in an epitaxial relation on the surface of low-index faces of a host crystal.

(1) Mechanism by three-dimensional nucleation on a host crystal. This is called the Volmer–Weber mechanism.
(2) Mechanism forming a thin film in an epitaxial relation through a monolayer spreading parallel to the interface. This is called the Frank–van der Merve mechanism.
(3) Intermediate mechanism between (1) and (2). At first, an epitaxial monolayer is formed, followed by three-dimensional nucleation. This is called the Stranski–Krastanov mechanism.

Table 7.1 *Three modes of growth mechanisms in epitaxial growth and corresponding conditions*[a]

Growth mechanism	*Energy conditions*	*Minimum $\Delta\mu$ for condensation to occur*
Volmer–Weber (weak adsorption) three-dimensional nucleation	$\Delta\alpha>0$ or $2\alpha>\alpha_s$	$\Delta\mu>0$
Frank–van der Merve (strong adsorption, saturated to bulk phase)	$\Delta\alpha<0$ or $2\alpha<\alpha_s$	$\varepsilon_1-\varepsilon_2<\Delta\mu<0$ (relatively under-perfect wetting)
Monolayer spreading parallel to the interface $-\Omega\lvert\Delta\alpha\rvert/a<\Delta\mu<0$ (thin film) $\Delta\mu>0$ (thick film)		
Stranski–Krastanov (intermediate) $-\Omega\lvert\Delta\alpha\rvert/a<\Delta\mu$ three-dimensional nucleation, $\Delta\mu>0$	first layer $\Delta\alpha<0$ monolayer spreading $\Delta\alpha>0$	$\Delta\mu<0$ (lateral spreading of first layer) $\Delta\mu>0$ (spreading of first layer and three-dimensional nucleation)

[a] Taken from ref. [14]

If we systematize the growth conditions under which the three modes of growth mechanisms operate and express them in a simple manner in relation to the work required to cause coagulation of the guest component onto the surface of the host, $\Delta\alpha=2\alpha-\alpha_s$, and the driving force $\Delta\mu=\mu_v-\mu_s$, we can formulate Table 7.1. This table is a simplified Chernov table [14], in which α is the host–guest interface; α_s is the adsorption energy of the guest; $\Delta\mu$ is the chemical potential difference between the vapor and solid phases; Ω is the molecular volume; and ε_1 and ε_2 are, respectively, the single bond energy and the bond energy between the host and guest molecules. We can predict that the epitaxial relation will collapse on increasing the driving force. If the epitaxial relation holds under a smaller driving force condition but not under a larger driving force, when the growth is governed by the Volmer–Weber mechanism, we can expect compact texture formation under small driving force conditions and coarse texture under large driving force conditions.

Table 7.2 *Regular intergrowth relations observed in minerals*

Terms	*Examples*
Coaxial intergrowth	amphibole and pyroxene (common *c*-axis)
	zircon and xenotime (common *c*-axis)
Oriented overgrowth (epitaxial growth)	rutile (100) on hematite (0001)
	chalcopyrite (112) on enargite (001)
	many other examples
Symplektitic intergrowth (including celyphite rim, corona, reaction rim)	quartz and feldspar
	plagioclase and magnetite
	garnet and quartz
Mylmekite intergrowth	quartz and plagioclase
Micrographic intergrowth	quartz and plagioclase
Topotaxy (syntaxy)	anatase, rutile
	pyrrhotite→marcasite + pyrite

This is an important relation in investigating the textures of crystal aggregates formed by physiological mechanisms.

7.5 Exsolution, precipitation, and spinodal decomposition

In addition to epitaxial relations, characteristic textures appear due to the intergrowth of crystals of two different species in a certain crystallographic relation. Various terms have been used in the mineralogical field to describe textures, as summarized in Table 7.2 [15], [16]. Observations of descriptive and taxonomy type have been accumulated, since they show the origin of rocks and ores, but understanding the mechanism of their formation still remains a future subject of research.

These intergrowth relations are formed through the processes of crystal growth, phase transformation or decomposition associated with a decrease in temperature and pressure, or metasomatism due to the supply of new components from outside.

Coaxial intergrowth is a paragenetic relation that describes crystals of two different species growing with a common axis; the misfit ratios between the two crystals in the direction of the common axis are small, without exception. The formation of coaxial intergrowth can be understood to be one crystal conjunct to the other in an epitaxial relation, where both continue to grow. If a liquid of eutectic A–B component is solidified from one side (unidirectional solidification), crystals of the two phases A and B precipitate in dotted, columnar or lamellar (with common axis) form, and show unique textures for unidirectional solidification. This is a well known phenomenon in metallurgy.

The solubility of a solid–solution component varies depending on temperature and pressure.

Even if crystals of a completely homogeneous single phase are formed at high temperature, a dissolved component is precipitated if the crystal experiences a decrease in temperature or an annealing condition. Many examples have been observed in metallurgy and mineralogy, and the terms "precipitation" and "exsolution" have been used in the respective fields. If aluminum alloy containing about 2% Cu (duralumin) is annealed, Cu precipitates in thin platy form parallel to a specific face of Al. Impurity nitrogen present in diamond in a substitutional form precipitates in the form of thin platelets of order 100 nm^2 parallel to the {100} plane of the host diamond (known as type I diamond). Since the presence of these platelets disturbs the movement of dislocations, duralumin and type I diamond containing impurity components have higher plastic strengths than the purer Al metal and type II diamond, which contains less nitrogen.

Exsolution is the term used to describe crystal growth occurring in a solid phase; it proceeds through nucleation and growth processes. Dislocations or inclusions, around which strain is concentrated, are highly likely to be sites for heterogeneous nucleation, and also the host and the exsolved phases exhibit a specific crystallographic orientation relation so as to minimize the excess energy induced by the interface between them. In the above examples, the exsolved phase appears as thin platelets parallel to a low-index face of the host, but arrangements of needles, dots, and stars aligned in specific directions controlled by the crystal structure of the host phase are also often encountered. Ti substitutionally present in corundum exsolves as needle crystals of rutile, TiO_2. In this case, exsolved TiO_2 needle crystals assemble in three directions on the {0001} face controlled by the oxygen close-packed structure of the host phase. Since numerous reflected light beams are focused by the curved surface of cabochon-cut crystals, which act as a lens, reflected light "strings" appear at right angles to the needles. This is the origin of the "stars" in star ruby and sapphire. Such an effect is called asterism, and is observed in many gemstones.

In the cases of silicate minerals, such as the feldspar group† and the pyroxene group,‡ the contents of the dissolved component are much higher than in the above cases. In such cases, even if (A, B) components completely dissolve to form a single solid–solution phase at high temperature, it is energetically favorable to be separated into two phases (A) and (B). Figure 7.14 is a phase diagram of a solid–solution of this type [17]. As a result, a texture consisting of alternating lamellae of two phases (A) and (B) is formed.

† Complex solid solution with $KAlSi_3O_8$ (Or), $NaAlSi_3O_8$ (Ab), and $CaAl_2Si_2O_8$ (An) as the major components, where Or is orthoclase, Ab is albite, and An is anorthite.

‡ Chain silicate minerals with $MgSiO_3$ (En), $FeSiO_3$ (Fs), and $CaSiO_3$ (Wo) as the major components, where En is enstatite, Fs is ferrosilite, and Wo is wollastonite.

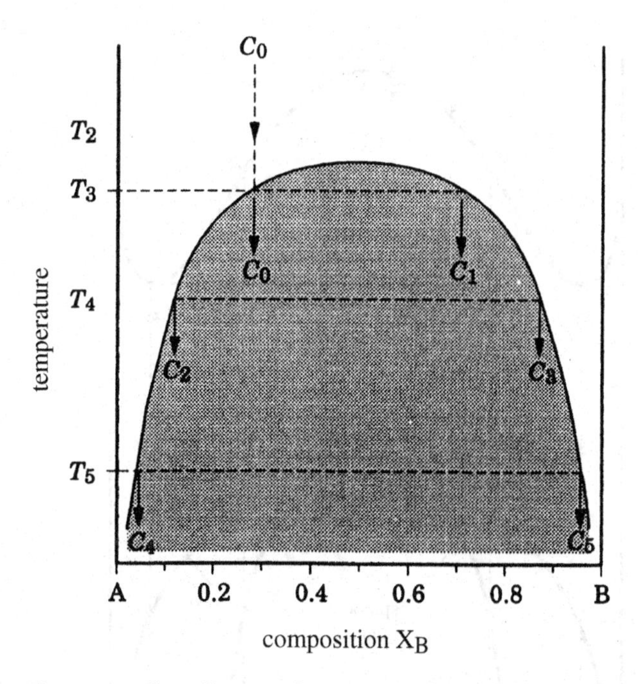

Figure 7.14. Phase diagram of a substance of (A, B) composition. In this system, one homogeneous solid–solution phase with composition C_0 is formed at high temperature. As the temperature decreases, it is energetically more stable for (A, B) to be separated into two phases (A) and (B). A solid–solution with composition C_0 will coexist as phases C_0 and C_1 at T_3 in equilibrium, and, on lowering the temperature further, phases $C_4 + C_5$ at T_5 will be in equilibrium.

The widths of these lamellae depend on the thermal history. Many examples have been observed in minerals, and these lamellae widths are used to analyze the thermal histories.

When the thicknesses of the lamellae are of the order of the wavelength of visible light, an alternating multiple lamellar structure with different refractive indices is formed, which causes interference of light beams, giving rise to color. Bright interference color shown by laboradorite ($Ab_{50}An_{50} \sim Ab_{30}An_{70}$, where $Ab = NaAlSi_3O_8$, $An = CaAl_2Si_2O_8$) is due to alternately exsolved lamellae of Ab and An components. Depending on the size or form of the exsolved lamellae, the structure can cause light scattering. Moonstones exhibiting what is known as the "moonstone effect" are such an example. In precipitation or exsolution, the formation of a new phase follows the form of crystal growth in a solid phase with a crystal structure, and involves nucleation and other growth processes that must overcome an energy barrier.

There are systems, however, in which the separation of a single phase into two phases as the temperature decreases is energetically favorable. Figure 7.15 indicates the phase diagram of such a phase [17].

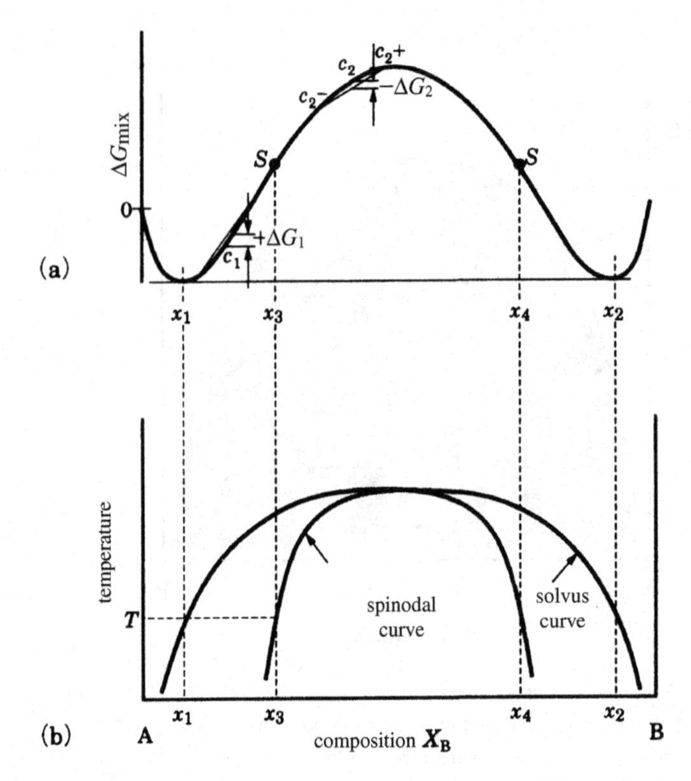

Figure 7.15. Phase diagram of a phase showing spinodal decomposition. (a) In a regular solution having positive mixing enthalpy, ΔG_{mix} has an opposite sign on the left and right sides of point *S*. So it is energetically favorable to have spinodal decomposition into x_1, x_3; x_2, x_4.

In this case, it is not necessary to overcome an energy barrier for nucleation to occur. Decomposition into two phases occurs without involving the process of nucleation. Separation into two phases in this type of system is called spinodal decomposition, and is seen among silicate minerals such as the pyroxene and feldspar groups. Generally, the boundaries between the two phases are wavy rather than flat, but the two phases are in a specific crystallographic relation.

Since the growth of a new phase is controlled by the structure of the host phase, the new phase should grow having a specific crystallographic relation with the host phase. This relation is applicable to the cases when a new phase, with composition of part of the host phase, is crystallized from the host phase by heating, or when a new phase is formed by the addition of another component. Since these processes involve the addition or departure of a part of the component, they are not the same as simple exsolution or spinodal decomposition. These phenomena are called syntaxy or topotaxy. This is because the new phase formation proceeds whilst keeping the basic structure of the host crystal, such as oxygen close-packing for example, unchanged. The term "topotaxy" is used to refer to this phenomenon; the term "syntaxy" is used to discuss the origin of the process.

References

1 Y. Takano, *Classification of Twins – Twin Groupoids*, Tokyo, Kokinshoin, 1973 (in Japanese)

2 G. Friedel, *Lecons de Cristallographie*, Paris, Berger-Lvzaült, 1926

3 M. L. Kronberg and F. H. Wilson, Secondary recrystallization in copper, *Trans. AIME*, **185**, 501–14

4 W. L. Bragg and G. F. Claringbull, *Crystal Structure of Minerals*, London, G. Bell & Sons, 1965

5 M. J. Buerger, The genesis of twin crystals, *Am. Min.*, **30**, 1945, 469–82

6 R. S. Wagner, On the growth of germanium dendrites, *Acta. Metall.*, **8**, 1960, 57–60

7 P. Hartman, On the morphology of growth twins, *Z. Krist.*, **107**, 1956, 225–37

8 M. Kitamura, S. Hosoya, and I. Sunagawa, Re-investigation of the re-entrant corner effect in twinned crystals, *J. Crystal Growth*, **47**, 1979, 93–9

9 Nai-ben Ming, K. Tsukamoto, I. Sunagawa, and A. A. Chernov, Stacking faults as self-perpetuating step sources, *J. Crystal Growth*, **91**, 1988, 11–19

10 Y. Aoki, Morphology of crystals grown from highly supersaturated solutions, *Mem. Sci., Kyushu Univ.*, Ser. D., **24**, 1979, 75–108

11 K. Maiwa, K. Tsukamoto, and I. Sunagawa, Activities of spiral growth hillocks on the (111) faces of barium nitrate crystals growing in an aqueous solution, *J. Crystal Growth*, **102**, 1990, 43–53

12 L. Royer, Recherches expermentales sur l'epitaxie ou orientation mutuelle de cristaux d'especes differents, *Bull. Soc. Franc. Min. Cristalle*, **51**, 1928, 7–159

13 P. Hartman, Epitaxial aspects of the atacamite twin, *Cursillos y Conferencias, C.S.I.C. (Espania)*, **7**, 1960, 15–18

14 A. A. Chernov, *Modern Crystallography III, Crystal Growth*, Berlin, Springer-Verlag, 1984

15 P. Ramdohr, *The Ore Minerals and Their Intergrowths*, New York, Pergamon Press, 1969

16 R. I. Santon, *Ore Petrology*, New York, McGraw-Hill, 1972

17 A. Putnis, *Introduction to Mineral Sciences*, Cambridge, Cambridge University Press, 1992

Suggested reading

P. Ramdohr, *The Ore Minerals and Their Intergrowths*, New York, Pergamon Press, 1969

A. Spray, *Metamorphic Textures*, New York, Pergamon Press, 1969

R. I. Santon, *Ore Petrology*, New York, McGraw-Hill, 1972

J. P. Bard, *Microtextures of Igneous and Metamorphic Rocks*, Dordrecht, D. Reidel, 1986

A. Putnis, *Introduction to Mineral Sciences*, Cambridge, Cambridge University Press, 1992

8

Forms and textures of polycrystalline aggregates

Spontaneously nucleated crystals can grow with various orientations. In such processes, surviving crystals are selected simply by their geometrical relation with the substrate surface. As a result, various textures of polycrystalline aggregates appear that are controlled by the form of the substrate surface. Spherulite, sheaf-like, and confeito-like* polycrystalline aggregates or the curved banding patterns seen in agate are all formed by a simple geometrical selection process. This principle has been practically utilized in single crystal synthesis and in epitaxial growth. It also acts in the formation of calculus in the internal organs of human bodies.

8.1 Geometrical selection

When nucleation occurs freely on a flat substrate surface under uncontrolled conditions, numerous crystals are formed at random orientations if there is no epitaxial relation between the substrate and the crystal. Crystals inclined to the substrate surface will make contact with crystals growing perpendicularly and stop their growth. If we assume the existence of equi-concentration lines parallel to the substrate surface, the growth of crystals perpendicular to the substrate will be promoted further, since their tips are in an ambient phase of higher driving force. In this way, only the crystals growing perpendicularly to the substrate surface will survive and continue to grow among many crystals formed on the substrate surface, and this is accompanied by a decrease in the number of individuals and appearance of textures consisting of many crystals aligned in a specific orientation. Since the selection of the surviving crystals is determined by geometrical

* Confeito is the name given to small pieces of candy made by crystallizing sugar around poppy-seed cores.

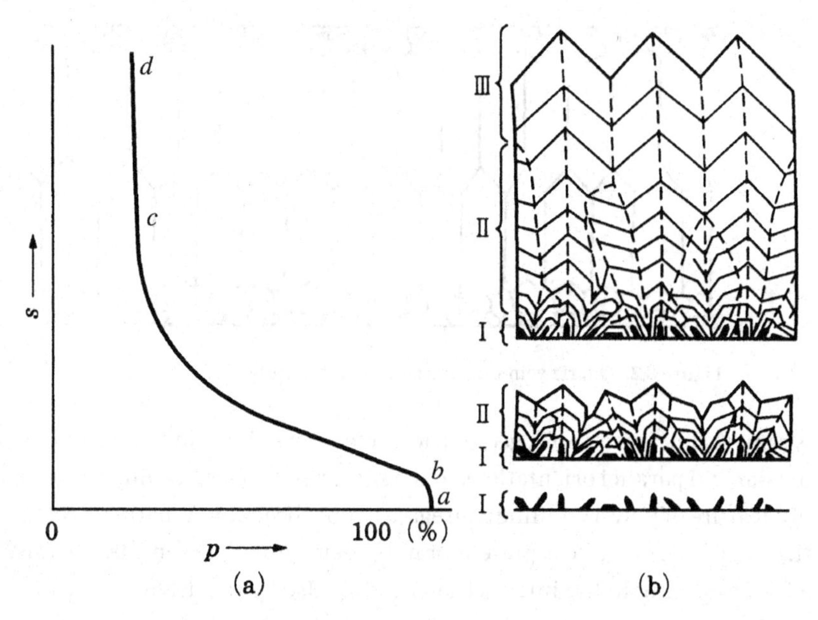

Figure 8.1. (a) Size s versus population density p. Points a, b, c, and d correspond to I (a, b), II (b, c), and III (d) in (b), which is a schematic illustration of geometrical selection.

orientation only, this is called geometrical selection. Figure 8.1 shows the concept schematically. Kolmogorov's theory is a probability theory solving the problem [1]; see ref. [14], Chapter 7. The population density n of surviving crystals by geometrical selection is expressed as follows. For the two-dimensional case,

$$n(h) \propto 1/h\,,$$

and, for the three-dimensional case,

$$n(h) \propto 1/\sqrt{h},$$

where h is the separation between neighboring crystals.

The textures of polycrystalline aggregates formed by geometrical selection have been widely observed in natural minerals, but the principle has also been actively used in various synthetic methods.

The most representative example is seen in the comb-like texture of quartz veins formed by the precipitation of quartz crystals on the wall of a vein from a hydrothermal solution that has entered into small fractures in the rocks (Fig. 8.2).

If there is no epitaxial (coherent) relation between the substrate and the growing crystals, and the nuclei formed initially have completely random orientation, and the growth rate is more or less isotropic, geometrical selection operates in one direction only: perpendicular to the substrate surface. When there is an epitaxial and coherent relation between the substrate and the crystals, and the growth rate

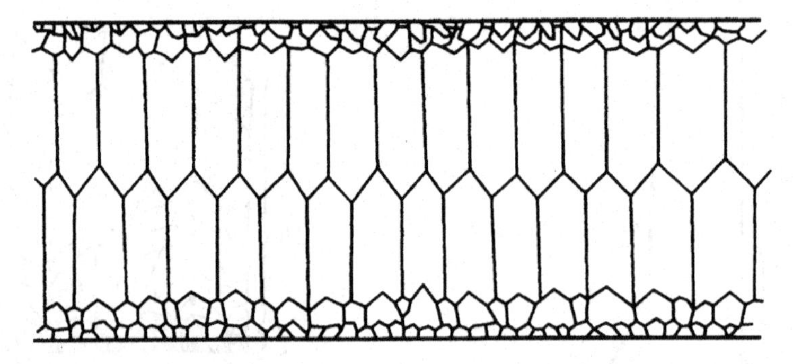

Figure 8.2. Quartz vein showing a comb-like texture.

is remarkably anisotropic, geometrical selection will be operative in both perpendicular and parallel orientations to the substrate surface, leading to the formation of a texture of polycrystalline aggregate in one direction. Whether a coherent relation with the substrate is present or not is determined not only by the misfit ratio, a factor relating to the interface energy, but also by the driving force (see Section 7.4). Even if a coherent relation is achieved under a low driving force condition, the relation will be violated under higher driving force conditions. Close-packed, densely arranged compact and hard textures will appear if a coherent relation is maintained, but the texture will be coarsely packed and soft if the coherency is violated. This relation is important for crystal growth in living bodies (see Chapter 14).

Geometrical selection is used practically in various methods of single crystal synthesis. In one such method from the melt phase, the Bridgman–Stockbarger method, crystallization is achieved by moving the melt zone, either by moving the heater or the crucible. To obtain large single crystals by this method, a technique is adopted in practice to design the crucible so as to encourage the survival of only a small number of crystals, by geometrical selection, out of the numerous crystals that are formed initially. In crystal growth from the vapor phase by the CVT method using a closed tube, a technique is adopted in practice in which the diameter of the tube is narrowed at the crystallization zone so that only a small number out of many crystals will survive and grow. To achieve an epitaxial relation between the substrate and the crystal with a large misfit ratio, a device that forms a buffered layer in between the substrate and the crystal is used; this may also be regarded as a method utilizing geometrical selection.

8.2 Formation of banding

When the substrate surface is curved or spherical, various textures of polycrystalline aggregate appear through geometrical selection. Spherulites will be formed when a sand grain or spherical polycrystalline aggregate formed at the early stage of nucleation acts as a substrate, and wavy banding parallel to the substrate surface will appear when the substrate surface is irregular. After one layer is completed via intermittent growth, and the process is repeated, a banding pattern will appear consisting of a unit layer of polycrystalline aggregate running perpendicularly to the substrate surface through geometrical selection, starting from minute crystallites of random orientation. For spherulites, concentric banding will appear in a similar manner.

The intermission and resumption of growth are inevitably involved in a system where there is an imbalance between the diffusion rate and the growth rate and a critical value such as the energy barrier is involved. When growth resumes, the particle size is small and the density is high, but the size increases and the density decreases as growth proceeds. Since the impurity concentration will also vary, the color intensity will be different.

Most mineral crystals formed under normal temperatures and pressures on or near the Earth's surface are minute, and often show a banding texture. Malachite, $Cu_2(OH)_2CO_3$, and azulite, $Cu_3(OH)_2(CO_3)_2$, are copper carbonate minerals resulting from secondary precipitation from an aqueous solution containing Cu ions and CO_2, which result from the dissolution of primary copper sulfide minerals in underground water. Due to the beauty of its banding pattern, consisting of green malachite or alternating layers of green malachite and indigo blue azulite, these minerals have been prized as ornamental materials since ancient times. Rhodochrosite, $MnCO_3$, is a mineral with a beautiful banding pattern, pink in color, formed from a low-temperature hydrothermal solution, and has been used as an ornamental material under the name Inca rose. The familiar banding pattern seen in agate is formed in a similar manner; agate will be considered in more detail in Chapter 10. Several examples of these banding patterns are shown in Fig. 8.3.

The free surface of these polycrystalline aggregates takes various forms depending on the degree of unevenness, and various terms including botryoidal, reniform, stalactite, and colloform have been used to describe them (see Fig. 8.4 and Table 8.1). In addition, the term oolitic is applied to an aggregate of equal sized spherulites. Terms used in mineralogy to express the characteristic forms of polycrystalline aggregates are given in Table 8.1. In most cases, the characteristic morphology is derived from geometrical selection. Other terms, such as bladed, foliated, and plumose, may be used to describe characteristic morphologies. These are, in most cases, forms shown by sub-parallel aggregation of thin platy crystals.

8.3 Spherulites

Spherulites are formed if geometrical selection takes place on a spherical substrate particle. Substrate particles may be a completely different material from those materials forming the spherulites, such as a sand grain, or a spherical particle of polycrystalline aggregate of the same species formed under a higher driving

(a)

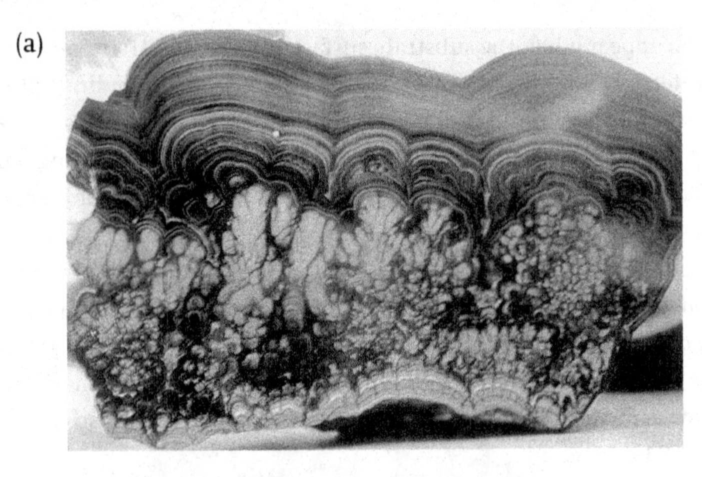

(b)

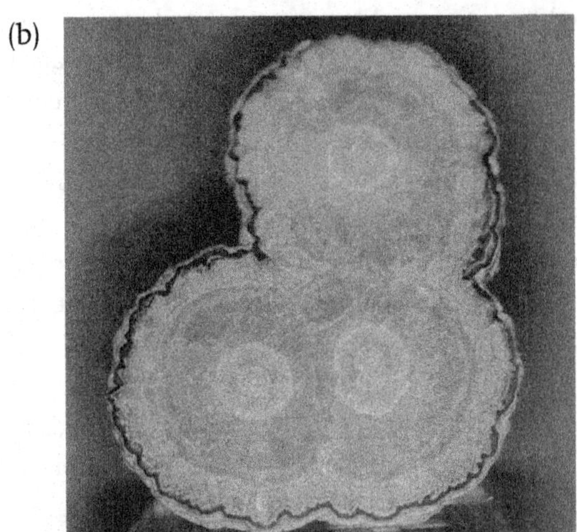

(c)

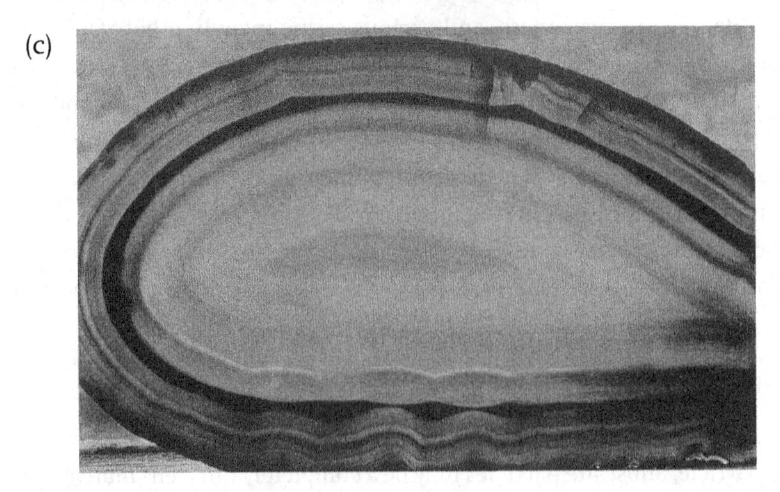

Figure 8.3. Examples of banding patterns formed by geometrical selection in (a) malachite; (b) rhodochrosite; (c) agate.

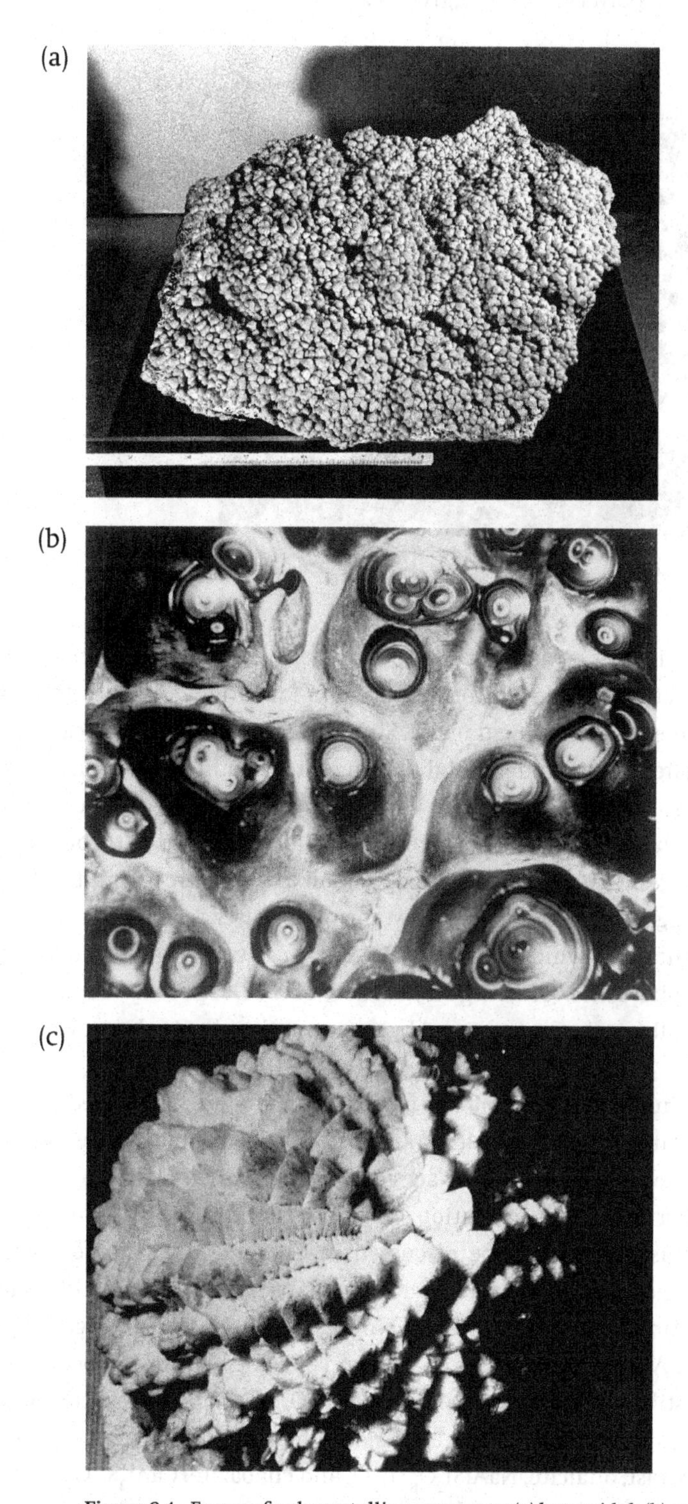

Figure 8.4. Forms of polycrystalline aggregate: (a) botryoidal; (b) mamillary; (c) spherulitic; (d) oolitic (see p. 156).

(d)

Figure 8.4 *(cont.)*

force condition, or it may be a polycrystalline aggregate of random orientation formed by agglutination of crystals formed under a low driving force condition through the movement of an ambient phase. Since the formation of spherical polycrystalline aggregates occurs under a high driving force condition in the absence of effect of flow, spherulites are a form of polycrystalline aggregate generally appearing under a high driving force condition. If there is a flow, or turbulence, the situation changes. Crystals formed under small driving force conditions in polyhedral form agglutinate by moving flow, forming polycrystalline aggregates with random orientation, which acts as a substrate for further growth to form spherulites. NaCl spherulites occurring on the shore of the Dead Sea have spherical nuclei of polycrystalline aggregate formed by agglutination due to the movement of seawater.

In the case of sodium uric acid, which is the cause of gout, needle-shaped single crystals are formed in the red corpuscles, transported by blood flow, and deposited at the knuckle of the big toe, where they accumulate, and this acts as a site for the nucleation of spherulites. The formation of kidney stones and gallstones is a similar process; this process will be discussed in Chapter 14. Compared with isotropic crystals, needle, long prismatic, and thin platy crystals are more likely to form spherulites. Among the zeolite group of minerals (silicates with a cage structure), natrolite, $Ba_2Al_2Si_3O_{10} \cdot 2H_2O$, whose *Habitus* is needle-like, always occurs as spherulites, and stilbite, $(Ca, Na_2, K_2)Al_2Si_7O_{18} \cdot 7H_2O$, with characteristic platy *Habitus* occur as sheaf-like or bladed aggregates, which correspond to uncompleted spherulites. In contrast, analcite, $NaAlSi_2O_6 \cdot H_2O$, and chabazite, $CaAl_2Si_4O_{17} \cdot H_2O$, whose *Habitus* are isotropic, occur only exceptionally as spherulites.

Table 8.1 *Terms used to describe textures of polycrystalline aggregates*

Terms	*Examples*
Granular	limestone[a]
Fibrous	asbestos[b]
Radiate	natrolite[b]
Bladed (wheatsheaf, bow-tie)	stilbite[b]
Globular	wavellite[c]
Oolitic	opal[b]
Botryoidal	chalcedony[b], limonite[d]
Mamillary	malachite[a]
Reniform	hematite[d]
Colloform	general term for botryoidal, mamillary, reniform with banded texture
Stalactite	calcite[a]
Foliate (micaceous, lamellar)	kyanite[b], mica[b]
Feathery, plumose	mica[b]
Reticulate	rutile[d]
Dendritic, arborescent	pyrolucite[d], native silver[e]
Spherulite	various minerals, polymer crystals

[a] Carbonates [b] Silicates [c] Phosphates [d] Oxides [e] Native elements

Polymer crystals whose *Habitus* is characteristically thin platy exclusively take spherulitic form under high driving force conditions. In crystals having this type of *Habitus*, crystals are often bent or twisted, and often sub-parallel growth results as growth proceeds, due to the strain induced into the growing crystal or introduced through the effect of precipitation of a foreign substance on the growing surface. As a result, instead of the formation of a perfect spherulite on the spherical substrate surface, various incomplete spherulites, such as those shown in Fig. 8.5(c), for example sheaf-like, bow-tie, and two-eye forms, appear.

Spherulite formation by geometrical selection may rarely be seen on crystals with isotropic *Habitus*. Native arsenic, As, occurs in a confeito-like form, and is a type of spherulite grown through the geometrical selection of rhombohedral crystals. Spherical aggregation of calcite crystals with nail-head *Habitus* is also observed. Semi-spherical aggregates of platy barite crystals known as desert rose are shown in Fig. 8.6.

8.4 Framboidal polycrystalline aggregation

There is a type of polycrystalline aggregate of pyrite crystals showing a framboidal appearance, known as framboidal pyrite. It occurs in sedimentary

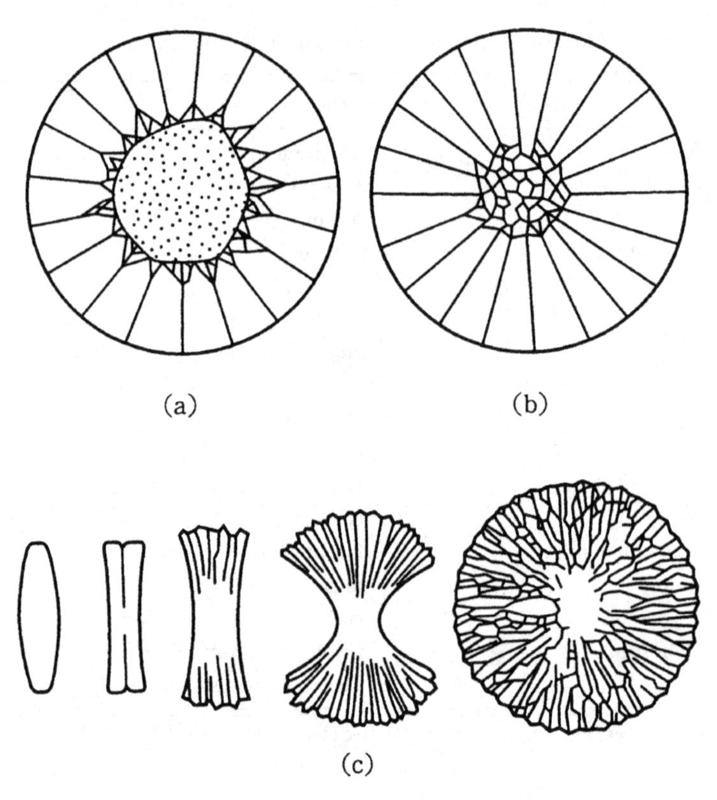

Figure 8.5. Modes of formation of spherulites [2], [3]. Spherulites formed (a) on a spherical particle of foreign material, (b) on a polycrystalline aggregate of the same species, and (c) by split growth.

rocks or as hot spring precipitates, and forms a framboidal aggregate of one micrometer to a few tens of micrometers order, consisting of close-packed idiomorphic crystals of less than 1 μm. Figure 8.7 shows a scanning electron microscope (SEM) photograph of a framboidal pyrite. Characteristic features are that pyrite crystals of framboidal appearance are idiomorphic crystals of equal size and the same *Habitus*. Various structures, such as cubes, octahedra, and pentagonal dodecahedra are known. Judging from the fact that these crystals are arranged in a close-packed manner, we may assume that geometrical selection does not operate in their formation.

Two different opinions have been put forward to explain the mechanism of formation of minute idiomorphic crystalline aggregation with equal size and the same *Habitus*. The first considers the aggregation of minute crystallites due to the stirring of an inorganic hydrothermal solution, and the second assumes it is due to living organisms such as bacteria. Hydrothermal synthetic experiments have demonstrated that the polycrystalline aggregate of pyrite, with the same texture

Figure 8.6. Semi-spherical aggregate of the platy barite crystals known as desert rose.

and structure as the natural crystal, could be reproduced, and it has been demonstrated that the formation of framboidal pyrite by an inorganic process, without the help of bacterial activity, is possible. However, the second argument is based on the observation of the modes of occurrence of this type of pyrite.

8.5 Texture formation in multi-component systems

The mechanisms operating in the formation of textures seen in polycrystalline aggregates of the same species have been discussed in Sections 8.1–8.4. This may correspond to the analysis of a mechanism controlling the so-called self-organization or self-assemblage. Other mechanisms are possible; for example, tiny spherical particles are assembled and a close-packed structure is formed due to surface tension. The formation of opal consisting of a close-packed structure of minute amorphous silica spheres may be such a case.

In this section, we shall analyze the factors controlling the formation of the texture and structure of polycrystalline aggregates of multiple phases seen in

Figure 8.7. Framboidal pyrite; SEM image.

rocks, ores, or ceramics when crystal growth takes place under entirely uncontrolled conditions. Since the population density and morphology of constituent particles determine the texture and structure of this type of system, various factors affect their formation. The texture and structure of the resulting form are determined through an inter-relation of the static and dynamic factors, such as temperature, pressure, driving force, chemical composition of the system, crystallization differentiation, the initial state, and movement.

Magma, which is a high-temperature solution, solidifies to form igneous rock. If magma cools down slowly in the depths of the Earth, the nucleation density is low and the crystal can grow until it comes into contact with others, after which a plutonic rock with holocrystalline texture appears. If magma containing growing crystals that nucleated earlier solidifies at an intermediate depth, a rock with porphyritic texture, consisting of large idiomorphic crystals (phenocrysts) and groundmass consisting of minute idiomorphic crystals, will result.

Depending on the solidification conditions, igneous rock with vitreous groundmass will be formed. If magma containing no phenocrysts solidifies rapidly, glassy igneous rocks are formed. When magma intrudes into cracks in a stratum forming a dyke, a quenched phase consisting of minute crystallites will appear because the surrounding stratum is quenched along the contact zone. Due to the uplifting movement of the magma, phenocrysts originally present in the magma are broken, and minute crystalline particles become detached; these act as secondary nucleation sites, resulting in the modification of the state of tiny crystallites in the groundmass. In laboratory experiments, in which cooling is performed in two stages (trying to grow phenocrysts in the first stage and tiny crystallites in groundmass in the second stage), it was impossible to reproduce the hyalopyritic texture commonly observed in natural basalt, which consists of phenocrysts and microcrystalline groundmass. The texture reproduced by the two-stage experiments consists of phenocrysts and the surrounding dendritic crystals, which is a result of the morphological variation due to the difference in supercooling. However, if the system is stirred, hyalopyritic textures may be successfully reproduced in the laboratory (see ref. [4], Chapter 3) because the nucleation density increases drastically by an order of magnitude as minute particles, detached from phenocrysts by stirring, act as secondary nucleation sites.

By consideration of these examples, we see that the knowledge accumulated in the industrial crystallization field is important and useful in understanding the crystallization process in magma. Since, in industrial crystallization, the nucleation, growth, and size distribution or morphological change of crystals proceed in a closed reaction vessel, it is necessary to evaluate the kinetics based on the final products. Also useful in the analysis of magmatic crystallization is the concept of crystal size distribution (CSD), which may be used to evaluate the nucleation rate and the growth rate from the resulting size distribution.

All apparatus used in industrial crystallization is designed to obtain crystal particles of uniform size, and so the CSD is chosen to facilitate this outcome. The evaluation of the peak in the CSD is made by using an average value of the CSD, or using the value of the dispersion coefficient using the central value (CV) and the standard deviation. When particle sizes are completely uniform, the CV equals zero, and in complete mixing CV = 0.5. In complete mixing, the density function of the CSD, $n(L)$ ($\mu m^{-1} m^{-3}$), or the logarithm (m^{-4}) and particle size L (μm) plot, is linear, and from its gradient the growth rate R ($\mu m/s$) and the stay period (τ (s)) are obtained. The section of L–0 is uniquely determined by the ratio of the nucleation rate to the growth rate. The distribution formula is

$$n(L) = (B/R)\exp(-L/R\tau).$$

Examples are shown in Figs. 8.8(a) and (b), in which the above relations obtained in the industrial crystallization field are applied to crystals in igneous rocks formed

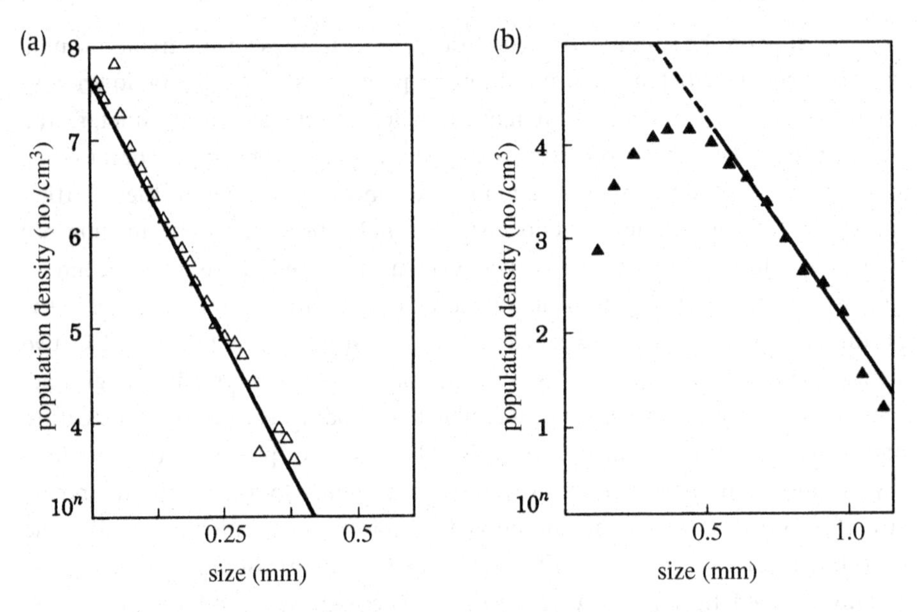

Figure 8.8. Crystal size distribution (CSD) plots of (a) plagioclase in igneous rock and (b) garnet porphyroblast in contact metamorphic rock [4].

by the solidification of magma, and those in metamorphic rocks formed by contact metasomatism. Figure 8.8(a) shows an example of the measurement on plagioclase in lava, and Fig. 8.8(b) illustrates garnet in contact metamorphic rocks. The horizontal axis is particle size, and the vertical axis is particle density/cm^3 [4]. From these results and geological data, the growth rate of plagioclase in the lava shown in Fig. 8.8(a) is estimated to be $10^{-10} \sim 10^{-11}$/(cm s); the nucleation rate is $10^{-2} \sim 10^{-3}$/(cm^3 s); the growth time is 10 ~ 100 years. In Fig. 8.8(b), the growth time is less than 100 years at $\Delta T = 10\,°C$, and the nucleation rate is $10^{-4} \sim 10^{-1}$/(cm^3 s). The difference in the curves in Figs. 8.8(a) and (b) is interpreted as due to the annealing effect involved in contact metasomatism.

Sedimentary rocks are composed of epigenetic minerals that have survived weathering and authigenic minerals formed *in situ* from low-temperature aqueous solution. Since authigenic minerals are formed from aqueous solution at low temperature, they are either amorphous or minute crystalline, and so the texture of sedimentary rocks is principally determined by the sizes and forms of epigenetic minerals.

In regional metamorphic rocks and contact metamorphic (metasomatic) rocks, new mineral crystals grow in solid rocks in which there was a change in conditions. However, the process is not the same as straightforward solid state growth or recrystallization. Since volatile components such as H_2O and CO_2, which were originally present in the rock, are involved, it is better to assume a process of dissolution–precipitation. In this respect, the crystallization resembles crystal growth taking place in sintering with a minor amount of solvent components. Since H_2O is released by reaction, Ostwald ripening has a significant effect. Actual examples will be analyzed in Chapter 13.

References

1 A. N. Kolmogorov, *Dokl. Akad. Nauk. SSSR*, **65**, 1949, 681 (in Russian)

2 D. P. Grigoriev, *Ontogeny of Minerals*, Jerusalem, Israel Program for Scientific Translation, 1961

3 G. G. Lemmlein, *Morphology and Genesis of Crystals*, Moscow, Nauk, 1973 (in Russian)

4 K. V. Cashman, in *Reviews in Mineralogy, vol. 24, Methods of Igneous Petrology*, eds. J. Nicholls and J. K. Russell, Washington, Mineralogical Society of America, p. 259

Suggested reading

J. P. Bard, *Microtextures of Igneous and Metamorphic Rocks*, Dordrecht, D. Reidel, 1986

R. Kretz, *Metamorphic Crystallization*, New York, John Wiley, 1994

S. Banno, M. Toriumi, M. Kobatake, and T. Nishiyama, *Dynamics of Rock Formation*, Tokyo, Tokyo University Press, 2000 (in Japanese)

PART II APPLICATION TO COMPLICATED AND COMPLEX SYSTEMS (CASE STUDIES)

In this second part of the book, we will present an explanation, using real examples, of how we can solve the problems involved in complicated and complex systems, based on the fundamental concepts explained in Part I and using crystal morphology as the key. Chapters 9 to 13 cover mineral crystals, and Chapter 14 discusses crystals formed in physiological activities. In particular, we discuss diamonds in Chapter 9 and quartz in Chapter 10; these are examples of minerals whose crystal morphologies have attracted deep interest from the earliest times. In Chapter 11, pyrite and calcite are selected as representative examples of mineral crystals that show elaborately varied *Habitus* and *Tracht*. Chapter 12 focuses on crystals grown from the vapor phase in free space, and the analysis will be principally based on the observations of surface microtopographs of crystal faces, imperfections, and inhomogeneities in crystals. Crystals formed in pegmatites or druses in volcanic rocks will be used as examples. In Chapter 13 our discussion will be centered on how crystal growth proceeds in the processes of metasomatism or metamorphism. In Chapter 14, biomineralization will be considered from the standpoint of crystal growth.

9

Diamond

There is no mineral, other than diamond, consisting of only sp^3 covalent bonds which fits perfectly the purpose of this book. Diamonds are formed deep in the Earth under high-temperature and high-pressure conditions, and are uplifted during geological events to the Earth's surface at a speed of 100 km/h. The history of natural diamond crystals, ranging from subduction of plates, growth of crystals in the upper mantle, plastic deformation associated with rapid uplifting, and partial dissolution experienced in the ascent process, is recorded within the crystal. Diamond crystals synthesized under high-temperature, high-pressure conditions, but from the solution phase with a different solvent component from that in natural crystallization, and those synthesized from the vapor phase under labile conditions, show different morphological characteristics, to the extent that a distinction between natural and synthetic diamonds can easily be made.

9.1 Structure, properties, and use

Diamond C is the high-pressure phase of carbon, and the C–C bonding is of sp^3 pure covalent nature. The structure has a three-dimensional framework as indicated in Fig. 9.1, and is different from the low-pressure phase, graphite, which has a sheet structure consisting of sp^2 covalent bonds and Van der Waals bonds connecting the sheets. Other polymorphs called lonsdalite, fullerene, and carbon nanotube, which consist of mixed sp^2 and sp^3 bonds, are also known.

In diamond C, it is the fact that the structure consists of sp^3 covalent bonds only that accounts for its unique physical properties, such as the highest hardness of any material (Mohs hardness 10), small compressibility (1.7×10^{-7}cm^2/kg), the highest elasticity among any known material ($4\sim6\times10^{12}$ dynes/cm^2, bulk elasticity), large thermal conductivity ($9\sim26$ W/(deg cm)), and small thermal expansion coefficient ($0.8\pm0.1\times10^{-6}$ at 20 °C, comparable to the value of invar). As a result,

(a)

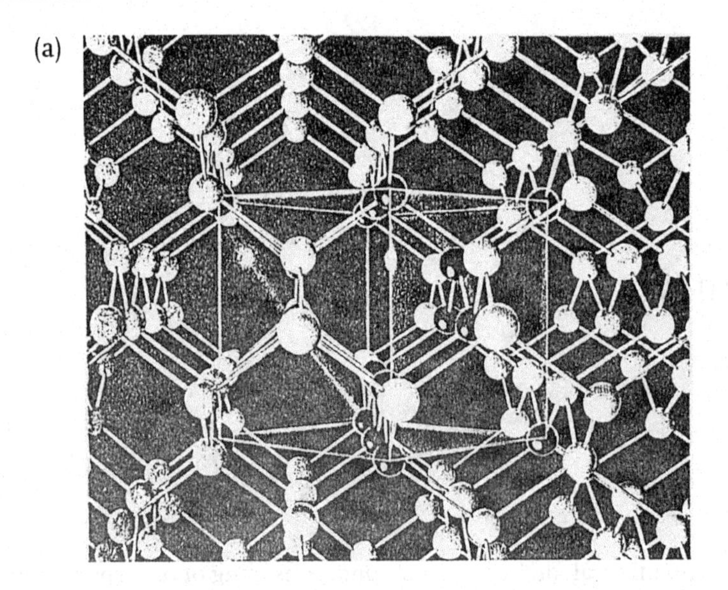

(b)

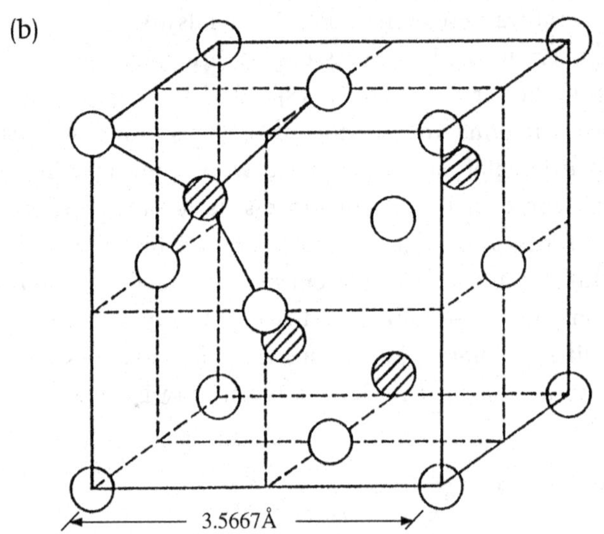

(c)

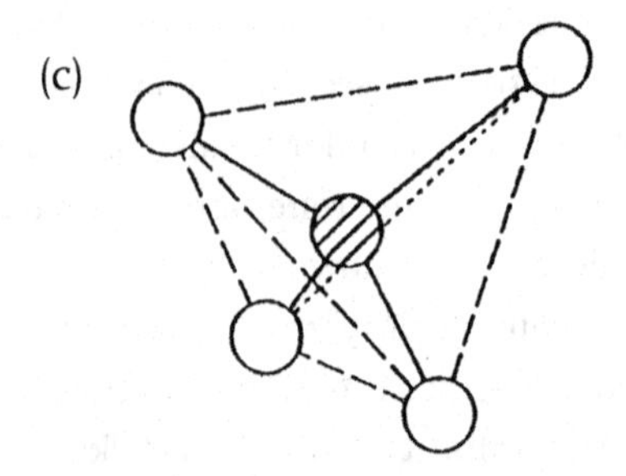

Figure 9.1. Crystal structure of diamond. (a) Three-dimensional representation; (b) unit cell; (c) bonding between carbon atoms by sp^3 covalent bond; and periodicity seen perpendicular to (d) the (111) direction and (e) the (100) direction.

(d)

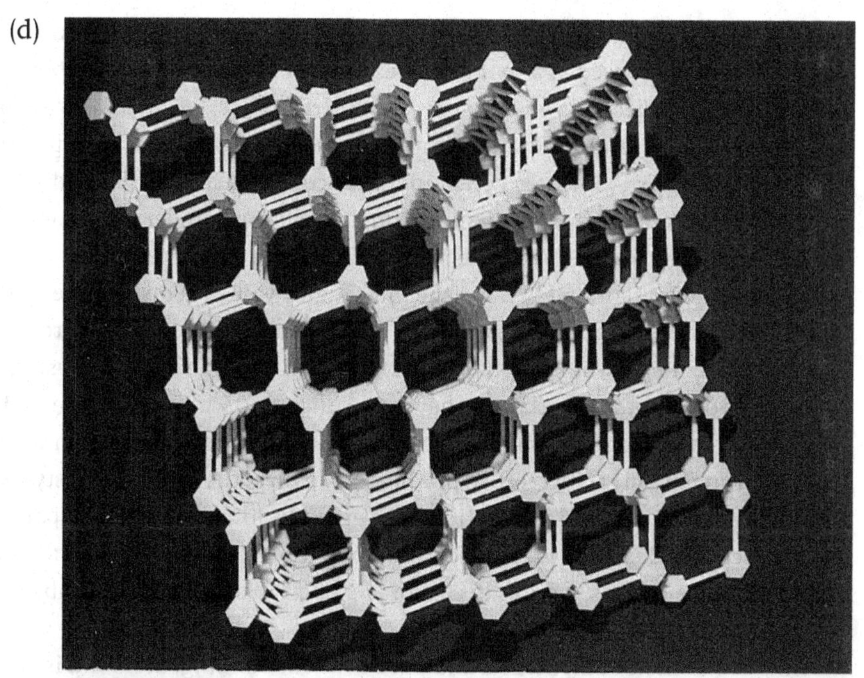

(e)

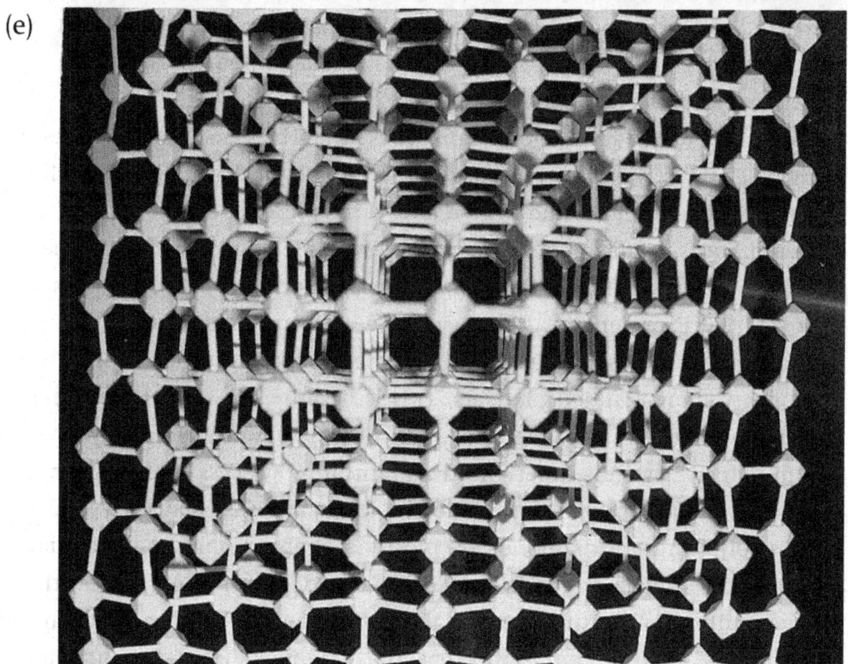

Figure 9.1 (*cont.*)

since the first diamond was found in India during the fourth to fifth centuries BC, diamonds have been used as a tool to process other hard materials, and even now they are used uniquely in a wide range of industrial processes, such as wire dies, tools to manufacture ultra-fine machining tools, surgical knives, high-pressure diamond anvils, heat sinks for compound semiconductors, optical windows for spacecraft, etc. Semiconductor diamond intended for use under extreme conditions is expected to become available in the near future.

The reasons why diamond became the most highly valued gemstone are (1) that the anisotropic properties in the hardness of diamond single crystals were recognized and a technique was created to polish diamond to a high brilliance using the powder, and (2) that the primary mother rock of diamond was found in the middle of the nineteenth century, which enabled the mining of the deposit on a large scale, resulting in a great increase in production. In turn, the brilliant-cut style was invented, which enhanced the optical properties (colorless, high transparency, high refractive index $n=2.144$), which promoted the position of diamond above that of other gemstones. Before this, colored stones, like emerald and ruby, were esteemed higher than diamond.

Owing to its single composition and pure covalent bonding, diamond is a standard solid material, and takes the role of the most appropriate sample in explaining the effects upon structure-sensitive properties when the structure of solid material deviates from the ideal state.

Natural diamonds are classified into Types I and II based on their transparency to ultra-violet rays. The two types have different physical properties, such as the absorption of ultra-violet and infra-red rays, the presence and absence of satellite reflections on X-ray diffraction, the roughness of cleavage surface, and the plastic deformation temperature; these differences arise principally from the amount of nitrogen that is present as an impurity and the state that the nitrogen takes. In Type I, which contains up to 0.2% N, the nitrogen usually presents itself in the form of clusters or precipitated platelets. This is because the crystal experienced an annealing process. In contrast, there are some diamonds, such as synthetic diamonds, which have not experienced an annealing process, and so the nitrogen appears in a substitutional form. The two forms are classified as Type Ia and Ib, respectively. Synthetic diamonds are mostly Type Ib, and most natural diamonds belong to the Type Ia group. Type II diamonds contain nitrogen at less than one part per million, and so they are crystals of purer carbon than Type I. When a small amount of Al or B is present in this type, the diamonds show semiconductor properties, and so are called Type IIb, discriminating them from the insulator Type IIa. Depending on the state of nitrogen clustering, Type I is further subdivided into Type Iaa, Type Iab, etc. Recently, the effect of another impurity, hydrogen, has attracted interest, and extensive studies have been made.

Diamonds are unique in existence as the standard sample for solid state physics, but they are also extremely valuable information transmitters of the conditions and events that have occurred in the depths of the Earth. There are many cases in which minerals formed in the mantle, deep within the Earth, are transported to the surface through global-scale movement such as plate tectonics, but most transform to other phases, decompose into more than two mineral species, or dissolve entirely. Although diamonds experience severe conditional changes, they do not transform into graphite, the low-pressure stable phase; instead, they survive in a metastable state, and thus can be a unique transmitter of information. In this way, information is recorded in the diamond crystals in the form of morphology, perfection, and homogeneity. It could be said that diamond is a unique letter sent from the depths of the Earth.

From geological and petrological investigations, and thermodynamic stability relations, it is believed that natural diamonds were formed deep under the Earth, under high-pressure, high-temperature conditions, then brought up to the Earth's surface through rapid ascent of kimberlite and lamproite magmas, and were finally quenched as a metastable phase by adiabatic expansion due to volcanic eruption. From investigations of solid state inclusions in natural diamonds, two kinds of environmental compositions, ultramafic and eclogitic, have been distinguished. In addition, diamond crystals of micrometer size are found in garnet or zircon crystals in ultra-high-pressure metamorphic rocks.

We will explain in the following sections what new information may be obtained from morphology, perfection, and homogeneity in diamond crystals, in addition to the geological, geochemical, petrological, and mineralogical data.

9.2 Growth versus dissolution

A total of 364 crystal figures of diamond are reproduced in Goldschmidt's *Atlas der Kristallformen* [1] published in 1913–23. The figures are beautiful sketches of crystals, and they even include surface microtopographs; some of these are shown in Fig. 9.2.

Although there are octahedral, dodecahedral, cubic, tetrahedral, elongated football-like forms, triangular platy forms, or star-like forms, the characteristic is that the forms are mostly rounded. Whether the rounded forms bounded by curved faces are due to growth or dissolution has been a subject of controversy since ancient times. Similar arguments arose about the origin of triangular depressions (trigons) universally observed on {111} faces with an opposite orientation to the triangle of the {111} faces, and on the origin of center-cross patterns seen in the center of octahedral crystals. The controversies surrounding these problems are summarized in Table 9.1.

The rounded forms of natural diamond crystals are commonly observed in crystals occurring both in alluvial deposits (secondary deposits) and in mother rocks

Figure 9.2. Sketches of natural diamond crystals extracted from the *Atlas der Kristallformen* [1].

Table 9.1 *Growth versus dissolution controversies surrounding natural diamond crystals*

Origin of rounded forms	
Growth model	solidification of liquid droplet of carbon, by Raman [3]
Dissolution model	Fersman and Goldschmidt [4], Williams [5], Seal [2]
Origin of trigons	
Growth model	Tolansky [6]; laboratory etching produces only etch pits of opposite orientation to trigons
Dissolution model	Frank [7]; etching in kimberlite powder produces trigons
Center-cross pattern	
Growth model	Frank [7]; combined form of smooth {111} and rough {100} faces
Deformation model	Tolansky [6]; due to plastic deformation

(primary deposits), and so cannot be due to simple abrasion in the transporting process by water, but must be due to geological processes experienced during their growth or post-growth history.

Among the controversies summarized in Table 9.1, the argument that the rounded forms are due to growth and represent the as-grown morphology has been completely rejected, and it is now believed that all these forms represent the forms appearing due to the dissolution process.

As-grown crystals of diamond are bounded by flat faces, sharp edges, and corners, similar to ordinary polyhedral crystals. Conclusive evidence to support the dissolution process was obtained by Seal [2], who observed the relation between the rounded surfaces and growth banding revealed in crystals by etching the sections. From the observations showing that the rounded external form cuts the straight growth banding parallel to {111} seen in the interior, it was evident that the original crystal had been larger than its present size, and had changed its form to the present rounded form because the surface became partially dissolved (Fig. 6.1(c)).

As discussed previously, natural diamond crystals formed in the mantle deep within the Earth under high-temperature and high-pressure conditions corresponding to its thermodynamically stable region; they were then brought up to the Earth's surface by magmatic movement of kimberlite and lamproite, and they eventually survived as a metastable phase due to adiabatic expansion by volcanic eruption. As diamonds ascend to the Earth's surface, they have to pass through temperature and pressure conditions that can cause instability in diamond at high temperature. In addition, oxidation–reduction conditions vary according to the compositional differences in H_2O and CO_2 in different magmas, and the degree of dissolution will change depending on the oxygen fugacity, f_{O_2}.

Figure 9.3 is a schematic diagram showing how the morphology of the crystal and the surface microtopographs of crystal faces change as dissolution proceeds, starting from an as-grown octahedral crystal. The corner and edges of an

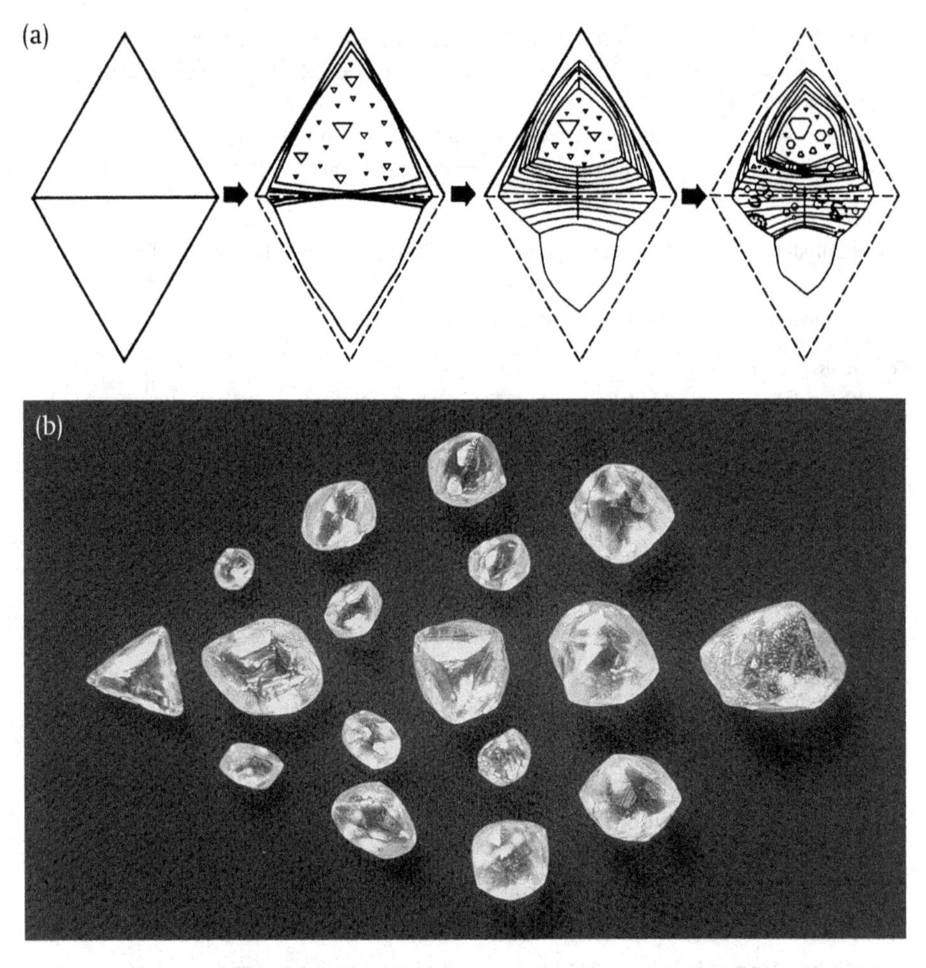

Figure 9.3. (a) Change of form as dissolution proceeds, starting from an octahedral crystal. (b) Rounded crystals.

octahedral crystal are rounded, resulting in a hexa-octahedral form bounded by forty-eight rounded faces consisting of {*hhl*} faces. Etching also preferentially takes place at outcrops of dislocations and point defects on {111} faces, forming point-bottomed (P-type) and flat-bottomed (F-type) etch pits (trigons) (Fig. 9.4).

Trigons are, in general, oppositely orientated to the triangular form of a {111} face. In etching experiments in the laboratory using an oxidizing agent such as KNO_3, it was only possible to produce etch pits with the same orientation as the triangular form of the {111} face. This led to arguments about whether the origin of trigons was due to growth or dissolution. It was shown experimentally that etch pits with the same orientation as naturally observed trigons were produced when diamonds were heated in powders of kimberlite. Later, it was demonstrated by Kanda *et al.* [8] that the orientations of etch pits of diamonds were reversed depending on the oxygen fugacity f_{O_2} and the temperature T (Fig. 9.5).

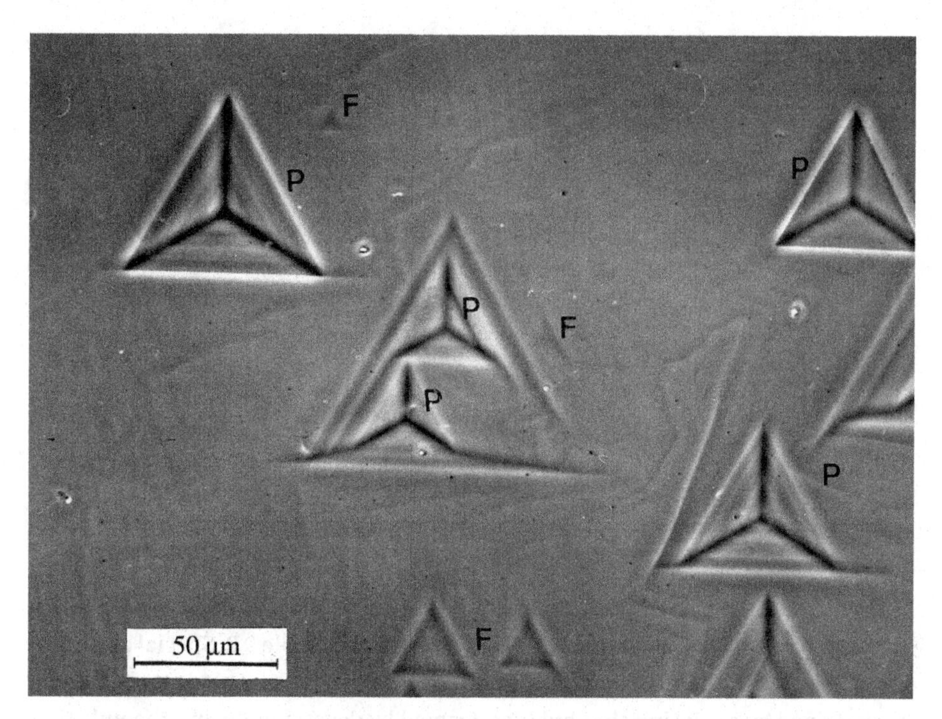

Figure 9.4. P-type and F-type trigons. Phase contrast photomicrograph.

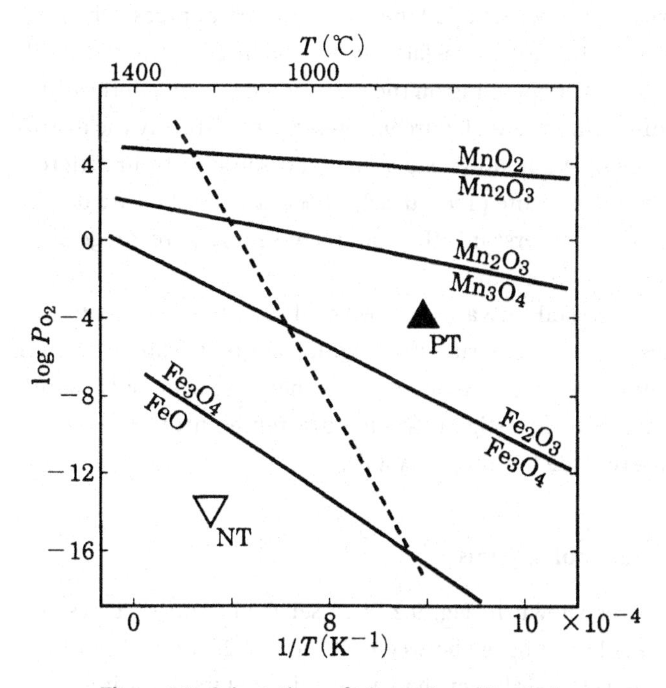

Figure 9.5. Orientations of trigons shown in relation to temperature T and oxygen fugacity f_{O_2} [8]. PT has the same orientation as the triangle form of the {111} face, and NT has the opposite orientation.

Table 9.2 *Morphological characteristics of natural diamond crystals showing that they were partially dissolved*

External forms	rounded corners and edges rounded faces rounded {*hll*}, {*hk*0} faces
{111} faces	trigons (P-type and F-type etch pits), mostly with opposite orientation to the triangular {111} face, rarely with the same orientation, or hexagonal pits
Surface of curved crystal faces	network of ditches superimposed circular ditches
Internal	straight growth banding cutting rounded external form banding pattern with irregular form

There is a great deal of evidence that demonstrates that natural diamond crystals were partially dissolved, and this is summarized in Table 9.2. Special attention should be paid to the superimposed circular ditches (Fig. 9.6). This patterning may be explained by assuming that bubbles, formed by degassing during the uplifting process of the magma, adhered on the crystal surface and resisted dissolution.

The magmas comprising kimberlite and lamproite cylindrical pipes, which act as carriers of diamonds to the Earth's surface, vary in their contents of volatile components like H_2O and CO_2, depending on the pipes, which results in the difference in the degree of dissolution rates. Figure 9.7 presents statistically the distribution of the morphology of diamond crystals made on samples from different kimberlite pipes. In some kimberlite pipes, nearly 75% of the crystals are octahedral with slightly rounded corners; in other pipes, the same percentage show strongly rounded forms.

Since natural diamond crystals are affected by the dissolution process to variable degrees, it is necessary to reconstruct the original, as-grown state. There are two methods of achieving this. One is to find crystals that have been only slightly dissolved, and the other is to investigate the texture representing the growth process, which might be recorded in single crystals.

9.3 Single crystals and polycrystals

The crystal figures shown in Fig. 9.2 were selected from sketches that appeared in a series of books published between 1913 and 1923, and all represent forms of single or twinned crystals larger than a few millimeters. They are mostly gem-quality diamonds. Other than these forms, there are translucent or opaque

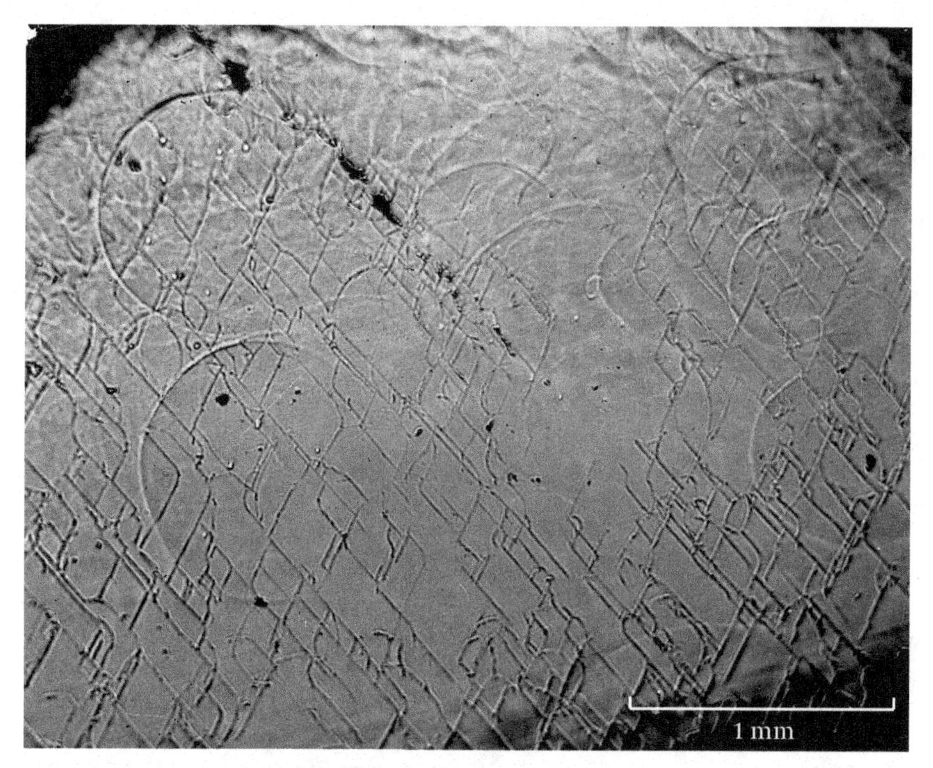

Figure 9.6. Superimposed circular ditches (etch patterns).

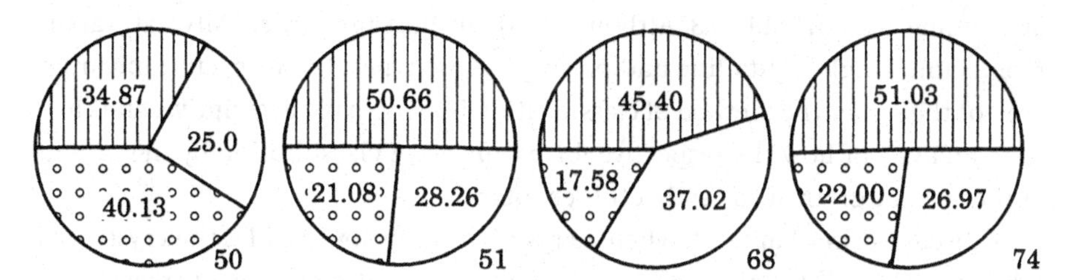

Figure 9.7. Statistics on diamond morphologies associated with different kimberlite pipes in Siberia. The areas with vertical lines represent an octahedral morphology with only a slight dissolution; those with the circles show crystals bounded by curved faces, which received heavier dissolution. Blank areas correspond to an intermediate type.

diamond crystals, which are mainly used for industrial purposes, either in single crystalline (but full of inclusions) or polycrystalline aggregates of minute crystals. The ratio of gem-quality to industrial-quality stones was, until recently, 1:4, but this has since dropped to 1:1. This is due to changes in the standards required for gem-quality diamonds.

The classification and naming by Dana [9], [10] and Orlov [11] of diamond forms

Table 9.3 *Morphology of diamond*

Classification by Dana and Orlov

Dana (1962) [9], [10]	*Orlov (1977) [11]*	
Single crystalline	*Single crystalline*	
Octahedral	*Variety*	
Cuboid	I	octahedral
Dodecahedral bounded by curved faces	II	cubic bounded by flat faces or cuboid, transparent faces
Tetrahedral	III	cuboid, translucent
Spinel twins, etc	IV	coated stone, clear core and milky coat
	V	coated stone, clear core and black coat
Polycrystalline	*Polycrystalline*	
Bort	*Variety*	
Framesite	VI	ballas, spherulite with fibrous radiating structure
Stuwartite	VII	aggregate of a small number of octahedral crystals
Short bort	VIII	bort, aggregate of idiomorphic crystals
Hailstone bort	IX	bort, aggregate of irregular grains
Ballas	X	carbonado, cryptocrystalline aggregate
Carbonado		

are summarized in Table 9.3. Although both authors broadly classify natural diamonds into single crystalline and polycrystalline types, the respective meanings are not the same. The method of classification is principally descriptive, and there is no analysis of how the respective forms appear. A classification by the present author will be given at the end of this chapter.

As discussed in Chapter 3, when a crystal grows below $\Delta\mu/kT^*$, the crystal will take a polyhedral form bounded by smooth interfaces, and on increasing the driving force the interface will transform into a rough interface, and the morphology changes to hopper, dendritic, and then to polycrystalline aggregate in spherulitic form. (See Fig. 3.21 for a schematic illustration of morphological changes depending on the driving force, assuming a crystal is bounded by {111} faces only.) Crystals grown under a small driving force condition grow as an octahedral single crystal bounded by flat faces, but those formed under higher driving force conditions will appear as polycrystalline aggregates such as spherulites.

Interface roughness varies depending on crystallographic direction (crystal faces). Therefore, on crystals growing under the same driving force condition, the roughness of the interfaces depends on crystallographic directions. In Section 9.4, we will analyze the morphology of diamond crystals, taking this into account.

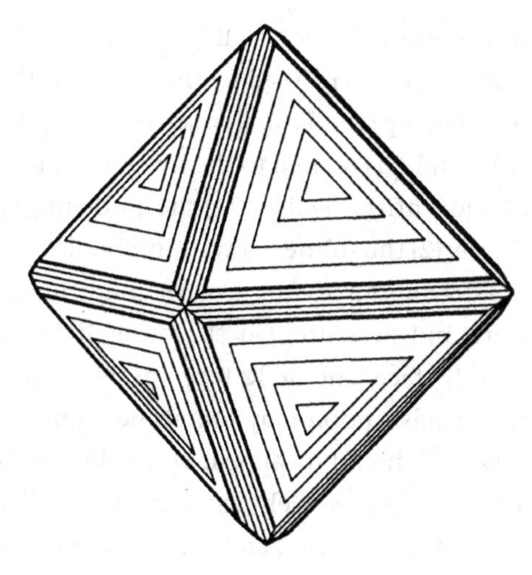

Figure 9.8. Structural form of diamond as predicted from PBC analysis.

9.4 Morphology of single crystals

9.4.1 *Structural form*

If PBC analysis (see Section 4.2) is performed based on the crystal structure of diamond (Fig. 9.1), we see that {111} is an F face containing three PBCs, and that {110} is an S face containing one PBC. On the other hand, since dangling bonds on the surface are perpendicularly oriented in successive atomic layers, {100} can be regarded as having a sort of double-decker structure, with no PBCs included, and so it should be classified as a K face. From this analysis, it is expected that {111} is the only smooth face on which layer or spiral growth can take place, whereas {100} corresponds to a rough interface which grows by an adhesive-type mechanism.

The {110} face is an S face, on which layer or spiral growth is not expected, and the face may appear by the piling up of the edges of the steps of the growth layers on {111} faces. Therefore, the morphology of a single crystal of diamond that can be expected based only on structural characteristics, entirely neglecting the effect of environmental conditions, is an octahedron bounded by largely developed {111} faces, accompanied by narrow, striated {110} faces, with no {100} faces appearing as crystal faces, as shown in Fig. 9.8. On {111} faces, growth or spiral layers with triangular form of the same orientation as the triangle of the face will be observed, whereas {110} faces are characterized by the development of striations, and {100} faces (if they appear) will exhibit rough and rugged surfaces.

Growth forms are the forms of a crystal determined by the structural characteristics and the effects due to environmental conditions. Assuming that a growing crystal is bounded by two F faces only, the *Tracht* of the crystal is determined by the relative normal growth rate of two F faces R, i.e. R_A:R_B. If both faces grow by the spiral growth mechanism, the R values are determined by the height of the spiral growth layer h, the advancing rate of the step v, and the step separation λ_0 in the following form: $R = hv/\lambda_0$. Since $\lambda_0 = 19r_c$, and r_c is related to the free energy of the step γ^2, and the driving force $\Delta\mu^3$, the factors affecting R are (1) the ambient phase, i.e. whether it is a vapor or solution phase; (2) the solute–solvent interaction energies, i.e. the solvent component; (3) the driving force and growth conditions; (4) the impurities; and (5) the temperature and pressure. On the other hand, when a growing crystal is bounded by an F face and a K face (or an S face), the solute–solvent interaction or impurity adsorption may sometimes suppress the R value of a K face, and the K face, which should disappear from the crystal, may develop as large as a habit-controlling face. Conversely, there is a possibility that a K face (or an S face) transforms into an F face, or that an F and an S face transform into a K face.

9.4.2 *Characteristics of {111}, {110}, and {100} faces*

The most commonly encountered forms of single crystals of diamond in nature are octahedral or forms bounded by curved faces originating therefrom. Although Seal concluded that all {110} faces appeared by dissolution [2], it is seen in one of Seal's own photographs (Fig. 6.1(c)) that a (110) face with straight striations appears by a series of alternating micro-facets of (111) and $(\bar{1}11)$. Although it must be accepted that the larger, rounded {110} faces (in reality, these are {*hhl*} faces) appear by dissolution, there are cases in which {110} faces appear due to growth. Such {110} faces are stepped faces characterized by striations only. {110} faces showing these types of characteristics (Fig. 9.9) are often encountered among micro-diamond crystals of micrometer size. In contrast to this, step patterns due to the development of growth layers have never been observed on {100} faces; only rugged surfaces are seen on this face, as shown clearly in Fig. 9.2. Since natural diamond crystals bounded by {100} faces always show the surface characteristics seen in Fig. 9.2, the term cuboid instead of cube is used.

On the {111} face of the diamond crystal from Siberia (Fig. 9.10), which is only partially dissolved, triangular growth hillocks with the same orientation as the triangle of the face are observed. The height of the steps of the growth layers consisting of the growth hillocks is less than 0.5 nm. It is confirmed that there is a P-type trigon at each summit of the respective growth hillocks, which correspond to the outcrops of screw dislocations, as identified by X-ray topography (Fig. 9.11), and so it can be concluded that the crystal grew by the spiral growth mechanism on {111} faces [12]. It is a characteristic of dislocations of this type that a bundle of dislocations originate from the center of a crystal and radiate nearly perpendicularly to the {111} surface; they have Burgers vector <110>.

Figure 9.9. SEM photograph of {110} faces appearing by growth, characterized by the development of straight striations parallel to the edge of {111}.

9.4.3 *Textures seen inside a single crystal*

Slight deviations in physical perfection and chemical homogeneity in nearly perfect single crystals can be visualized by etching, X-ray topography, cathodoluminescence methods, etc. These deviations provide superb records, revealing fluctuations in growth rate and partitioning of impurity elements through the growth process, as already discussed in Chapter 6.

As discussed above, most natural diamond crystals are characterized by dislocation bundles originating from the center of a crystal and running nearly perpendicularly to the {111} surface with a growth banding pattern parallel to {111}. There are, however, crystals showing complicated curved banding patterns, the origin of which will be a subject for future study.

Some crystals exhibit a texture called a center-cross pattern, the origin of which was, at one time, a subject of controversy, as to whether the origin was by growth or by plastic deformation. The center-cross pattern is schematically illustrated in Fig. 9.12; it corresponds to a texture shown by the growth sectors of two coexisting crystal faces, {111} and {100}. This pattern indicates that the arms of the cross correspond to the growth sectors of {100}, which disappear at the later stage of

Figure 9.10. Phase contrast photomicrograph of a (111) face of an octahedral diamond crystal from Siberia. Note the triangular pyramidal growth hillocks and small trigons at the summits of the respective growth hillocks. Photographed by K. Tsukamoto [12].

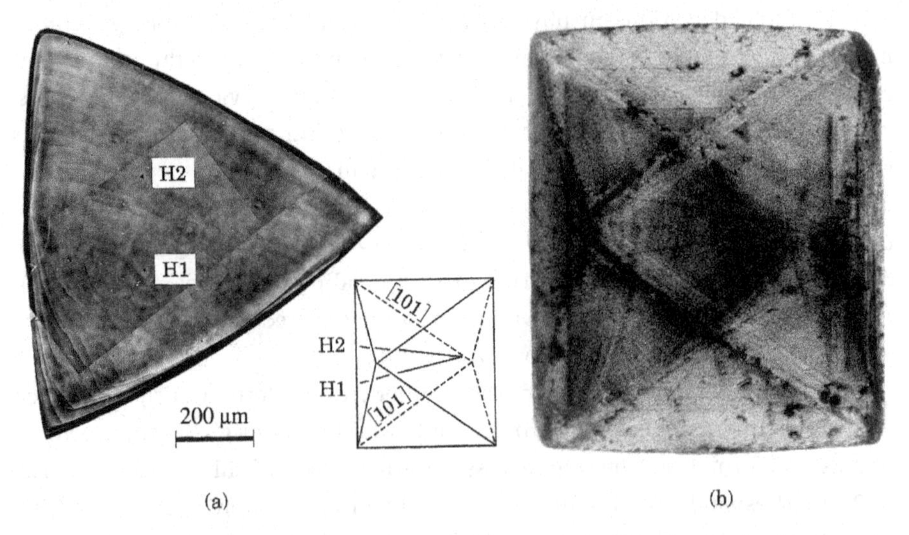

Figure 9.11. (a) Surface microtopograph (phase contrast photograph taken by K. Tsukamoto) and (b) X-ray topograph (taken by T. Yasuda [12]) of a (111) face. Steps with height of less than 0.5 nm spread from the outcrops of dislocations H1 and H2.

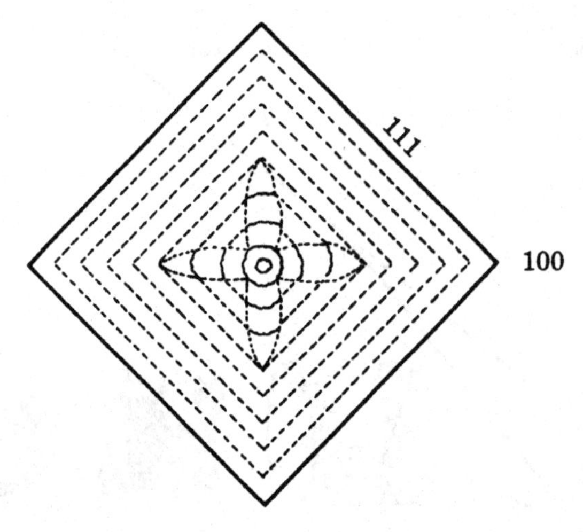

Figure 9.12. Growth sector showing a center-cross pattern formed by the growth of smooth {111} and rough {100} interfaces.

growth, and the crystal eventually takes a simple octahedral form. Whereas the growth banding in the {111} growth sectors is straight and parallel to {111}, that in the {100} growth sectors is wavy and curved, like a hammock. This indicates that on the same crystal {111} behaves as a smooth interface (F face), whereas {100} behaves as a rough interface (K face). In natural diamond crystals, {100} behaves exclusively as a rough interface, and no evidence has been obtained so far to show that {100} may sometimes behave as a smooth interface.

From these observations, we may conclude that, in the growth of natural diamond crystals, three faces, {111}, {110}, and {100}, behave and show characteristics completely in agreement with the characteristics expected from PBC analysis. Therefore, we may conclude that, under the environmental conditions of natural diamond growth (principally in the silicate solution phase), {111} always behaves as a smooth interface under $\Delta\mu/kT^*$ conditions, whereas the $\Delta\mu/kT^{**}$ of {110}, and particularly of {100}, stays close to the origin under any conditions, and these faces behave exclusively as rough interfaces.

Figure 9.13 shows morphological changes of diamond crystals on a diagram of growth rate versus driving force relation (see Fig. 3.15), based on the relation of the ** positions of the {111} and {100} faces. From this, it is expected that spherulitic forms and cuboids occur under higher driving force conditions, whereas octahedral crystals are expected under lower driving force conditions.

9.4.4 *Different solvents (synthetic diamond)*

In the case of synthetic diamond, grown under high-temperature, high-pressure conditions from a high-temperature solution with metal or alloy as the solvent, diamond crystals exhibit a cubo-octahedral *Tracht* bounded by {100} and

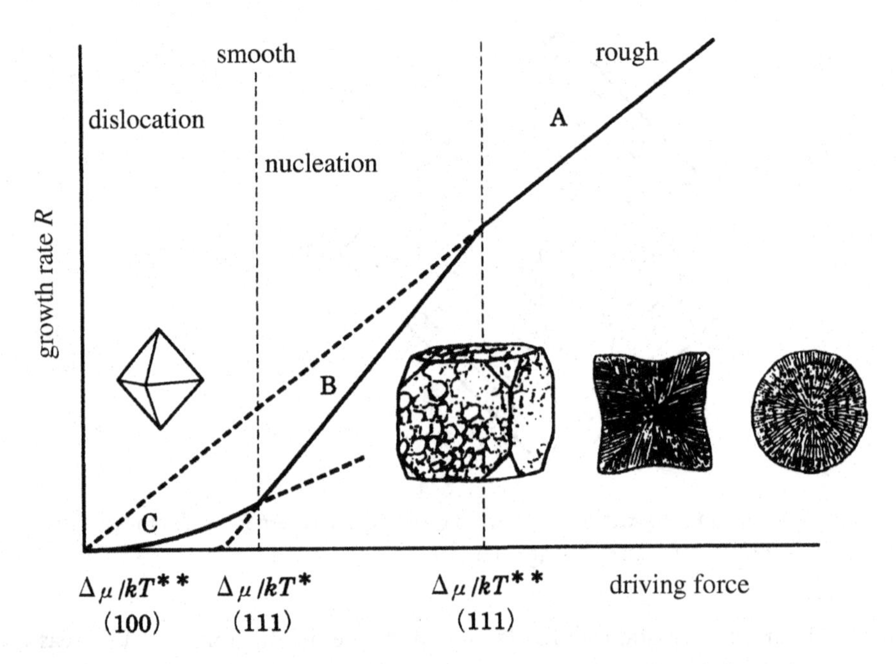

Figure 9.13. Morphology of natural diamond crystals expected in relation to the driving force. The morphological change is as expected assuming that the positions of $\Delta\mu/kT^*$ and $\Delta\mu/kT^{**}$ are different for {111} and {100}.

{111}. Although the development of these two faces may vary depending on the driving force, both show evidence of spiral growth, as can be seen in Fig. 9.14. This indicates that {100} transforms from a rough K face to a smooth F face in a metallic solution. A transformation from a K face to an F face requires a change in the surface structure of the {100} face (i.e. surface reconstruction should take place), resulting in the same effect as introducing a new PBC. In the growth of natural diamond in a high-temperature solution phase with silicate as the solvent, surface reconstruction does not occur, whereas it does occur if a metal solvent is used. Surface reconstruction occurs in a solution in which the metallic element has a small ionic radius as the solvent, whereas it does not occur in a solvent with large ionic radius.

Recently, diamond synthesis has been successfully performed under high-temperature, high-pressure conditions in a system using kimberlite powder, various carbonates, sulphates or water as the solvent [13], [14]. Higher pressure and temperature conditions are required in a non-metallic solution than in a metallic solution, and the crystals obtained are mainly simple octahedral, differing from those observed in crystals grown from metallic solutions. Crystals synthesized in a non-metallic solution show the same characteristics as natural diamond *Tracht*. These observations indicate that the solvent components have a definitive effect upon surface reconstruction, and thus on the morphology of the crystals.

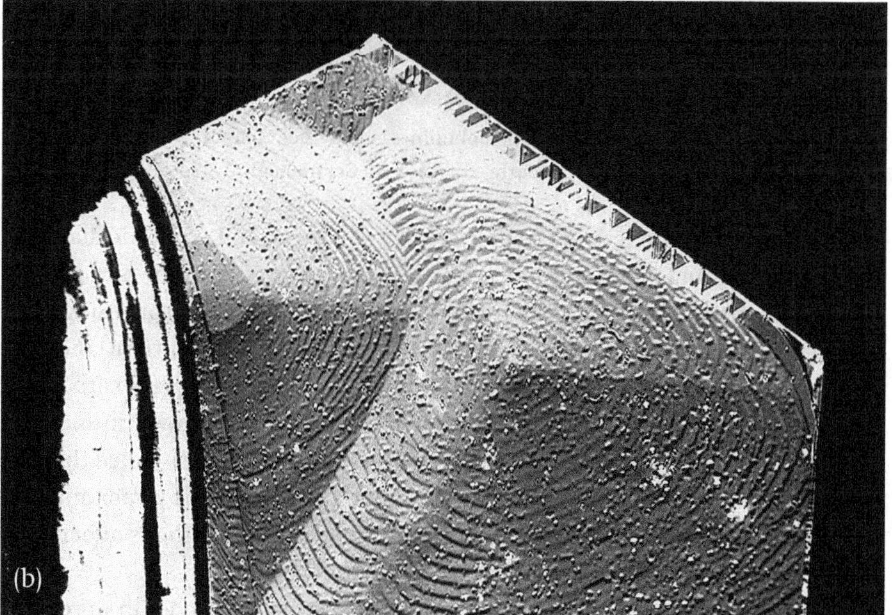

Figure 9.14. Spiral growth layers observed on (a) {100} and (b) {111} faces of diamond crystals synthesized under high-pressure and high-temperature conditions.

(a)

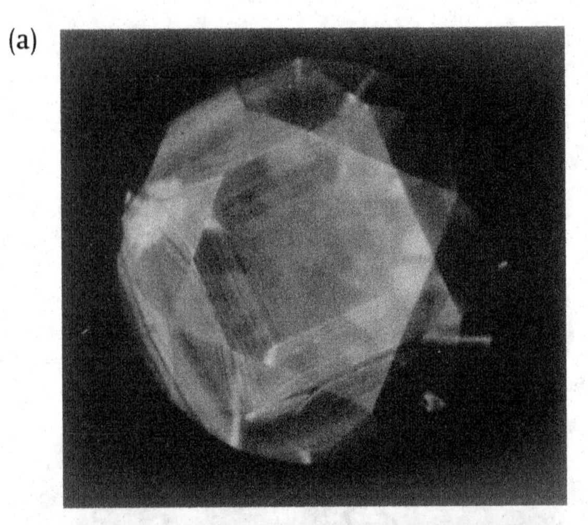

(b)

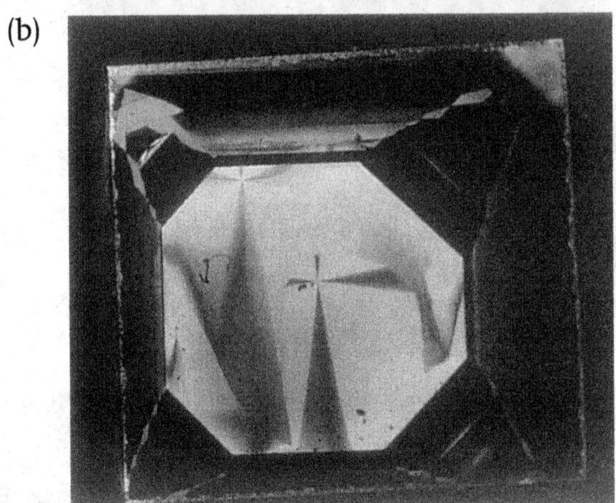

Figure 9.15. Comparison of cathodoluminescence tomographs of (a) natural and (b) synthetic diamonds. (By courtesy of All Japan Gemmological Association [15].)

The resulting differences in the morphology of natural and synthetic diamond single crystals using metal solvents is also recorded in patterns shown by growth sectors and growth banding. If we reveal inhomogeneities of this sort using such techniques as cathodoluminescence, the differences between natural and synthetic diamonds are easily distinguishable. Figure 9.15 is an example of cathodoluminescence photographs of natural and synthetic diamonds grown in metallic solution, and the difference between the two is very evident [15]. It is noted that the intensity of the cathodoluminescence is remarkably different depending on growth sector. Figure 9.15 also indicates that element partitioning is affected by crystallographic directions, and therefore by kinetics.

Vapor grown diamonds are synthesized by the CVD method under low-pressure conditions, for which diamond becomes unstable, and it is possible to obtain single crystals of micrometer size. Frequent occurrences of multiply twinned particles (refer to Section 7.2), cubo-octahedral *Tracht* bounded by {111} and {100} (similar to HPHT synthetic diamond), and the universal occurrence of spiral growth layers on the {100} faces are the morphological characteristics shown by these CVD diamonds. However, CVD diamonds show a distinct difference from high-pressure, high-temperature synthetic diamonds in the relative relation between the morphological importance of {111} and {100}. In both natural and high-pressure, high-temperature synthetic diamonds, the order of morphological importance is {111} > {100}, whereas this relation is reversed in CVD diamonds, and {100} behaves as the morphologically more important face. {111} faces show a skeletal form on those crystals whose {100} faces show spiral growth layers. Also, it is observed that the orientation of spiral growth layers on {111} is reversed, and is opposite to the orientation observed on crystals formed under high-pressure, high-temperature conditions.

The {111} face of diamond contains three PBCs, but {100} contains a maximum of two PBCs, even if surface reconstruction takes place. Therefore, the characteristics observed in CVD diamonds cannot be accounted for in terms of surface reconstruction due to the difference in solvents. If the surface energy term can be modified for any other reason, it is possible that the order of morphological importance may be reversed. A possible reason may be the surface adsorption of H_2 molecules on the surface of a growing CVD diamond. It has been calculated that H_2 molecules adsorbed on the surface can drastically modify the surface energy state of diamond.

9.4.5 *Twins*

As can be seen from the lower part of Fig. 9.2, diamonds occur as spinel twins with (111) as the twin plane (= composition plane) and <111> as the twin axis. Crystals twinned according to the spinel twin law exclusively take triangular platy forms. This morphology is due to the pseudo re-entrant corner effect (see Section 7.2). Cyclic twins showing apparent five-fold symmetry elements due to repeated twinning are occasionally reported. Only one locality is known to produce cyclic twins of this type in nature, but in the synthesis of diamond under high-pressure, high-temperature conditions it is possible to grow them at will by providing high driving force conditions. Multiple twin particles have a similar twin relation as this, and are found universally in CVD diamonds synthesized from the vapor phase under low-pressure (1 Pa) conditions. An interesting point is that the distinct pseudo re-entrant corner effect is not seen in either cyclic twin or multiply twinned particles.

9.4.6 *Coated diamond and cuboid form*

Natural crystal growth occurs under non-controlled conditions, which may vary greatly during the crystal growth processes. In such a case, growth may be initiated under higher driving force conditions than $\Delta\mu/kT^{**}$, and the driving force gradually diminishes to lower than $\Delta\mu/kT^{*}$, which leads to the formation of a single crystal in appearance, with a central core portion formed by dendritic growth, surrounded by mantle sections formed by layer growth. If magma containing polyhedral crystals grown under $\Delta\mu/kT^{*}$ is uplifted, the driving force increases above $\Delta\mu/kT^{**}$, and dendritic growth will occur on the surface of preformed single crystals. As a result, a crystal appears with a clear core, surrounded by mantles with fibrous texture formed by dendritic growth. Depending on the growth history, variations of this type may occur repeatedly, resulting in more complicated textures; for a discussion, see Section 3.10. Conditional changes of this type are recorded as internal textures in diamond single crystals.

There are two types of internal textures seen in natural diamond crystals that show conditional changes. Although Dana [9], [10] and Orlov [11] distinguished between single crystals and polycrystalline aggregates, they did not put crystals that had experienced the two conditions into different categories.

The first and the most typical type is called coated stone, which has a clear, gem-quality internal portion, coated by a mantle portion with fibrous texture. In appearance, this type of diamond is opaque and of industrial quality, but if a window is made the clear, gem-quality interior is discernible. Coated stones have the following growth history.

If magma containing polyhedral single crystals of diamond formed under a driving force condition below $\Delta\mu/kT^{*}$ is uplifted and the diamonds are placed under a higher driving force condition, dendritic growth takes place on the substrate of a polyhedral crystal, thus forming a mantle portion surrounding a clear single crystalline core. Crystals showing similar textures to coated stones are universally observed among various phenocrysts of rock-forming minerals in volcanic rocks, and diamond is not an exception.

There is a group of natural diamonds called cuboids, which exhibit a cubic or cuboid morphology. These occur frequently in a particular locality, such as in the Republic of the Congo, and are mostly of industrial quality. Cuboids are not bounded by crystallographically flat {100} faces, but instead show exclusively rugged surfaces. According to X-ray topographic investigations [16], [17], cuboid diamonds are characterized by a columnar (thick fibrous) texture developing from the center in the <111> and <100> directions, as schematically shown in Fig. 9.16. Sometimes a central clear core portion is detected; sometimes it is not. Therefore, the formation of cuboids is essentially the same as that of the mantle portion of coated stones, and the difference between the two types is simply the difference in the thickness of the mantle portion.

In the second type (layers), the cuboid takes the role of a seed, as opposed to the above case, in which a single crystalline diamond grows under the condition below

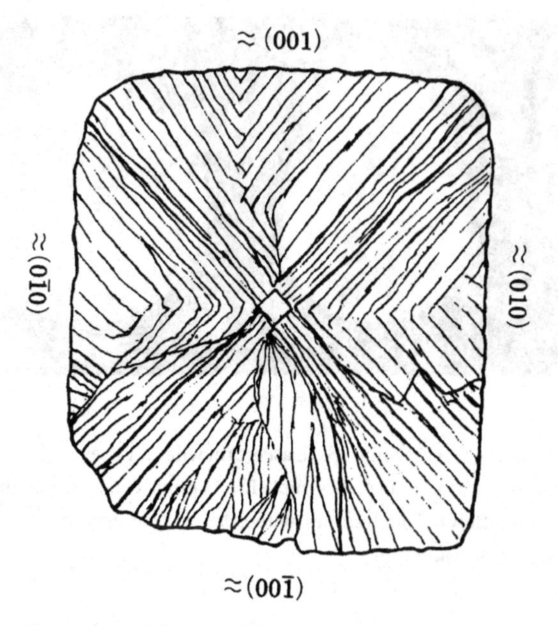

Figure 9.16. Schematic illustration of the internal texture of a cuboid, based on an X-ray topograph [16], [17].

$\Delta\mu/kT^*$. In recent investigations on cut stones, an example showing, for the first time, the presence of a seed in the crystal growth of natural diamonds has been discovered [18]. It is also the first example to show the relation between diamonds formed by ultra-high-pressure metamorphic rocks and those formed in ultramafic magma.

9.4.7 *Origin of seed crystals*

Gem-quality diamond is usually cut along a line parallel to the (100) direction of an octahedral crystal to obtain two cut stones. Two such samples, A and B, shown in Fig. 9.17, were compared to see if it is possible to identify whether the two stones came from the same rough or not. X-ray topography indicated that sample A contains a small number of dislocation bundles radiating from the center to {111}, and the Burgers vector is exclusively along <110>. Growth banding that is straight and parallel to {111} is observed in sample A. These features are commonly observed in gem-quality octahedral crystals. In contrast, in sample B, a square core portion is observed at the center, and dislocation bundles are generated principally from the surface of the core, and the Burgers vector is along <100>. The nature of the dislocations is entirely different in the two samples, and it is clear that samples A and B came from different rough stones. A further sample, C, was then compared to A and B. X-ray topographs of sample B and C match perfectly, proving that the two stones came from the same rough (Fig. 9.18).

Figure 9.17. Three cut stones were compared to ascertain whether they originated from the same rough stone. (a) Sample A; (b) sample B; (c) sample C. Samples A and B were seen to come from different rough stones, whereas it was confirmed that samples B and C came from the same rough [18].

The two stones B and C show hitherto unknown features [18], as follows. There is a core portion with a square outline in cross-section and cuboid form in three dimensions, and all dislocation bundles with Burgers vector $<100>$ generate from the surface of the core portion (Fig. 9.19). This implies that the core portion was formed somewhere else; it was then trapped in a different environmental phase and acted as a seed under the new conditions, after which the major part of the crystal was formed. This was the first piece of evidence to prove the presence of seed crystals in the growth of natural diamond.

The size and morphology of the core portion provide us with useful information that allows us to deduce where the diamond crystal that acted as the seed was formed. The core portion is cuboid in form, indicating that it was formed under a driving condition above the $\Delta\mu/kT^{**}$ of the $\{100\}$ face. In contrast to this, the growth of the major portion of the crystal, which grew around the seed, took place under conditions below the $\Delta\mu/kT^{*}$ of the $\{111\}$ face. The dislocation directions are $<100>$, which are distinctly different from the directions of $<110>$ generally observed in gem-quality diamonds. However, on X-ray topographs and cathodoluminescence tomographs of the samples, it is observed that, starting from the seed, micro-facets of $\{111\}$ appear, which gradually change to larger $\{111\}$ faces.

There are three types of rocks that are known to fulfil the conditions for diamond growth, and, in fact, diamonds are found in all of them.

(1) Ultra-high-pressure metamorphic rocks. Diamonds occur sporadically in crystals of garnet or zircon in various ultra-high-pressure metamorphic rocks formed in deep subduction zones. Crystals are of micrometer size, and the morphology is mostly spherulitic or cuboid, but octahedral is also

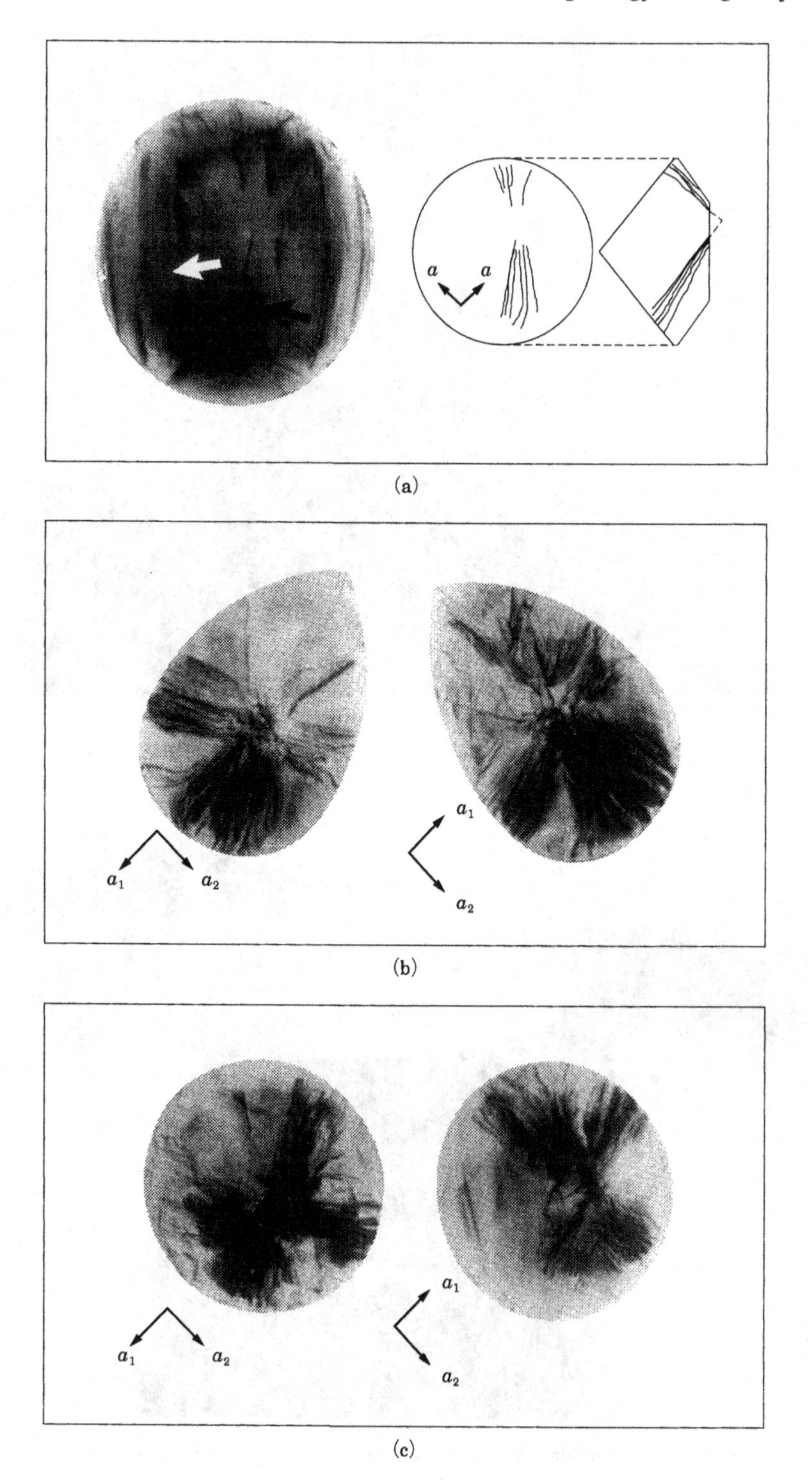

Figure 9.18. X-ray topographs of the three cut stones shown in Fig. 9.17. (a) Sample A; (b) sample B; (c) sample C. X-ray topographs taken by T. Yasuda.

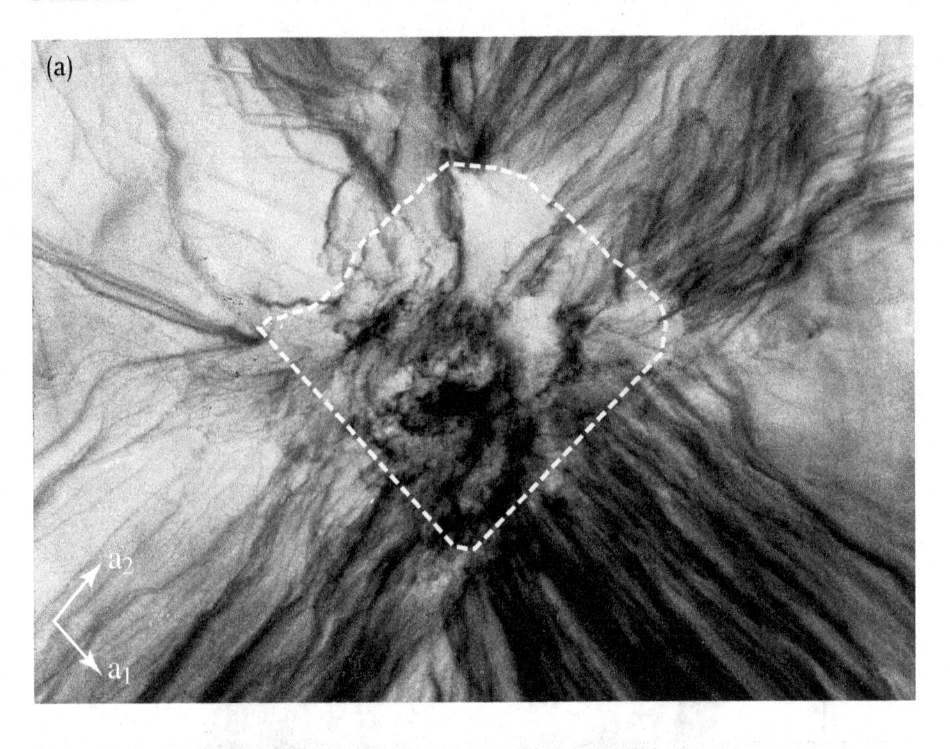

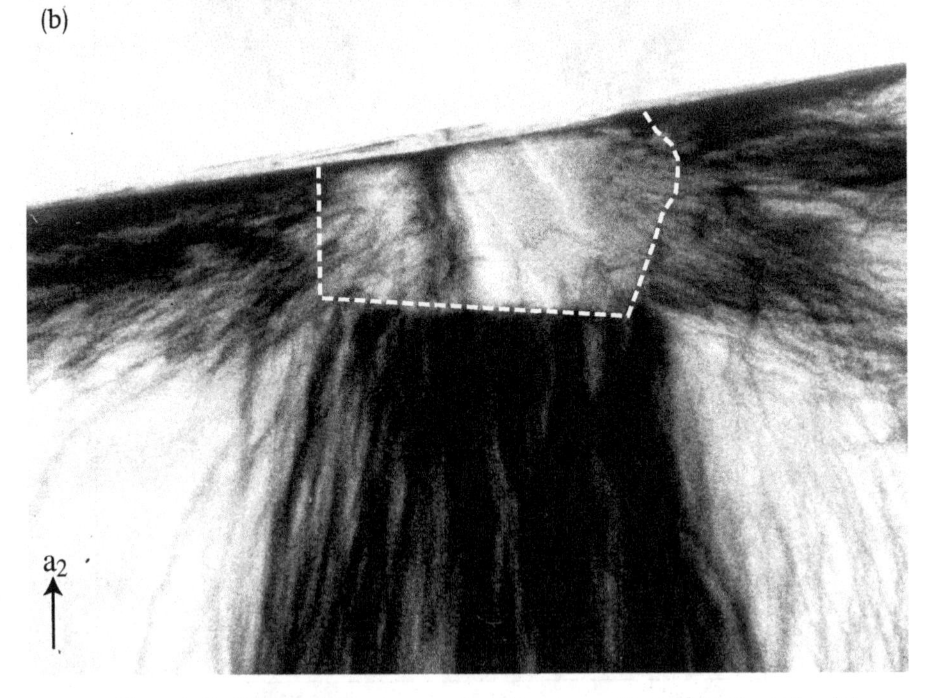

Figure 9.19. Magnified X-ray topographs of the core portions of cut stones B and C. (a) Stone B, cut parallel to the table facet; (b) stone C, cut perpendicular to the table facet. X-ray topographs taken by T. Yasuda.

found. The content of diamonds is extraordinary high, up to 2%, which is the characteristic feature of this mode of occurrence. Considering that the original source of ultra-high-pressure metamorphic rocks was oceanic sediments subducted deep into the Earth, the high content of diamond is understandable. This suggests that these diamonds grew in silicate liquid droplets formed by the partial melting of mother rocks under the condition above $\Delta\mu/kT^{**}$ for diamond growth, since carbon content can be very high due to subducted carbon of organic origin.

(2) Eclogite. This is found as xenolith in ultramafic rocks (see (3) below), and consists mainly of garnet. The origin has not been clarified, and so it is not known yet whether it is of ultra-high-pressure metamorphic origin or of magmatic origin. According to Orlov [11], the proportion of polycrystalline versus single crystalline diamond in eclogite is much higher than in ultramafic rocks.

(3) Ultramafic rocks. Diamonds are found and are considered to have been grown in ultramafic magma (which has a high content of Mg and Fe, and a low SiO_2 content). The content of diamonds is extremely low, 0.2 g/tonne; the proportion of single crystals within this group is high. It is assumed that diamonds were grown under conditions below $\Delta\mu/kT^*$.

There is a clear difference in syngenetic inclusions in diamond crystals between (2) and (3), indicating a difference in the chemical environments of diamond formation.

Many of the diamonds in types (2) and (3) were trapped in the uplifting process of kimberlite and lamproite magma and were brought up to the Earth's surface, whereas it is thought that those in type (1) have been brought up by a reverse subduction movement.

Considering the sites of diamond formation and the subsequent uplifting movement of magma, we may suggest the following scenario from the discovery of cuboid seeds of type (2) (growth of diamond under conditions below $\Delta\mu/kT^*$ on a cuboid seed). Namely, diamond crystals of micrometer size formed by ultra-high-pressure metamorphism due to subduction are further transported to the depths of the earth and incorporated into ultramafic magma, at which time they acted as seeds, and growth of a diamond proceeded below $\Delta\mu/kT^*$. It is surprising to see within such a small diamond crystal a record of the vast movement and history of the Earth's hidden depths.

Although Dana [9], [10] and Orlov [11] classified diamond crystals broadly into two types, single and polycrystalline, we can also construct a classification (given in Table 9.4) based on the preceding analysis. By this classification we are able to correlate the morphology of diamonds with their growth conditions and growth histories.

Table 9.4 *Morphology of diamond*
Sunagawa's classification

Single crystalline	octahedral, tetrahedral, dodecahedral with curved faces, twins
Polycrystalline	ballas, bort, short bort, hailstone bort, framesite, stuwartite, carbonado
Two-stage growth	(a) coated stone, cuboid
	(b) octahedral growth on cuboid seed

9.4.8 *Type II crystals showing irregular forms*

Natural diamonds grew in the mantle deep under the Earth and were brought up to the surface by the rapidly ascending movement of kimberlite or lamproite magma. During this process, crystals were partially dissolved and also experienced plastic deformation due to stress associated with the rapid ascending movement. The evidence supporting this is recorded in the form of tangled dislocations observed in single crystals, the occurrence of slip lines, and bent crystals.

Type I diamond has a high content of nitrogen, which occurs in the form of precipitated nitrogen platelets or aggregates. In contrast to the fact that Type I crystals may be regarded as a C–N alloy, Type II diamond contains nitrogen at less than part per million order, and can be regarded as a purer carbon crystal. As can be seen from the large differences in critical shear stress that causes plastic deformation between pure aluminum and duralumin, an Al alloy containing less than 2% Cu, Type II (corresponding to pure metal) is plastically weaker than Type I (corresponding to an alloy), and thus it is expected that Type II diamond will be plastically more deformed and will eventually fracture before Type I crystals under the same applied shear stress.

External forms of diamond crystals are used as a criterion in sorting rough stones of diamond. Rounded crystals, octahedrons, and cuboids are all grouped as Type I crystals, and are further subdivided by color and forms on a fashioning basis. In contrast, crystals showing irregular or platy forms without distinct crystal faces are all classified as Type II. This empirical classification has been found to be essentially correct on checking using the transmittance of ultra-violet rays. Namely, Type I crystals exhibit rounded, polyhedral forms, and Type II crystals are characterized by irregular and platy forms, not bounded by crystallographic faces. The following reasons have been suggested as explanations [19].

(1) Irregular forms of Type II crystals may represent broken forms of what were originally polyhedral forms in the Earth and became irregular either (i) by the uplifting process or (ii) due to shock received during or after mining operations.
(2) The remarkable anisotropy in the dissolution in the magma may have contributed to the irregularity.

Figure 9.20. Type II diamond showing irregular forms.

(3) The growth of diamond in interstices of solid particles may be an explanation.

(4) Polyhedral (principally octahedral) crystals of Type I may have appeared due to selective adsorption of impurities to suppress the growth rate of {111}, and therefore pure Type II grew without such an effect, thus resulting in irregular forms.

The last explanation, i.e. the development of {111} due to impurity adsorption, can be easily refuted on the basis of the anisotropy involved in diamond structure (PBC analysis). The likelihood of (2) and (3) being true is also remote.

As seen from Fig. 9.20, Type II crystals show irregular forms and their surfaces show minute undulation, indicating that the surface was etched after the irregular forms appeared. Under a polarization microscope, Type II crystals show a characteristic tatami-mat pattern [20] formed by crossing slip lines parallel to {111} or they universally show strain birefringence. X-ray topographs of Type II crystals consist of irregular areas showing contrast images, and those are entirely out of contrast, which indicates that the crystal is bent. The X-ray topographic characteristics are very different from those seen in Type I crystals. All this indicates that Type II crystals are plastically more heavily deformed than Type I crystals. It is anticipated that the probability of Type II crystals taking irregular forms is much higher than for Type I crystals, if deformation proceeds further and crystals are broken. Judging from the observation that Type II crystals show surface etching, we may conclude that the morphological characteristics of irregular or platy forms

Figure 9.21. Micro-diamond photographed under an ultra-violet filter of 2250 Å. The transparent crystals are Type II; the opaque ones are Type I. There is no difference in morphology between the two. Note the high proportion of Type II crystals. In macro-diamond, the ratio is much lower because the Type I and II layers are alternately stacked.

of Type II crystals appeared in the ascending process from the Earth's depth to the surface, and not during the mining process.

In the morphology of as-grown micro-diamond crystals, no essential difference was detected between Type I and II crystals, which both take octahedral forms. This is clearly shown in Fig. 9.21 [21], in which the two types are compared by the transmittance of ultra-violet rays. There is no essential difference in morphology between Type II, which is transparent under the ultra-violet ray used, and Type I, which is opaque to the same wavelength.

References

1 V. Goldschmidt, *Atlas der Kristallformen*, B.I–B.IX, Heidelberg, Carl Winters Universitatsbuchhandlung, 1913–23
2 M. Seal, Structure in diamond as revealed by etching, *Am. Min.*, **50**, 1965, 105–23
3 C. V. Raman and S. Ramaseshan, The crystal forms of diamond and their significance, *Proc. Ind. Acad. Sci.*, **A24**, 1946, 1–24
4 A. Fersman and V. Goldschmidt, *Der Diamant*, Heidelberg, Winter, 1911

5 A. F. Williams, *Genesis of the Diamond*, London, Benn, 1932

6 S. Tolansky, *The Microstructures of Diamond Surfaces*, London, N. A. G. Press, 1955

7 F. C. Frank, Defects in diamond, in *Science and Technology of Industrial Diamonds*, ed. J. Burls, London, IDIB, 1967, pp. 119–35

8 H. Kanda, S. Yamaoka, N. Setaka, and H. Komatsu, Etching of diamond octahedron by high pressure water, *J. Crystal Growth*, **38**, 1977, 1–7

9 E. S. Dana, *The System of Mineralogy*, 6th edn, New York, John Wiley & Sons, 1892–1915

10 C. Palache, H. Berman, and C. Frondel, *The System of Mineralogy*, 7th edn, vol. 1, New York, John Wiley & Sons, 1944

11 Yu. L. Orlov, *The Mineralogy of The Diamond*, New York, John Wiley, 1977

12 I. Sunagawa, K. Tsukamoto, and T. Yasuda, Surface microtopographic and X-ray topographic study of octahedral crystals of natural diamond from Siberia, in *Materials Science of the Earth's Interior*, ed. I. Sunagawa, Dordrecht, D. Reidel, 1984

13 M. Arima, Experimental study of growth and resorption of diamond in kimberlitic melts at high pressures and temperatures, in *Advanced Materials '96*, Tsukuba, NIRIM, 1996, pp. 223–8

14 H. Yamaoka, M. Akaishi, and S. Yamaoka, Diamond formation in the graphite-MgO-H_2O system, *Advanced Materials '96*, Tsukuba, NIRIM, 1996, pp. 245–50

15 I. Sunagawa, The distinction of natural from synthetic diamonds, *J. Gemmol.*, **24**, 1995, 489–99

16 F. C. Frank and A. R. Lang, X-ray topography of diamond, in *Physical Properties of Diamond*, ed. R. Berman, Oxford, Clarendon Press, 1965, pp. 69–115

17 A. R. Lang, Internal structure, in *The Properties of Diamond*, ed. J. E. Field, London, Academic Press, 1979, pp. 425–69

18 I. Sunagawa, T. Yasuda, and H. Fukushima, Fingerprinting of two diamonds cut from the same rough, *Gem and Gemmology*, Winter Issue, 1998, 270–80

19 I. Sunagawa, A discussion on the origin of irregular shapes of Type II diamonds, *J. Gemmol.*, **27**, 2001, 417–25

20 M. Takagi and A. R. Lang, X-ray Bragg reflexion, "spike" and ultra-violet adsorption topography of diamonds, *Proc. Roy. Soc.*, **A 281**, 1964, 310–22

21 S. Tolansky and H. Komatsu, Abundance of type II diamonds, *Science*, **157**, 1967, 1173–5

Suggested reading

I. Sunagawa, *Talks on Diamond*, Tokyo, Iwanami Pub. Co., 1964 (in Japanese)

R. Berman (ed.), *Physical Properties of Diamond*, Oxford, Clarendon Press, 1965

J. Burls (ed.), *Science and Technology of Industrial Diamond*, London, IDIB, 1967

I. Sunagawa, *Diamonds, Their Genesis and Properties*, Tokyo, Ratis, 1969 (in Japanese)

Yu. L. Orlov, *The Mineralogy of The Diamond*, New York, John Wiley, 1977

J. E. Field (ed.), *The Properties of Diamond*, London, Academic Press, 1979

E. Wilks and J. Wilks, *Properties and Applications of Diamonds*, London, Butterworth-Heinemann, 1991

J. E. Field (ed.), *The Properties of Natural and Synthetic Diamond*, London, Academic Press, 1992

10

Rock-crystal (quartz)

Single crystals of quartz (SiO_2) showing euhedral forms are traditionally called rock-crystals, and interest in their form has continued since the time of Steno. Hexagonal prismatic crystals with two types of rhombohedral faces, $\{10\bar{1}1\}$ and $\{01\bar{1}1\}$, at the tip show short-prismatic to long-prismatic forms or tapered forms determined by the relative growth rates between the rhombohedral and the prismatic faces, as well as various intergrowth forms such as sub-parallel or scepter intergrowth. Twins growing according to the Japan law or Brazil law exhibit characteristic forms (see Sections 10.6.1–10.6.3). It is the aim of this chapter to analyze how various forms of rock-crystals appear. Chalcedony and agate, two polycrystalline aggregates of quartz crystals of micrometer size, exhibit characteristic textures. How these textures are formed will also be explained in this chapter.

10.1 Silica minerals

Rock-crystal is a typical mineral crystal, which, in its regular geometric hexagonal prismatic form, has attracted interest since the earliest times. Its chemical composition is SiO_2, and its mineral name is quartz. More than six polymorphs are known among minerals of chemical composition SiO_2. Within this group, the polymorph belonging to the hexagonal system, crystal group 622, and space group $P6_222$ or $P6_422$, which is stable above 573 °C under 1 atmosphere pressure, is called high-temperature quartz, and that belonging to the trigonal system, crystal group 32, is called low-temperature quartz. In low-temperature quartz, there are two types of structures, with space groups $P3_121$ and $P3_221$, called right-handed quartz and left-handed quartz, respectively. In addition to these, high-temperature polymorphs called tridymite and cristobalite and high-pressure polymorphs called coesite and stishovite (as well as a couple of other phases) are known. The general name "rock-crystal" has been traditionally applied to clear single crystals of quartz attaining sizes discernible to the naked eye.

Various names, such as amethyst, citrine, and smoky or black quartz, have been used for colored rock-crystal, whereas for the cryptocrystalline aggregate of quartz, names such as chalcedony and jasper are used. Agate and cornelian, for example, are types of chalcedony that have specific textures or colors. In this chapter, we analyze how a variety of morphologies of high-temperature and low-temperature quartz appear, and how textures of polycrystalline aggregate seen in agate and other crystals are formed.

In both high- and low-temperature quartz, the unit of construction is SiO_4 (a tetrahedron consisting of one silicon and four oxygen atoms) (Fig. 10.1). A three-dimensional network structure is constructed by the sharing of all oxygen atoms at the summits of the SiO_4 tetrahedra. Viewed from the *c*-axis direction, the symmetry axes constructed by connecting SiO_4 are either 6_2 or 6_4 screw axes in high-temperature quartz, and either 3_1 or 3_2 screw axes in low-temperature quartz. We see that slightly changing the angles that are formed on connecting the SiO_4 tetrahedra causes the difference between a crystal belonging to a hexagonal system, crystal group 622, and a crystal belonging to a trigonal system, crystal group 32.

Since only an angular adjustment is required to cause the transition between the two phases, it occurs sharply at 573 °C under 1 atmosphere pressure. However, we should expect a transitional (precursor) state, corresponding to the energy required for the angular adjustment, to be present during the transition. Since two possible orientations are associated with this transition, Dauphiné twinning occurs. Dauphiné twinning may also occur because of mechanical or electrical stress. In the phase transition from tridymite or cristobalite to low-temperature quartz, structural rearrangement is required, and so the transition is not sharp and these two forms may remain as metastable phases at low temperatures.

The principal form of single crystals of natural rock-crystal is hexagonal prismatic bounded by six prisms *m* $\{10\bar{1}0\}$, terminated alternately by three major *r* $\{10\bar{1}1\}$ and three minor *z* $\{01\bar{1}1\}$ rhombohedral faces. The difference between right-handed and left-handed quartz appears in the position of the *s* $\{11\bar{2}1\}$ and *x* $\{51\bar{6}1\}$ faces. It is exceptionally rare to observe the appearance of a basal plane $\{0001\}$ in natural rock-crystal, but the faces appear universally on synthetic quartz grown on seeds. In natural rock-crystal, the *r* and *z* faces are flat surfaces on which growth hillocks with triangular pyramidal forms are universally observed, whereas the *m* face is characterized by the development of striations parallel to the edges with the *r* and *z* faces. On synthetic quartz, conical growth hillocks are observed on the *r* and *z* faces, whereas polygonal growth hillocks are commonly seen on the *m* faces. Only when synthesis by NaCl aqueous solution is performed do striation patterns similar to those seen on natural crystals appear.

Figure 10.1. Crystal structure of low-temperature quartz.

Although the principal morphology of rock-crystal is a hexagonal prismatic *Habitus*, natural crystals may deviate from this. In Goldschmidt's *Atlas der Kristallformen* (see ref. [1], Chapter 9), 855 crystal figures are compiled in 54 plates. A few examples are shown in Fig. 10.2, in which various forms are observed, such as malformed hexagonal prisms, tapered prisms, platy, and scepter forms. (See also Fig. 1.1.)

In contrast to the hexagonal prismatic morphology of low-temperature quartz, it has been assumed that the characteristic morphology of high-temperature quartz is hexagonal bipyramidal where no $\{10\bar{1}0\}$ faces appear on the crystal

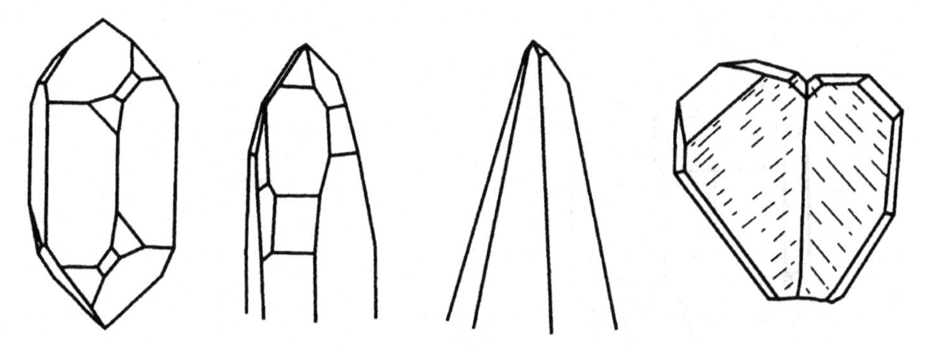

Figure 10.2. Various forms of rock-crystal. Selected from *Atlas der Kristallformen* (ref. [1], Chapter 9).

which is bounded by $\{10\bar{1}1\}$ faces. In the following sections we will analyze the origin of the malformation seen in low-temperature quartz and discuss why high-temperature quartz takes on a hexagonal bipyramidal form.

We shall start from an analysis of what sort of morphology we should expect for quartz crystals if we entirely neglect the effect of environmental conditions.

10.2 Structural form

If the structural form of rock-crystal is predicted based on the Bravais–Friedel (BF) law, a form is obtained that is entirely different from the naturally observed *Habitus*, a polyhedral form bounded by nearly equally developed $\{10\bar{1}1\}$, $\{01\bar{1}1\}$, and $\{0001\}$ faces, as explained in Section 4.2. From the Donnay–Harker (DH) law, in which a three-fold screw axis in the *c*-axis direction is taken into consideration, $\{0001\}$ should be $\{0003\}$, and the reticular density reduces to one-third of that predicted by the BF law, leading to the prediction that it is not necessary for $\{0001\}$ to appear as a *Habitus*-controlling face. The structural form of rock-crystal based on the DH law is hexagonal prismatic with alternately appearing r $\{10\bar{1}1\}$ and z $\{01\bar{1}1\}$ faces at the termination, which is in good agreement with the observed growth forms (see Fig. 4.2). Since there is not a big difference between the structures of the high- and low-temperature forms of quartz, we may expect similar hexagonal prismatic growth form bounded by $\{10\bar{1}0\}$, $\{10\bar{1}1\}$, and $\{01\bar{1}1\}$ for high-temperature quartz also. The hexagonal, bipyramidal form with no association of $\{10\bar{1}0\}$, which is commonly seen as the growth form (*Habitus*) of high-temperature quartz, is therefore not expected from a structural point of view.

In the PBC analysis of Hartman–Perdok theory, $\{10\bar{1}0\}$, $\{10\bar{1}1\}$, and $\{10\bar{1}0\}$ are F faces (see ref. [3], Chapter 4). The predicted structural form of rock-crystal by PBC analysis is hexagonal prismatic (Fig. 10.3). Therefore, using this morphology as a

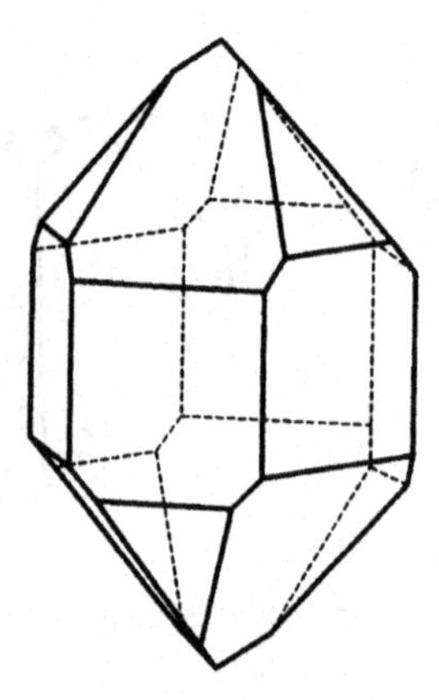

Figure 10.3. The structural form of low-temperature quartz predicted from PBC analysis (see ref. [3], Chapter 4).

criterion, we may analyze the effect of growth conditions on the growth forms of rock-crystal. This analysis will be useful in understanding various geoscientific problems and in identifying appropriate growth conditions in quartz synthesis.

10.3 Growth forms

The growth forms of a crystal are determined by the anisotropy involved in the growth rate, and thus the difference in aspect ratio, i.e. long-prismatic or short-prismatic, is determined by the relative ratio of growth rates $R_m{:}R_{r,z}$, i.e. the relative normal growth rates of m $\{10\bar{1}0\}$, r $\{10\bar{1}1\}$, and z $\{01\bar{1}1\}$ faces. When R_m is much smaller than $R_{r,z}$ the crystal takes a long-prismatic *Habitus*. When $R_r \leq R_z$, six well developed pyramidal faces, r being slightly larger than z, will appear, but crystals with $R_r \ll R_z$ will take a triangular prismatic form with only three r faces at the termination.

All m, r, and z faces grow by the spiral growth mechanism, and so the growth rate R is determined by the height of the spiral growth layers, their advancing rates, and the step separation; thus, factors influencing these values are the same factors which affect the growth forms of rock-crystal.

Low-temperature quartz, irrespective of whether it is natural or synthetic, grows in hydrothermal solution. In addition to H_2O, minor amounts of NaCl, NaOH, NaF_2, and Na_2CO_3 (in synthesis, these are called mineralizers and increase the solubility of quartz) play important roles as solvent components.

The anisotropy in the growth rates of the synthesis of rock-crystal at an industrial scale is depicted in Figure 10.4 [1].* R_m is the smallest, the order being

* Original figures were published in ref. [1], and in most of the later publications Fig. 10.4 is cited. Details are explained in ref. [2]. New data have been published in ref. [3].

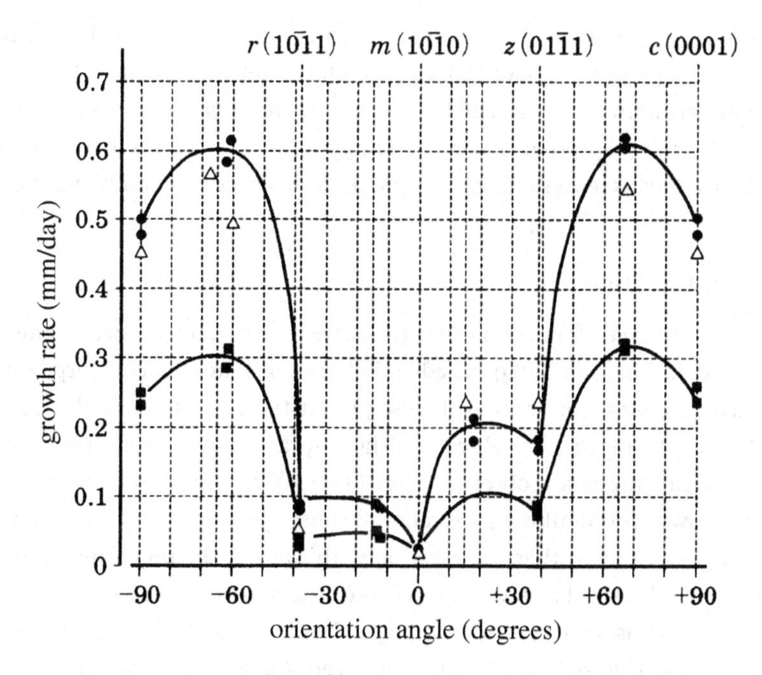

Figure 10.4. Growth rates of crystal faces in synthetic quartz [1]–[3]; ■ = normal growth rate of crystal faces; ● = growth rate in weight; △ = reference data.

$R_m < R_r < R_z \ll R_{0001}$. However, the difference between R_r and R_z is small, and the relation may vary depending on temperature difference and the strength of convection. It has been empirically shown that if these values are large $R_r = R_z$, and if they are small $R_r \ll R_z$. In the latter case, spontaneously grown crystals, grown without seed in an autoclave, exhibit triangular prismatic form with only *r* termination, and the *z* faces disappear.

Whether low-temperature quartz crystals take long- or short-prismatic forms depends on the modes of occurrence (the growth conditions). Compared with the long-prismatic or needle-like forms with an aspect ratio exceeding ten shown by spontaneously grown synthetic quartz crystals, the aspect ratio of quartz formed under high-temperature conditions and occurring in pegmatite is two to three at maximum; for example, that of amethyst formed in geodes at lower temperatures is in the range one to two, and it exhibits short-prismatic *Habitus*. Crystals showing an aspect ratio as high as spontaneously grown synthetic crystals are found only exceptionally in nature; however, very rarely a case in which needle crystals coexist with crystals having a smaller aspect ratio occurs in a druse. In many cases, crystals formed at the later stages take $r \gg z$.

Since high-temperature quartz occurring as phenocrysts in acidic igneous rocks (igneous rocks with around 70% SiO_2 content) exclusively take hexagonal bipyramidal forms with no prismatic faces, it has been assumed that this is the typical form of high-temperature quartz. When quartz crystals are hydrothermally synthesized at temperatures above 573 °C, prism faces appear and the crystals take a hexagonal prismatic form [4]. The hexagonal bipyramidal form of high-temperature quartz simply represents the morphology formed in acidic igneous magmas, and is thus a *Habitus* representing the growth in this particular ambient phase.

10.4 Striated faces

The prismatic faces of natural rock-crystal are characterized by the development of striations parallel to the edges between *m*, *r*, and *z* faces (perpendicular to the *c*-axis). Natural rock-crystal showing no distinct striations is almost exceptional. In industrially mass-produced synthetic quartz using NaOH or KOH as mineralizers, no striations are observable on $\{10\bar{1}0\}$ faces. As shown in Fig. 10.5(a), five-sided growth spiral hillocks are generally observed. However, if quartz crystals are synthesized in hydrothermal solution with NaCl as the mineralizer, the prismatic faces exhibit similar striations to those observed on natural crystals [5].

The ratio $R_{r,z}$:R_m is up to ten times higher in NaCl solution than the ratios seen in NaOH or KOH solutions. From this, it is deduced that the striations are due to the remarkable anisotropy involved in the step advancing rate of the growth spirals developing on the *m* faces The main reason why this anisotropy occurs is understood to be due to the NaCl, which is added as a mineralizer in H_2O. The hydrothermal solution in which natural rock-crystal grows is, in general, NaCl aqueous solution.

10.5 Growth forms of single crystals

10.5.1 *Seed crystals and forms*

Seed crystals are always used in the synthesis of quartz. Various seed orientations are used depending on the individual requirements. In growing synthetic quartz for industrial purposes, a seed bar parallel to the *y*-axis (a *y*-bar) is generally used, whereas for colored quartz, such as amethyst, platy seed parallel to $\{10\bar{1}1\}$ is sometimes used. (This effectively produces color intensity in the growth sectors, taking into account the difference in partitioning of the impurity element, for example Fe, between the *r* and *z* faces.) Experiments using circular disks, holed circular disks, and spheres as seed crystals are also reported [6]. If prolonged growth is achieved on a seed crystal of any form, the crystal will eventually take a polyhedral form in short- to long-hexagonal-prismatic forms, with the aspect ratio determined by $R_{r,z}$:R_m corresponding to the growth condition. The forms can be predicted by computer experiments. Figure 10.6 shows a morphodrom prepared by Iwasaki *et al.* [7]. However, in real synthesis, growth is

(a)

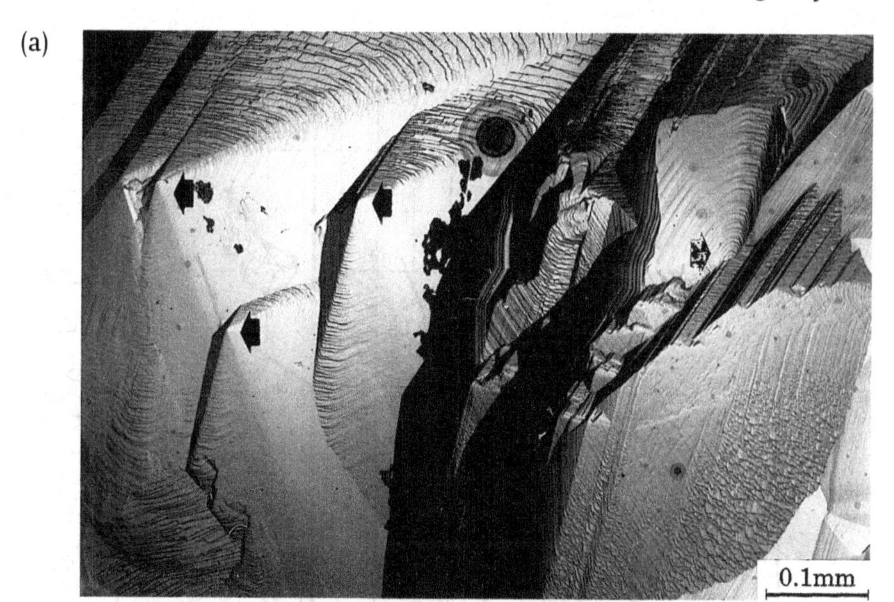

(b)

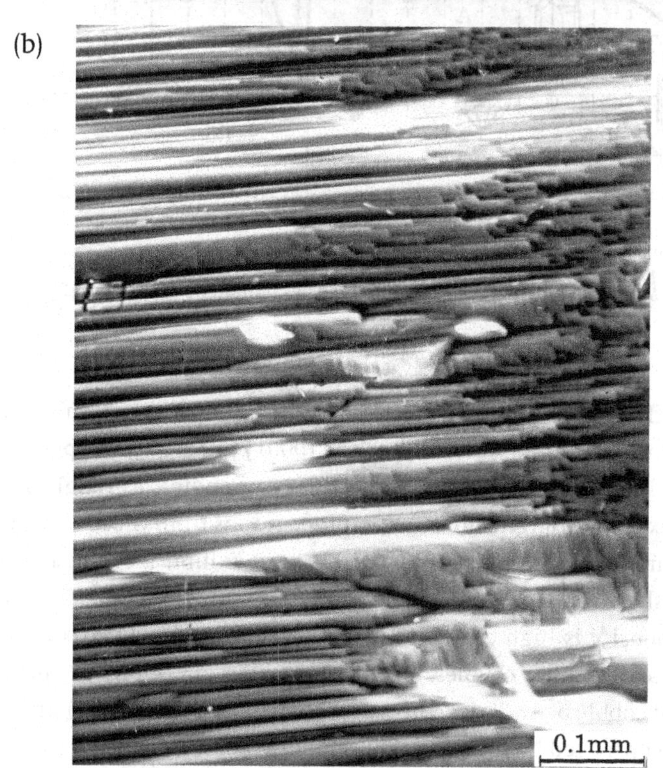

Figure 10.5. (a) Polygonal spiral growth hillocks universally observed on $\{10\bar{1}0\}$ faces of synthetic quartz. (The arrows indicate the summits of the growth hillocks.) (b) Striation patterns commonly seen on the $\{10\bar{1}0\}$ faces of natural and synthetic quartz grown in NaCl aqueous solution.

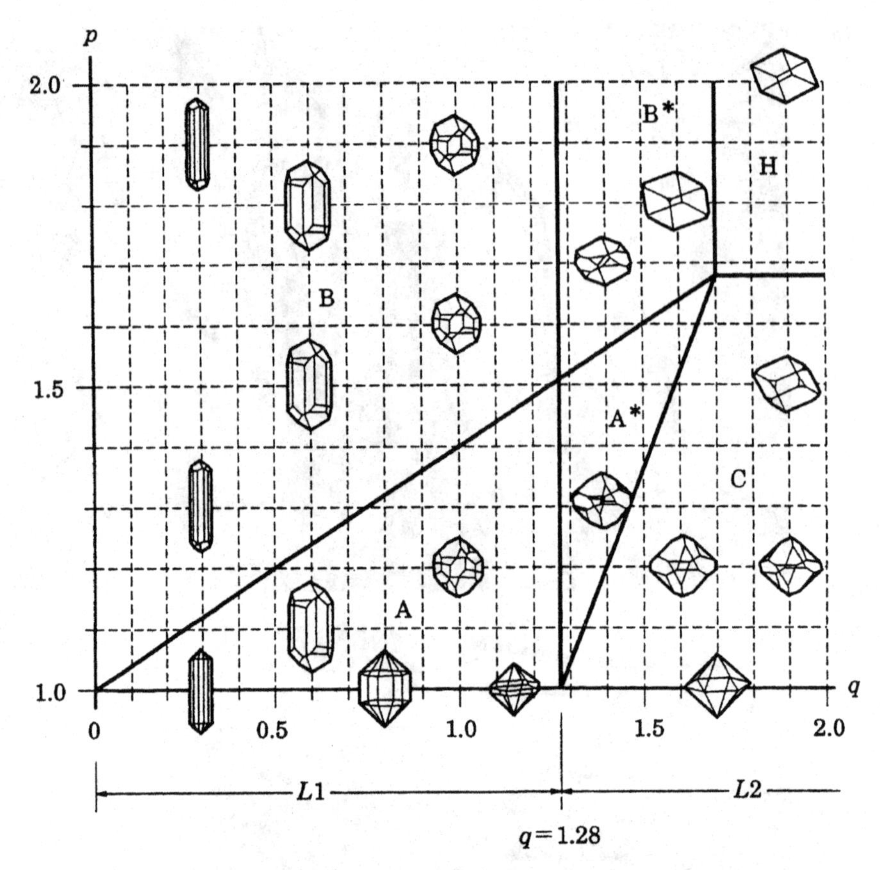

Figure 10.6. Morphodrom of rock-crystal obtained by computer experiments [7]; $p = R_z/R_r$, $q = R_m/R_r$.

terminated before the crystals reach the form shown in Fig. 10.3. As a result, faces that do not appear on natural rock-crystal, such as {0001} or $\{11\bar{2}0\}$, do appear on synthetic crystals. These are structurally K faces, corresponding to rough interfaces, and so they exhibit different surface microtopographs from those of $\{10\bar{1}0\}$ or $\{10\bar{1}1\}$, which are F faces. For example, {0001} faces exhibit a cobbled structure, such as that shown in Fig. 10.7. Cobbles are growth hillocks, but not spiral hillocks, and the density is high close to the seed and decreases as growth proceeds by coalescence of smaller cobbles. Since dislocations and impurities principally concentrate at cobble boundaries, the defect structures in the {0001} growth sectors are mainly controlled by the state of development of the cobbles.

Since *r*, *z*, and *m* faces grow by the spiral growth mechanism, spiral growth hillocks or spiral growth steps showing respective characteristic forms can be observed on their surfaces. The $\{10\bar{1}0\}$ faces, having the smallest normal growth rate, exhibit polygonal forms, whereas on $\{10\bar{1}1\}$ and $\{01\bar{1}1\}$ (faces having larger normal growth rate than the former) circular spiral growth layers are observed.

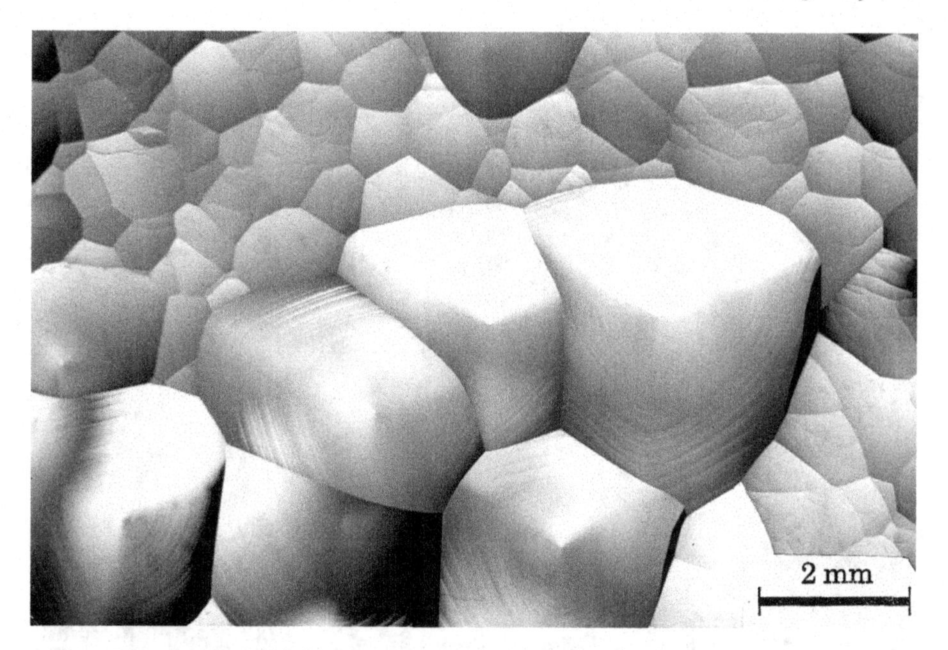

Figure 10.7. Cobbled structure observed on {0001} face of synthetic quartz. Photographed by M. Kawasaki.

10.5.2 *Effect of impurities*

If we consider NaOH, KOH, and NaF_2, which act as mineralizers, to be the solvent components, and other minor amounts of elements such as Fe^{3+} and Al^{3+} to be impurity elements, then the partitioning of these impurity elements is controlled principally by kinetics. The impurity partitioning is related to color, or radiation-induced color, and crystal morphology.

The violet color of amethyst is due to the presence of a color center that is induced by the irradiation of point defects caused by the impurity FeOOH, and the smoky color in smoky quartz and black quartz is due to the presence of a color center related to Al, which susbstitutes Si. Since impurity partitioning is affected by growth sectors and growth rates, different color intensities will appear in different growth sectors and will be affected by growth banding. Amethyst is a deeper color in the *r* growth sector, and a paler color in the *z* growth sector, resulting in the appearance of a triangular pattern with alternating intense and pale color. Variation in color intensity may also be observed in association with growth banding. Tapered crystals are examples showing the effect of impurity adsorption upon crystal morphology (see Section 10.5.3).

10.5.3 *Tapered crystals*

Some rock-crystals exhibit dog-tooth forms as hexagonal prismatic forms taper off towards the tip of the prism. Some examples are shown in Fig. 10.8. This

Figure 10.8. Tapered rock-crystals.

phenomenon is called tapering, and has been observed in crystals of ADP and KDP growing from an aqueous solution for piezoelectric materials, and is the subject of extensive investigation [8]–[10]. Crystals of ADP and KDP are bounded by {100} and {111} faces, and tapering occurs in the *c*-axis direction. It has been understood that tapering occurs because trivalent ions, such as Fe^{3+}, which are present as impurities, are selectively adsorbed in a particular direction that is related to the growth layers on the {100} face, thus retarding the advancing rate of the steps.

The origin of tapering observed in rock-crystal is basically the same as in the cases of ADP and KDP [11]. In the case of natural rock-crystal, it is suggested that the precipitation of clusters formed in the solution on the growing surface has a similar effect to that of impurity Fe^{3+} ions.

10.5.4 *Solution flow*

Rock-crystal occurring in vein-type ore deposits grows in ascending hydrothermal solution through cracks in the strata. The flow of solution causes the solute component to be supplied to crystals growing inclined to or perpendicular to the wall of the crack. In laminar flow, the growth rate of the side facing the flow increases compared with the opposite side. In turbulent flow, the situation will be reversed.

As a result, the section of hexagonal prismatic crystal changes from regular hexagonal (expected when the crystal grows in an isotropic environment) to malformed hexagonal. This variation is also recorded in the crystal as a directional variation in the separation of growth banding. This is utilized in the analysis of Earth science problems, such as evaluating environmental conditions, analyzing the effect of solution flow upon growth rates, and estimating the flow direction of hydrothermal solution, based on the analysis of changes in the separation of growth banding in crystallographically equivalent growth sectors and their fluctuations in the same direction. Based on this viewpoint, a difference is noted between hydrothermal veins, in which one-directional flow of hydrothermal solution is assumed, and pegmatite formed in a void in solidifying magma. In the former, one-directional flow of solution is clearly seen, whereas in the latter the flow directions change during growth.

To demonstrate the extent to which the flow of solution affects the crystal morphology, we describe an experiment performed by Balitzky (ref. [14], Chapter 4). Balitsky investigated the effect of solution flow on the growth rates of the *r* and *z* faces of quartz crystal in an autoclave, by reversing the position of the seed. As the supply of solute component changes depending on whether we have up-flow or down-flow, the growth rates and impurity partitioning can change according to the positioning of the seed. Although the general tendency is for larger development of the *r* face, since $R_r < R_z$ when crystals grow in an isotropic environment, the *r* and *z* faces develop nearly equally because R_r is promoted and $R_z = R_r$, if the *z* face is faced up-current.

This experiment demonstrates that a distinction between the *r* and *z* faces based on the size difference becomes impossible. Also, the growth sector of the *z* face takes on a violet color. In an isotropic environment, the violet coloring appears selectively in the *r* growth sectors, and the *z* growth sectors are colorless. This experiment also demonstrates that the partitioning of impurity Fe is affected by the growth rate.

10.6 Twins

10.6.1 *Types of twins*

Several different types of twinning relations have been observed in quartz, such as: the Dauphiné law, a twin relation between two right-handed or two left-handed structures with the *c*-axis as the twin axis; the Brazil law, a twin relation between a right-handed and a left-handed structure; the Japan law, with the *c*-axes forming an angle of 84°34′ with the composition plane $(11\bar{2}2)$; and others. (It should be noted that all these types might be described differently in different localities.) Dauphiné twins are formed in growth, in transformation, and by external factors such as mechanical, thermal, and electrical forces, and do not differ greatly from the morphology of single crystals. Compared with the positions of the *s* $\{11\bar{2}1\}$ and *x* $\{51\bar{6}1\}$ faces, those appearing on an *m* face will differ according to whether they are left- or right-handed; the *s* $\{11\bar{2}1\}$ and *x* $\{51\bar{6}1\}$ faces appear on

Figure 10.9. Various forms of rock-crystal twinned according to the Japan law from Narushima Island, Nagasaki Prefecture, Japan. The evolution of crystal forms in the order *a*→*b*→*c* is expected as growth proceeds, due to the re-entrant corner or pseudo re-entrant corner effect.

right and left corners of an *m* face in a Dauphiné twin. Therefore, we shall pay particular attention in this section to the explanation of why a characteristic morphology appears in growth twins according to the Japan law and the Brazil law. It usually happens that two individuals twinned according to the Japan law also contain Dauphiné twins and Brazil twins.

10.6.2 *Japanese twins*

Japanese twins of quartz have attracted interest since ancient times because they exhibit a remarkably platy V-shape, in contrast to the hexagonal prismatic morphology of coexisting single crystals. Since they grow on substrate, the V-shape was assumed to represent the upper half of an X-shape [12], which implies that the Japanese twins are penetration twins. The platy form of Japanese twins has been explained as being due to preferential growth at a re-entrant corner formed by two individuals. If the platy form is indeed simply due to the re-entrant corner effect, we should expect a variation of forms, from V-shape to fan-shape, as the effect proceeds, as shown in Fig. 10.9. If, however, it represents the upper half of an X-shape, Japanese twins should be penetration twins, not contact twins.

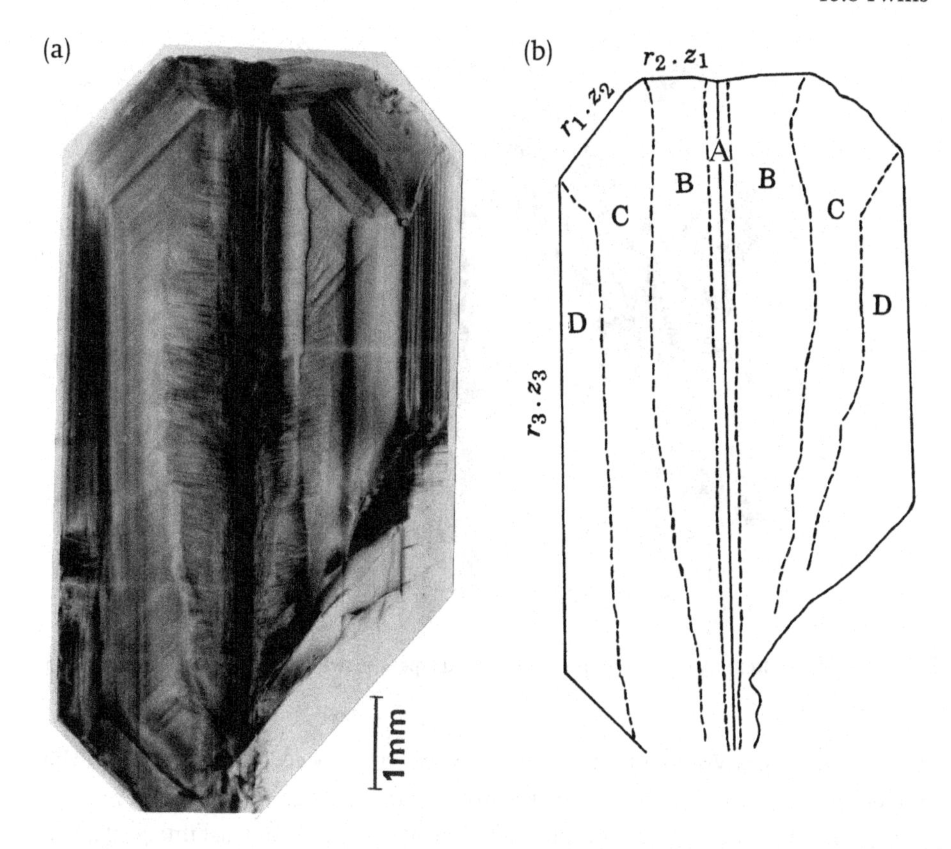

Figure 10.10. (a) X-ray topograph and (b) the corresponding sketch of a fan-shaped Japanese twin. X-ray topograph by T. Yasuda [13].

As already explained in Section 7.2, the re-entrant corner effect in its original sense can be expected only when two individuals are perfect crystals, containing no dislocations either in the individual crystals or in the composition plane. In real crystals, the pseudo re-entrant corner effect, by the mediation of dislocations, creates changes in the morphology of the twinned crystal.

Figure 10.10 shows an X-ray topograph of a fan-shaped Japanese twin. It is observed that (i) dislocations concentrate in the composition plane of two individuals (A), and (ii) the spacing of the growth banding in the r or z growth sectors adjacent to the composition plane (B) is approximately twice as wide as that in the growth sectors of crystallographically equivalent r and z growth sectors a distance away from the composition plane (C, D). Growth banding in the former sectors is wavy, whereas in the latter it is straight. This observation demonstrates that the growth is promoted due to preferential growth by concentrated dislocations in the composition plane, and platy or fan-like morphology is due to the pseudo re-entrant corner effect [13], [14]. The morphology of a Japanese twin should

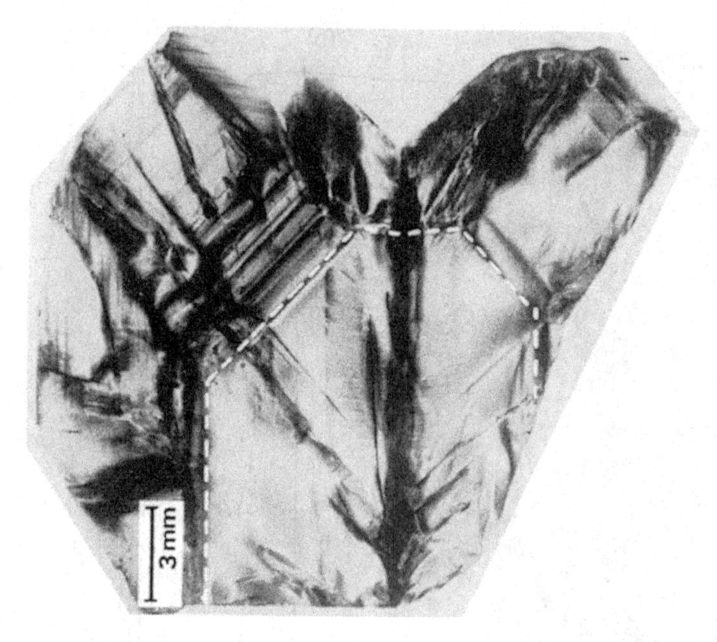

Figure 10.11. X-ray topograph of a V-shaped Japanese twin. X-ray topograph by T. Yasuda [14].

transform from a V-shape to a fan-shape as growth proceeds, irrespective of the effects of the original or the pseudo re-entrant corner effect.

Figure 10.11 is an X-ray topograph of a V-shaped sample. A contrast image of a fan-shape is discernible in the V-shape. Contrary to expectations, the morphology of the Japanese twin transforms from a fan-shape to a V-shape as growth proceeds. Whether a V-shape or a fan-shape results is dependent on whether $\{10\bar{1}0\}$ faces appear at the contact point of two individuals (V-shape), or whether $\{10\bar{1}1\}$ faces appear (fan-shape). The $\{10\bar{1}1\}$ faces are nearly perpendicular to the dislocations concentrated in the composition plane, whereas $\{10\bar{1}0\}$ faces are inclined at an angle. This creates the difference in the step heights of the growth layers that have originated from the dislocations, and affects the growth rates. This observation demonstrates that if $\{10\bar{1}1\}$ faces are present in the composition plane, the pseudo re-entrant corner effect is at maximum effect, but once $\{10\bar{1}0\}$ appears, the effect diminishes sharply [14]. Furthermore, we may expect that at the nucleation stage of a Japanese twin, the quartz crystals should take ditrigonal dipyramidal form, with no $\{10\bar{1}0\}$ faces present.

If two individuals conjugate on an *r* or a *z* face, a nucleus of a Japanese twin with $\{11\bar{2}2\}$ as the composition plane is formed. This indicates that a Japanese twin is not a penetration twin (i.e. not the upper half of an X-shape), but a contact twin (i.e. the upper half of a Y-shape). Horizontal banding in geode agate (see Section 10.9) appears through grain size variation due to gravitational sedimentation, and con-

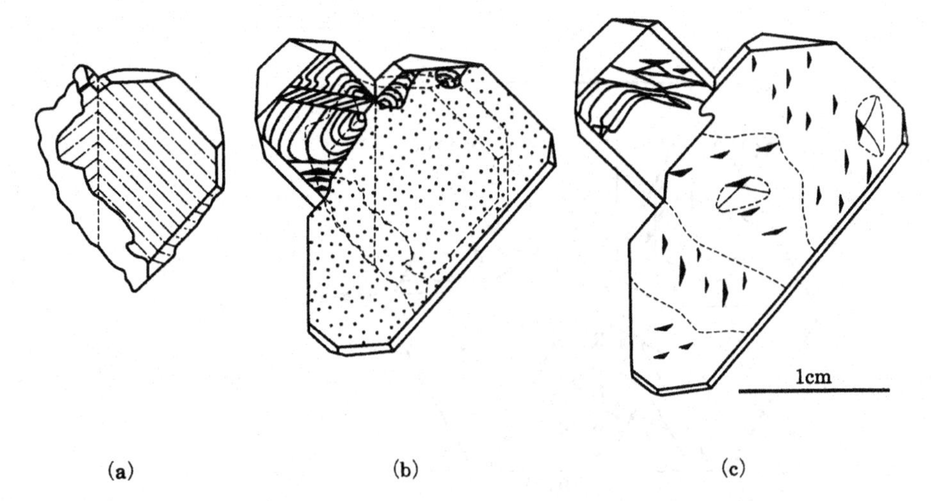

Figure 10.12. Change in form and surface microtopographs observed in regrowth experiments on a Japanese twinned sample used as a seed [15]. When the Japanese twin grows freely, the form changes from a V-shape to a Y-shape.

sists of idiomorphic crystals of quartz less than 1 μm. On quartz crystals of this size, $\{10\bar{1}0\}$ faces do not appear, and the crystals take on a ditrigonal dipyramidal form bounded by *r* and *z* faces. A Japanese twin is considered to have been formed if two individuals conjugate on well developed *r* and *z* faces. The introduction of $\{11\bar{2}2\}$ as the composition plane is simply due to the geometrical relation between the two individuals.

Figure 10.12 shows a series of sketches showing the variation in morphology and surface microtopography as growth proceeds, starting from a natural Japanese twin as seed grown in an industrial autoclave for quartz synthesis [15]. It is clearly seen that, starting from a partly broken Japanese twin with a V-shape, the form changes to a Y-shape when growth occurs in an open space. This experiment proves that the V-shape does not represent the upper half of an X-shape but that of a Y-shape, and therefore that the Japanese twin is not a penetration twin but a contact twin. It was later observed that natural Japanese twins occasionally exhibit a Y-shape when they grow in open spaces. Figure 10.12 also vividly demonstrates the sorts of changes that occur in the process of transformation from a rough to a smooth interface, and it may also be seen that two individuals are Dauphiné twinned.

10.6.3 *Brazil twins*

The Brazil law is used to describe an intergrowth of right- and left-handed structures with the *c*-axis as the twin axis. Since this is a relation between different structures, strictly speaking this does not belong to the twin category, but it has always been treated as an important twinning phenomenon. In Brazil twins, two

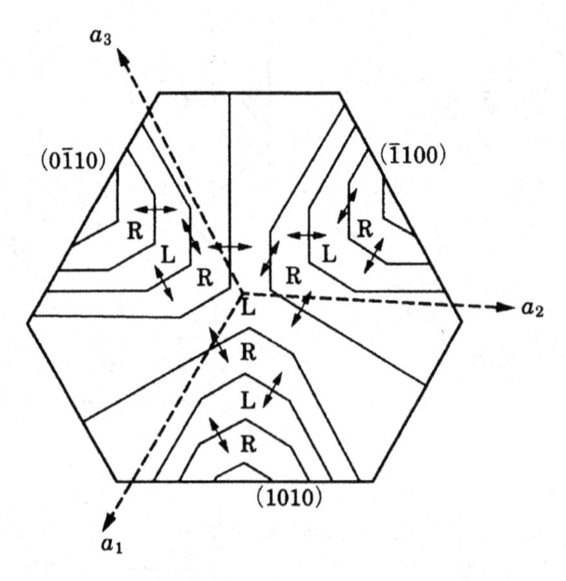

Figure 10.13. Brewster fringes frequently seen in amethyst crystals. L and R indicate left-handed and right-handed structures, respectively.

individuals are in a reflection relation on $\{11\bar{2}0\}$, and occur as twin sectors in lamellar form bounded by $\{10\bar{1}0\}$, $\{10\bar{1}1\}$, and $\{01\bar{1}1\}$. Brazil twins occur universally in quartz crystals formed at lower temperatures, as in amethyst occurring in geode, and are only exceptionally found in quartz crystals formed at higher temperatures, such as quartz crystals occurring in pegmatites. One report [16] claims that Brazil twin lamellae are found in quartz particles constituting agate in geode exceeding 10 nm, and the widths of the lamellae range from nanometer to micrometer order. The so-called Brewster fringes often observed in amethyst (Fig. 10.13) are the boundaries of Brazil twin lamellae formed by the coagulation of numerous Brazil twin lamellae.

Considering the fact that the lamellae widths are of nanometer to micrometer order and that a Brazil twin is formed by combining right-handed and left-handed structures, we have to assume the presence of clusters with right-handed and left-handed structures in the ambient phase, with the Brazil twin being formed by the joining together of these clusters. A Brazil twin is not formed by transformation from left-handed to right-handed structures while a crystal is growing by incorporation of ionic entities or SiO_4 tetrahedra as the growth unit. This is in agreement with the observation that Brazil twins are universally observed in quartz crystals formed at lower temperature rather than at higher temperature, since the probability of cluster formation is higher at lower temperatures.

In amethyst crystals showing Brewster fringes, a pattern such as that shown in Fig. 10.14 is observed in the growth sectors of or on the surface of *r* faces, and not in

(a)

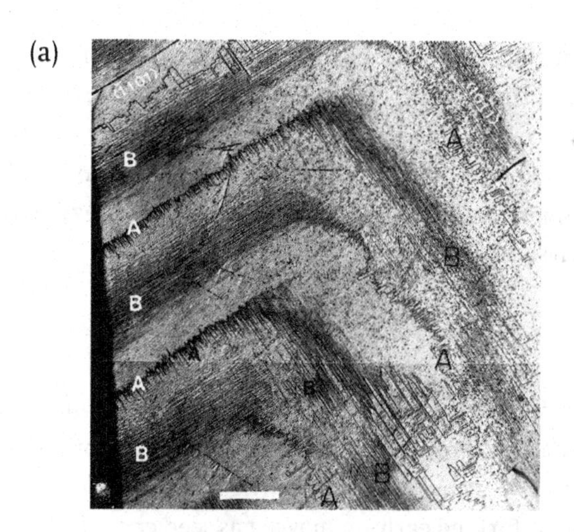

(b)

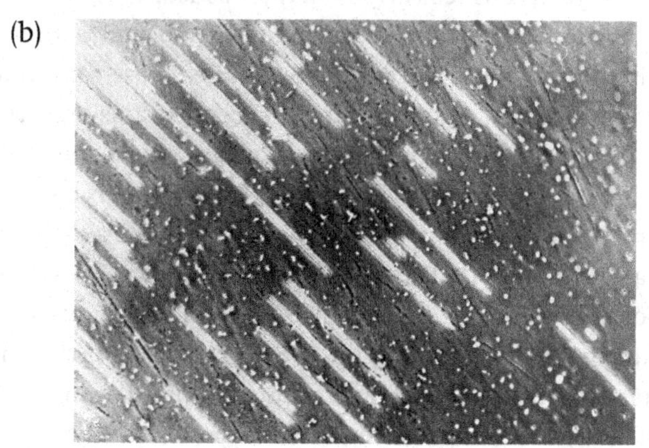

(c)

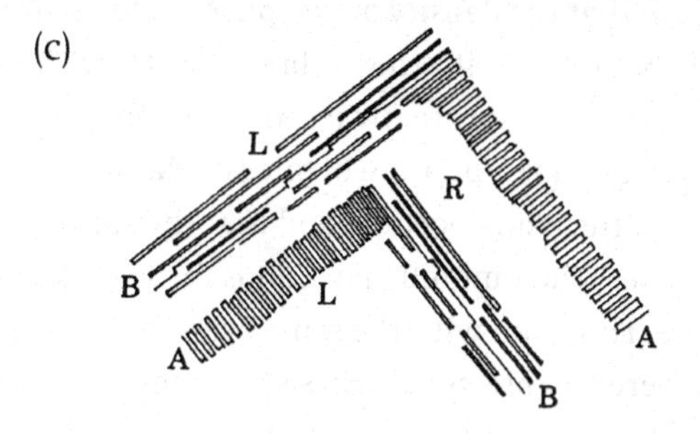

Figure 10.14. (a), (b) Reflection photomicrographs of weakly etched surface, and (c) the structure of Brewster fringes on an *r* face (model by Lu and Sunagawa [17]). Zigzag (A in (a) and (c)) and closed lamellae (B in (a) and (c)) appear alternately. L = left-handed quartz; R = right-handed quartz.

the growth sectors of or on the surface of *z* faces. In the banding pattern of *r* faces, Brazil twin lamellae continue in a zigzag form on one side of triangular banding, whereas on the other side they occur as an alignment of numerous rectangular closed lamellae, as can be seen in Fig. 10.14. In a section parallel to the *c*-axis, it is observed that Brewster fringes appear by the conjugation of numerous Brazil twin lamellae as the growth of amethyst proceeds.

10.7 Scepter quartz

Since the growth conditions are uncontrolled in the growth of natural quartz crystals, it is probable that the growing crystals experience major conditional changes, such as cessation of growth and regrowth from new solution. Under such circumstances, crystals formed early on may act as seed crystals.

Scepter quartz, which is composed of a trunk portion and a thicker umbrella (cap) portion, is an example of a crystal grown under these conditions (see Fig. 10.15). It is generally observed among groups of quartz crystals occurring in a druse that only the larger crystals show a scepter form, whereas coexisting smaller or shorter crystals do not. In some crystals, the trunk portion is colorless and the umbrella portion is violet amethyst, or smoky, and in other cases the relation is reversed. It is clear either way that the conditions changed at some time between the formation of the trunk portion and the formation of the umbrella portion. It is also generally observed (i) that the surface of the trunk portion and the coexisting non-scepter quartz are covered by precipitation of other minerals, such as mica or clay minerals, whereas a coating of this type is not seen on the surface of the umbrella portion, and (ii) that the density of two-phase inclusions is high in the trunk portion near to the boundary with the umbrella portion. These facts indicate that the growth conditions changed greatly between the formation of the trunk and the umbrella portions, that precipitation of mica or clay minerals occurs at the latest stage of formation of the trunk portion, and that the surface of the trunk is covered by these precipitates. The umbrella portion is formed from a purer hydrothermal solution than the trunk, but with the same orientation, after the surface of the trunk portion is covered by precipitate. Based on these observations, it is understood that the growth of the umbrella portion began from that portion not covered by foreign minerals and that it retains an epitaxial relation with the trunk portion, and overgrows the area covered by foreign minerals. This mechanism is exactly the same as ELO (epitaxial lateral overgrowth) [18], [19], explained in Section 7.4, which is used as a method of dislocation control in liquid phase epitaxy.

Figure 10.16 indicates why both scepter quartz and common quartz coexist in a druse.When quartz crystals are synthesized hydrothermally using a seed, a part of

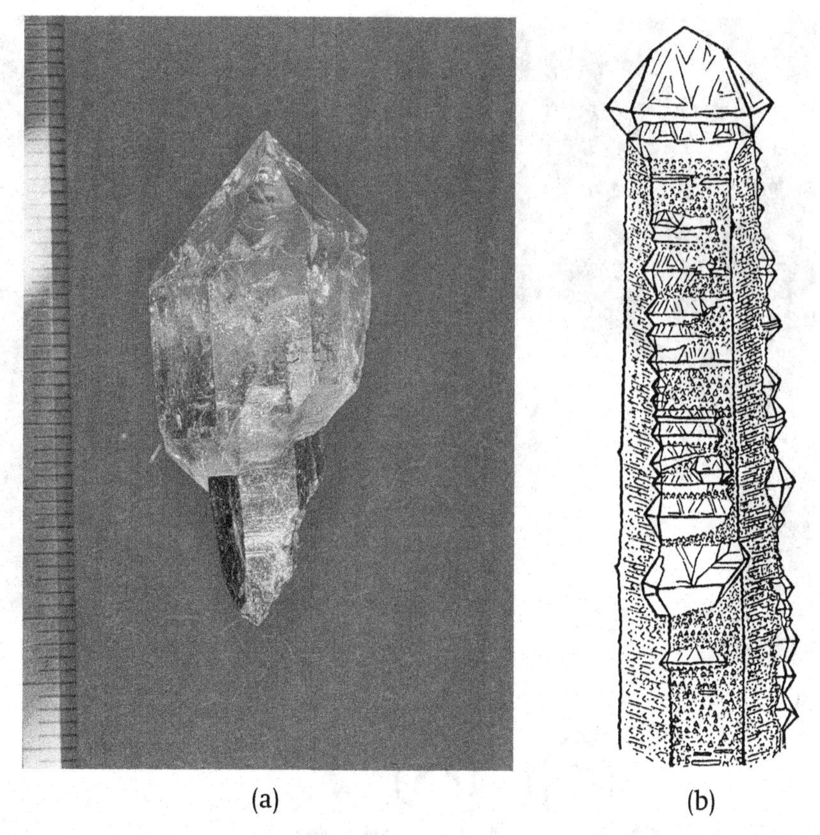

Figure 10.15. (a) Ordinary scepter quartz with umbrella portion only at the top of the trunk portion. (b) Scepter quartz with many umbrella portions at the top and along the edges of the trunk.

which is masked by silver clay, quartz growing from an unmasked part grows laterally over the surface of the silver clay, and scepter quartz is formed artificially [20]. If the seed surface is covered with outcrops of dislocations and the area has no dislocation outcrops unmasked, allowing ELO to take place, synthesis of dislocation-free quartz crystals may become possible.

10.8 Thin platy crystals and curved crystals

There may be crystals of other minerals already present in a vein before the growth of quartz crystals begins from a hydrothermal solution. These crystals may act as heterogeneous nucleation sites for later-forming quartz crystals. In particular, the edges and corners of earlier-formed crystals play an important role as heterogeneous nucleation sites. As a result, many quartz crystals formed later are aligned almost parallel to each other due to preferential nucleation on edges and

Figure 10.16. (a) An example of a geode with only part of the crystals showing scepter form; (b) schematic explanation.

corners, or quartz crystals in sub-parallel intergrowth may be produced, showing macroscopically curved surfaces, controlled by the curved forms of crystal aggregates that are already present. If crystals of other minerals which acted as seeds are dissolved after the growth of quartz crystals is complete, the quartz crystals will show markedly different forms from those of spontaneously formed quartz crystals; for example, see Fig. 10.17. Quartz crystals showing morphologies of this type often occur in Alpine-type mineral fissures, and are referred to as Gwindel-type crystals. The above mechanism was accounted for by Bonev [21].

10.9 Agate

Variety names, such as chalcedony and jasper, are given to minute crystalline aggregates of low-temperature quartz. Those showing distinct banding pat-

Figure 10.17. Gwindel-type quartz.

terns are called agate, which has been used widely as an ornamental stone for centuries. Agate occurs as a lining in the walls of voids in lava flows, or as nodules in sedimentary rocks. Two controversial suggestions have been put forward to account for the genesis of agate in lavas: one assumes that agate banding is due to the so-called Liesegang phenomenon in the crystallization of silica liquid droplets formed by the melting of siliceous rock fragments occluded in the magma; the other explains that the precipitation of quartz from underground water containing SiO_2 intruded into the voids of solidified lava. The former hypothesis assumes high-temperature conditions above 700 °C, but the second assumption, of precipitation from aqueous solution at temperatures as low as 50 °C, is now widely accepted.

Agate that fills the voids in igneous rocks commonly exhibits the following four types of textures; see also Fig. 10.18.

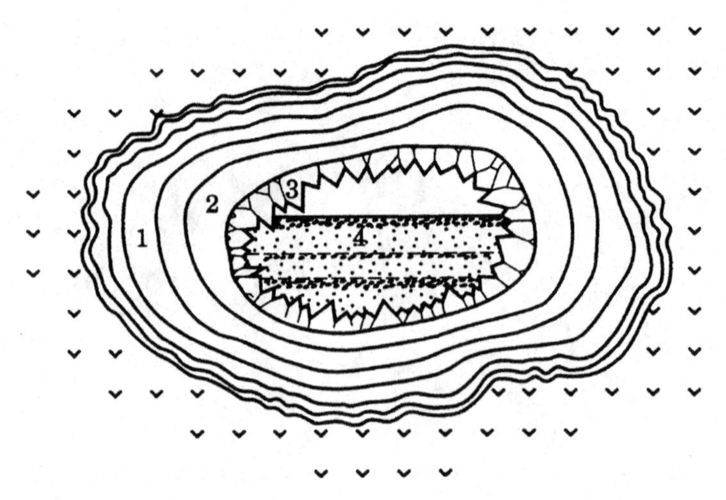

Figure 10.18. Schematic illustration of textures seen in a geode. (1) Agate banding (including (2) Rundel bundlung); (3) larger crystals; (4) Uruguay band.

(1) Those showing irregularly curved banding parallel to the wall surface of a geode (Fig. 10.19). This is called agate banding. The bands vary irregularly in width, but in general they become gradually narrower towards the inner portion, and eventually they have uniform and very narrow width. Such portions with narrow and uniform banding are called Rundel bundlung; this is explained in (2). Agate banding is composed of a fibrous texture running perpendicularly to the banding. A single fiber with a thickness of micrometer order, which is observable under an optical microscope, is composed of several tens of fibers with width of nanometer order, observable under an electron microscope. A single fiber is composed of an alignment of short-prismatic quartz crystals of nanometer size, whose *c*-axis is perpendicular to the elongation of the fiber. Fibers of optical microscopic order may sometimes show sheaf, semi-spherical, and spherulitic form, and tend to occur more frequently as the wall of the void is approached.

(2) Uniformly spaced banding of Rundel bundlung, which appears more frequently as the center of a void is approached (Fig. 10.20), is a banding pattern due to a repeated cycle of size distribution of quartz grains (from smaller to larger). The large quartz crystal developing towards the void center corresponds to the final stage of the cycle. Namely, changes from (1) to (3) can be regarded as representing a series of events.

(3) Large short-prismatic quartz crystals develop toward the center of a void. In many cases, these crystals are amethyst, in which Brazil twins and Brewster fringes are universally observed.

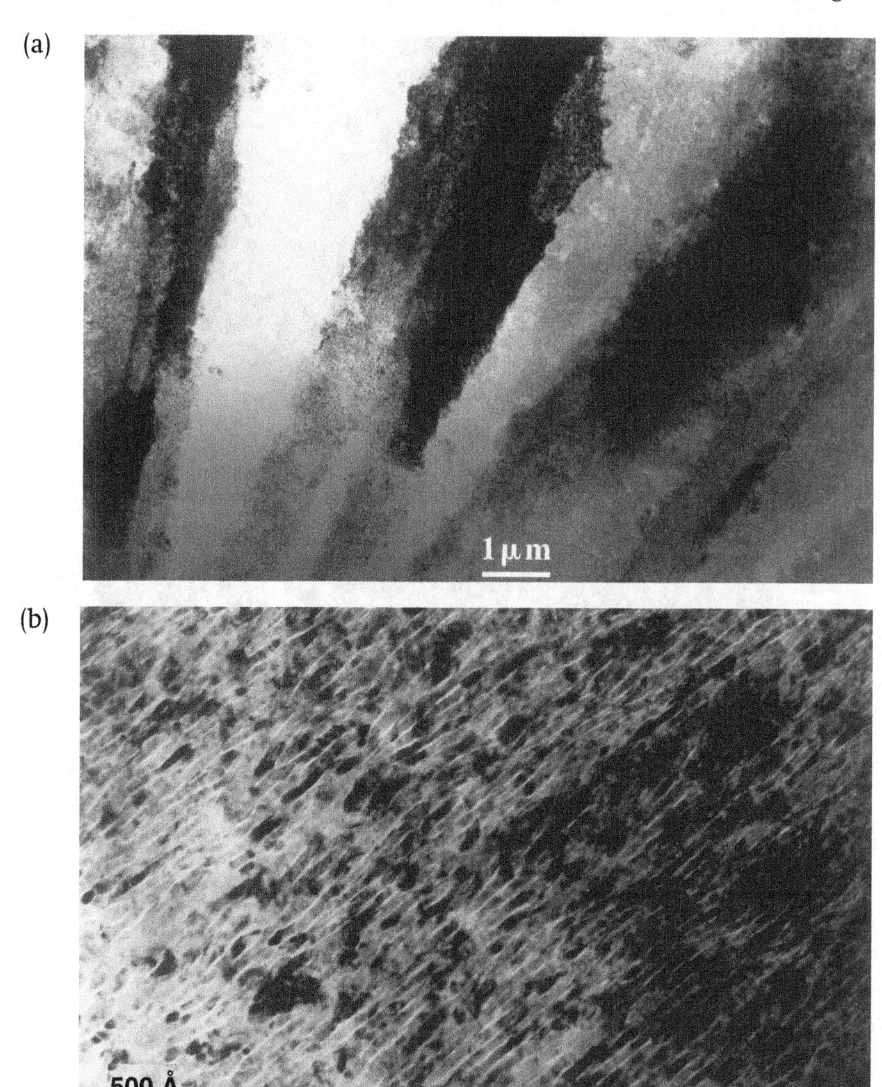

Figure 10.19. (a) Polarizing photomicrograph of agate banding. (b) Transmission-type electron photomicrograph [16].

(4) In the center of a void after formation of (1)–(3), a pattern with horizontal banding appears, which is called a Uruguay band. This band appears due to grain size variation of minute quartz crystallites of sub-micrometer order, and is formed by gravitational settlement. Quartz particles consisting of this band are idiomorphic, and show ditrigonal dipyramidal form bounded by *r* and *z* faces, with no *m* faces (Fig. 10.21). Brazil twinning is universally observed in these quartz particles.

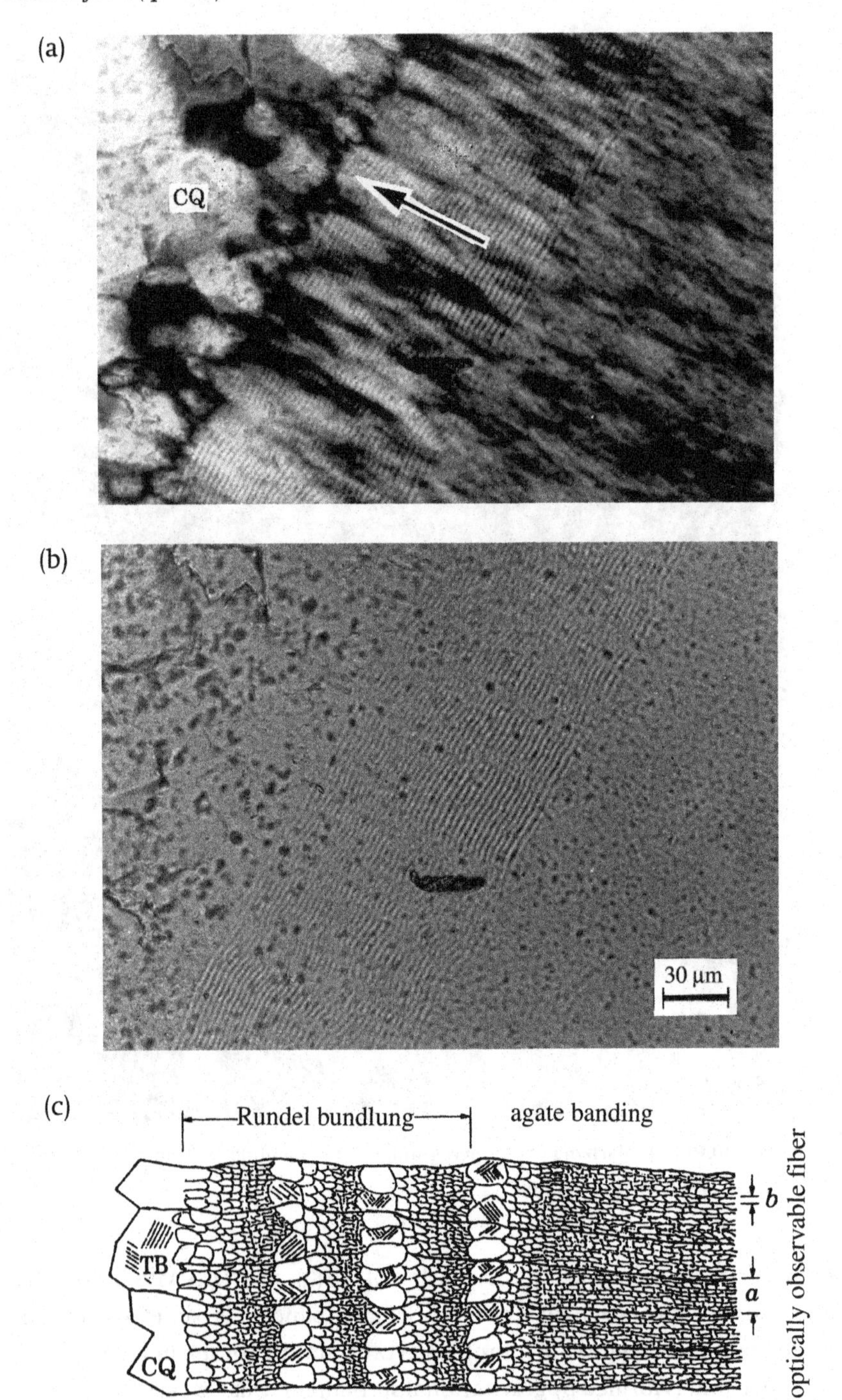

Figure 10.20. (a), (b) Photomicrographs showing textural changes from agate banding, Rundel bundlung, and large crystals (CQ); (c) schematic diagram [16].

Figure 10.21. Scanning electron micrograph of Uruguay band.

The order of formation of the banding pattern is (1)→(4), and there can be cases that all four stages are present, or one part may be absent, or, in some cases, the cycle repeats.

From these observations of the texture, it is considered that agate precipitates from an aqueous solution containing SiO_2, which intruded into a geode, and that in the aqueous solution clusters having right-handed and left-handed structures were already formed. When these clusters aggregate through electromigration due to the wall surface, the banding patterns discussed in (1) and (2) and large quartz crystals, (3), are formed. However, Uruguay banding is principally formed by gravitational settlement once the wall surface effect has diminished.

References

1 R. Bechman and D. R. Hale, Electronic grade synthetic quartz, *Brush Strokes*, September, 1955, 1–7

2 F. Iwasaki and H. Iwasaki, Industrialization of synthetic quartz in the States and Japan, *J. Japan. Assoc. Crystal Growth*, **25**, 1988, 247–50 (in Japanese with English abstract)

3 F. Iwasaki, H. Iwasaki, and Y. Okabe, Growth rate anisotropy of synthetic quartz grown in Na_2CO_3 solution, *J. Crystal Growth*, **178**, 1999, 648–52

4 M. Hosaka, T. Miyata, and I. Sunagawa, Growth and morphology of quartz crystals synthesized above the transition temperature, *J. Crystal Growth*, **152**, 1995, 300–6

5 M. Hosaka and S. Taki, Hydrothermal growth of quartz crystals in NaCl solution, *J. Crystal Growth*, **51**, 1981, 589–600

6 W. Zhong, *Synthetic Quartz*, 2nd edn, Beijing, Scientific Pub., 1994 (in Chinese)
7 H. Iwasaki and F. Iwasaki, Morphological variations of quartz crystals as deduced from computer experiments, *J. Crystal Growth*, **151**, 1995, 348–58
8 J. W. Mullin, A. Amatavivadhana, and M. Chakraborty, *J. Appl. Chem.*, **20**, 1970, 153
9 A. A. Chernov, L. N. Roshkovich, and M. M. Mkrtchan, *J. Crystal Growth*, **74**, 1986, 101
10 J. Owezarek and K. Sangwal, *J. Crystal Growth*, **102**, 1990, 547
11 Lu Taijing, R. B. Yallee, C. K. Ong, and I. Sunagawa, Formation mechanism of tapering of crystals: A comparative study between potassium dihydrogen phosphate crystals and natural quartz crystals, *J. Crystal Growth*, **151**, 1995, 342–7
12 C. Frondel, *The System of Mineralogy*, 7th edn, vol. 3, New York, John Wiley, 1962
13 T. Yasuda and I. Sunagawa, X-ray topographic study of quartz crystals twinned according to Japan twin law, *Phys. Chem. Min.*, **8**, 1982, 121–7
14 I. Sunagawa and T. Yasuda, Apparent re-entrant corner effect upon the morphologies of twinned crystals; A case study of quartz twinned according to Japanese twin law, *J. Crystal Growth*, **65**, 1983, 43–9
15 I. Sunagawa, J. Takahashi, K. Aonuma, and M. Takahashi, Growth of quartz crystals twinned after Japan law, *Phys. Chem. Min.*, **5**, 1979, 53–63
16 Lu Taijing and I. Sunagawa, Texture formation of agate in geode, *Min. J.*, **17**, 1994, 53–76
17 Lu Taijing and I. Sunagawa, Structure of Brazil twin boundaries in amethyst showing Brewster fringes, *Phys. Chem. Min.*, **17**, 1990, 207–11
18 T. Nishinaga, T. Nakano, and S. Zhang, Epitaxial lateral overgrowth of GaAs by LPE, *Jpn. J. Appl. Phys.*, **27**, 1988, L964–7
19 T. Nishinaga, Microchannel epitaxy: An overview, *J. Crystal Growth*, **237–9**, 2002, 1410–17
20 Y. Takahashi, H. Imai, M. Hosaka, and I. Sunagawa, Genesis of scepter quartz, *Collected Abstracts*, 2000 Annual Meeting of Miner. Soc. Japan, 2000, p. 97 (in Japanese)
21 I. K. Bonev, Nature and origin of the twisted quartz crystals (“Gwindels”) and of quartz with white stripes (“Fadenquartz”), *Collected Abstracts*, IMA–17, Toronto, A84, 1998

11

Pyrite and calcite

Pyrite and calcite are mineral crystals that represent a wide *Habitus* and *Tracht* variation. Pyrite is the most persistent mineral among sulfide minerals, occurring in a wide range of modes, including inorganic processes and bacterial action, and it can also be synthesized by hydrothermal or chemical vapor transport methods.

Pyrite crystals exhibit a wide range of *Tracht* and *Habitus*, and also occur in unusual forms of polycrystalline aggregate, such as framboidal pyrite. Although numerous crystal faces have been reported, the most important ones are {100}, {111}, and {210}. Calcite also exhibits a variety of *Tracht* and *Habitus*, such as platy, nail-head, prismatic, or dog-tooth forms, but $\{10\bar{1}1\}$ is the only F face. In this chapter, we focus our attention on the factors controlling the observed variations in *Tracht* and *Habitus* of pyrite and calcite.

11.1 Pyrite

11.1.1 Tracht *and* Habitus

Pyrite is the most common mineral among sulfides. It occurs not only as a major mineral of sulfide ore deposits of base metals, such as Cu, Pb, Zn, in vein-type, massive-replacement type, kuroko-type* deposits, etc., but also sporadically as an accessory mineral in volcanic, sedimentary, and metamorphic rocks. It also occurs as a precipitate in hot springs, and it may be formed by bacterial action. Pyrite itself is not an ore of Fe, though it contains iron, and at best may have economic value as an ore to obtain sulfuric acid. However, due to its occurrence in and

* Kuroko deposits are massive-type ore deposits formed by the deposition of various sulfide minerals around submarine fumaroles. They are major ore deposits of Cu, Pb, and Zn in Japan.

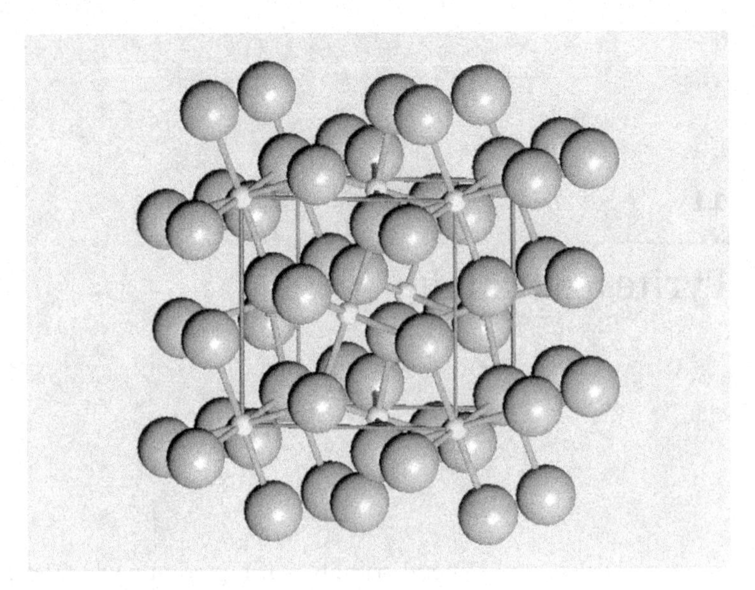

Figure 11.1. Crystal structure of pyrite.

around economically viable ore deposits, many investigations on its *Tracht* and *Habitus* were made, particularly in the former Soviet regions, in the hope that the results would assist the ore prospecting process. Single crystal synthesis has been carried out by CVT and hydrothermal methods, to utilize the semiconductor properties of pyrite, such as thermoelectromotive force.

Reflecting its wide mode of occurrence, pyrite is observed to be extremely rich in variations of *Tracht* and *Habitus*, and more than 460 crystal faces have been reported. In addition to pyrite's equi-dimensional polyhedral *Habitus*, such as cubic, octahedral, or pentagonal dodecahedral, *Habitus* such as remarkably flattened, malformed or kinked whiskers are reported. Crystal sizes range from sub-micrometer order to more than 10 cm along an edge. Polycrystalline aggregate, in the forms of framboidal pyrite (see Section 8.4), nodules, spherulitic, and irregular and granular aggregates have been reported. Although so many crystal faces and a wide variety of *Habitus* and *Tracht* are known, the three major crystal faces determining the morphology of pyrite crystals are {100}, {111}, and {210}, and the *Tracht* variation seen in Fig. 2.5 appears as combinations of these three faces. Although many crystal faces , such as {*hhl*}, {*hkl*}, and {*hk*0}, are known, {*hhl*} and {*hkl*} appear due to the growth of {111} faces, and {*hk*0} appear due to that of {100} and {210}. The reason why {210} appears as the major crystal face among {*hk*0} faces is due to the presence of a glide plane in the symmetry elements involved (see Section 4.2).

In the crystal structure of pyrite (Fig. 11.1), S_2 molecules with dumbbell form and

Figure 11.2. Characteristics of surface microtopographs of three major faces: {100} ((a), (b), (d), (e)); {111} ((d), (e)); {210} ((b), (c), (f)).

Fe are arranged alternately. Since the dumbbells are aslant in opposite orientations, pyrite belongs to the hemihedral crystal group m3, and the space group is Pa3. The reticular density of {110} in the Bravais–Friedel law should be calculated on the basis of {220} in the Donnay–Harker law, and the order of morphological importance becomes {210}≫{110}. According to PBC analysis on Hartman–Perdok theory, {100}, {111}, and {210} are F faces.

11.1.2 *Characteristics of surface microtopographs*

The surface microtopographs of the three major faces of natural pyrite crystals have the following characteristics (see Fig. 11.2 for a schematic representation).

(1) {100} faces are characterized by step patterns with elongated rectangular form in one axis, or elongated rectangular form with truncated corners parallel to the edge with {111}. Such step patterns are universally observed on {100} faces of pyrite crystals formed under any conditions, indicating that this face always grows either by two-dimensional nucleation or the spiral growth mechanism. The direction of elongation of the rectangular form follows the symmetry of point group m3, and is thus perpendicularly oriented on the neighboring {100} faces.

(2) Triangular step patterns are observed on {111} faces. {111} faces also grow either by two-dimensional nucleation or the spiral growth mechanism. However, triangular step patterns are oppositely oriented and inclined to the triangle of the {111} face. It should be noted that the orientation and inclination of triangular growth layers on the {111} face of pyrite differ from those of regular triangular growth hillocks observed on the {111} faces of diamond, which belongs to the holohedral crystal group. For diamond, the step patterns on the {111} face are triangular with the same orientation as the triangle of the face.

(3) Two entirely different surface microtopographs are observed on {210} faces depending on the locality where the crystal grew [1]. The most commonly observed has striations parallel to the edge with the neighboring {100} faces, showing no step patterns. This characteristic corresponds to that of an S face in PBC analysis. The {*hk*0} faces of crystals showing characteristics of this type are usually associated with other {*hk*0} faces, such as {430}, {410}, and {310} in addition to {210}. On the other hand, in the {210} faces of pyrite crystals from a specific locality, such as Elba, the observed striations are perpendicular to the edge with the neighboring {100} face, and the striations represent the edges of step patterns of narrow, elongated growth layers on the {210} face. In this case, {210} faces are grown as F faces. Similar situations have been observed on the $\{10\bar{1}0\}$ face of quartz (see Chapter 10), and in both cases the same crystal face behaves either as an F face or an S face, depending on the growth conditions.

From the surface microtopographic characteristics summarized above, it is concluded that the order of morphological importance of pyrite is $\{100\} > \{111\} > \{210\}$. This order is in agreement with the results of PBC analysis.

11.1.3 *Growth conditions and* Tracht

If a statistical analysis is performed on the appearance of different *Tracht* dependent on grain size of a few hundred pyrite crystallites occurring in a handful of clay in a kuroko ore deposit, it is universally observed that, as crystal sizes

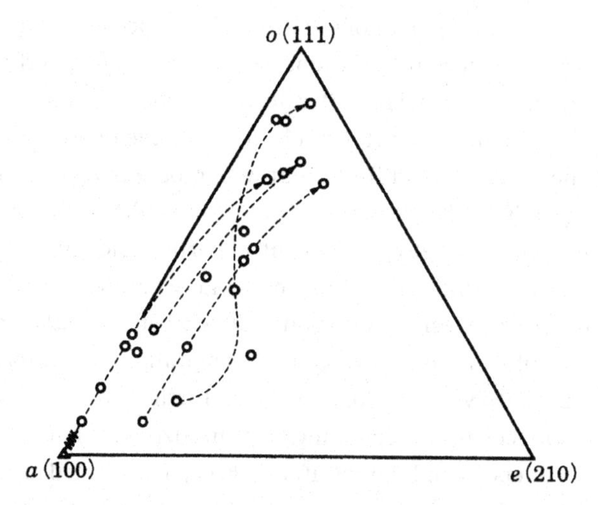

Figure 11.3. Difference in *Tracht* changes of pyrite crystals from the Kanbe Mine depending on grain size between ore deposits (**o**), weakly altered zone (**×**), and in country rock (▲) (see ref. [2], Chapter 2).

decrease, crystals mostly exhibit cubic *Tracht* bounded by {100}, and, as the grain sizes increase, {210} faces start to appear, and the frequency of appearance of pentagonal dodecahedral *Tracht* bounded by {210} faces increases. It was explained that this variation is due to *Tracht* variation associated with growth (see Section 4.4). The degree of change from cubic to pentagonal dodecahedral or to octahedral within the same grain size range varies from the center of an ore deposit to the outer region.

The change is most remarkable at the center of an ore deposit; as the region outside the deposit is approached, practically no *Tracht* change is observable. In other words, *Tracht* variation associated with growth is most remarkable in the region with a high grade of sulfide mineralization, which includes pyrite, where *Trachts* exhibiting well developed {210} or {111} faces appear, whereas pyrite crystals formed in a weakly mineralized zone, for example outside an ore deposit, do not show distinct change and remain as cubic even if their sizes increase. In Fig. 11.3, the statistical results obtained within ore deposits and outside are compared.

A similar tendency is observed between vein and country rock of vein deposits; in country rocks, pyrite generally takes cubic *Tracht*. This tendency is recognized in many observations on typomorphism reported in the former Soviet countries [2].

If growth layers spreading on {100} faces either by two-dimensional nucleation or the spiral growth mechanism reach the edge of the face, the *Tracht* remains as simple cubic. If steps bunch together to form macro-steps as they spread, {*hk*0} faces consisting of striations will appear as vicinal faces through the piling up of macro-steps. Faces having higher stability than others will increase in size, but the face is characterized by the development of striations since layer growth does not occur on the face. A {210} face showing striations only is a crystal face of this kind. In such a crystal, the larger the normal growth rate, or the smaller the step separation, the more {*hk*0} or {210} faces develop, and the larger the degree of change from cubic to pentagonal dodecahedral *Tracht* as growth proceeds. A larger driving force condition corresponds to this situation. The observed difference in *Tracht* variation depending on grain size between the central portion and outer region of a deposit appears to reflect this difference. When pyrite grows under favorable conditions, such as high sulfur fugacity, pentagonal dodecahedral *Tracht* appears, whereas under less favorable conditions pyrite crystals maintain a cubic form. It is probably for this reason that pyrite crystals occurring further away from an ore deposit do not show distinct changes depending on grain size, and that the *Trachts* are different between ore deposit and the country rock, crystals formed in country rock taking exclusively simple cubic form.

Although crystallographically equivalent, the {210} faces of pyrite from a particular locality are characterized by striations perpendicular to the edge with {100}. These striations are different from the striations described above and appear due to narrow growth layers on the {210} face, indicating that, under this condition, {210} behaves as an F face. It is still to be discovered why {210} faces behave differently depending on the growth conditions.

As compared to the relation between {210} and {100}, the appearance of {111} is slightly different. It is known that octahedral crystals contain more arsenic (As) than cubic crystals in the same locality, and there is a possibility that {111} faces develop due to an impurity adsorption of As. It is observed that the internal texture of this type of crystal shows a drastic change from cubic to octahedral in association with precipitation of minerals containing As.

11.1.4 *Polycrystalline aggregates*

In sedimentary rocks, pyrite often occurs as nodular, spherical, irregular, massive, framboidal polycrystalline aggregates of minute crystals. It often occurs in fossils, without modifying the external fossil form, and a bacterial origin has been suggested for pyrite aggregates of this type. Among such polycrystalline aggregates, framboidal pyrites have attracted particular interest. They are spherical, they measure a few tens of micrometers across, and one framboid consists of close-packed, equal-sized idiomorphic pyrite crystals with the same *Tracht*. Cubic, octahedral, pentagonal dodecahedral *Trachts*, etc. are known. One framboid consists of a unique *Tracht*.

Framboidal pyrite occurs, for example, in sedimentary rocks, muddy sediments, and precipitates in hot springs; two controversial origins have been suggested, one bacterial and the other relating to agitation in hydrothermal solution. Framboidal pyrite may also be synthesized by hydrothermal experiments. The genesis of framboidal pyrite remains a stimulating problem. It remains for us to resolve the problem of why the close-packed aggregation of equal-sized polyhedral crystals exists with the same *Tracht*.

11.2 Calcite

11.2.1 *Habitus*

The main polymorphs of $CaCO_3$ are calcite (trigonal system, $\bar{3}m$ (low-pressure phase)), and aragonite (orthorhombic system, mmm), but vaterite (hexagonal system, 6/mmm) is also known. In biological activity, both aragonite and calcite are formed as major minerals constituting shells and exo-skeletons under 1 atmospheric pressure, but, after some time of precipitation and deposition, aragonite transforms into calcite, the stable phase that exists below 5000 Pa. As a result, limestone, which is a sedimentary rock of biological origin, consists only of calcite. When a granitic magma intrudes into limestone, grain growth occurs and marble is formed by a contact metamorphic (metasomatic) process. Since there is free space in druses or veins formed during this process, idiomorphic crystals of calcite are formed. Idiomorphic crystals of calcite also occur in a wide temperature range, from hydrothermal veins to limestone eroded by underground water. They also occur in druses of volcanic rocks. The *Habitus* and *Tracht* show the greatest variations among mineral crystals. The *Habitus* may be thin platy, thick platy, short-prismatic, long-prismatic, rhombohedral, dog-tooth (scalenohedral), nail-head, etc. Many crystal faces are reported, but the only one corresponding to an F face is $\{10\bar{1}1\}$, which is a cleavage face. If $\{10\bar{1}1\}$ is the only F face, the representative *Habitus* of calcite should be rhombohedral, but real crystals show astonishingly variable *Habitus*. On the other hand, when calcite crystals are synthesized from pure hydrothermal solution, only rhombohedral *Habitus* is obtainable, whereas when polymer is added to the solution as an impurity component, dog-tooth *Habitus* is obtained. In natural calcite crystals, those formed under relatively high-temperature conditions, such as those occurring in contact metasomatic deposits, exhibit platy or nail-head *Habitus*, and, on decreasing the temperature, prismatic or dog-tooth *Habitus* occur. That *Habitus* changes with decreasing temperature has been shown to be a general tendency.

Figure 11.4 shows the various *Habitus* of calcite crystals observed in nature. To demonstrate *Habitus* changes at different orders of crystallization, or those depending on crystallization temperatures, three examples of intergrowths of two different *Habitus* of calcite are shown in Fig. 11.5 [3], see also Fig. 7.10. Figure 11.5(a) shows that later-grown calcite crystals on earlier-formed prismatic crystals of calcite show a different prismatic to a dog-tooth *Habitus* from the host crystal.

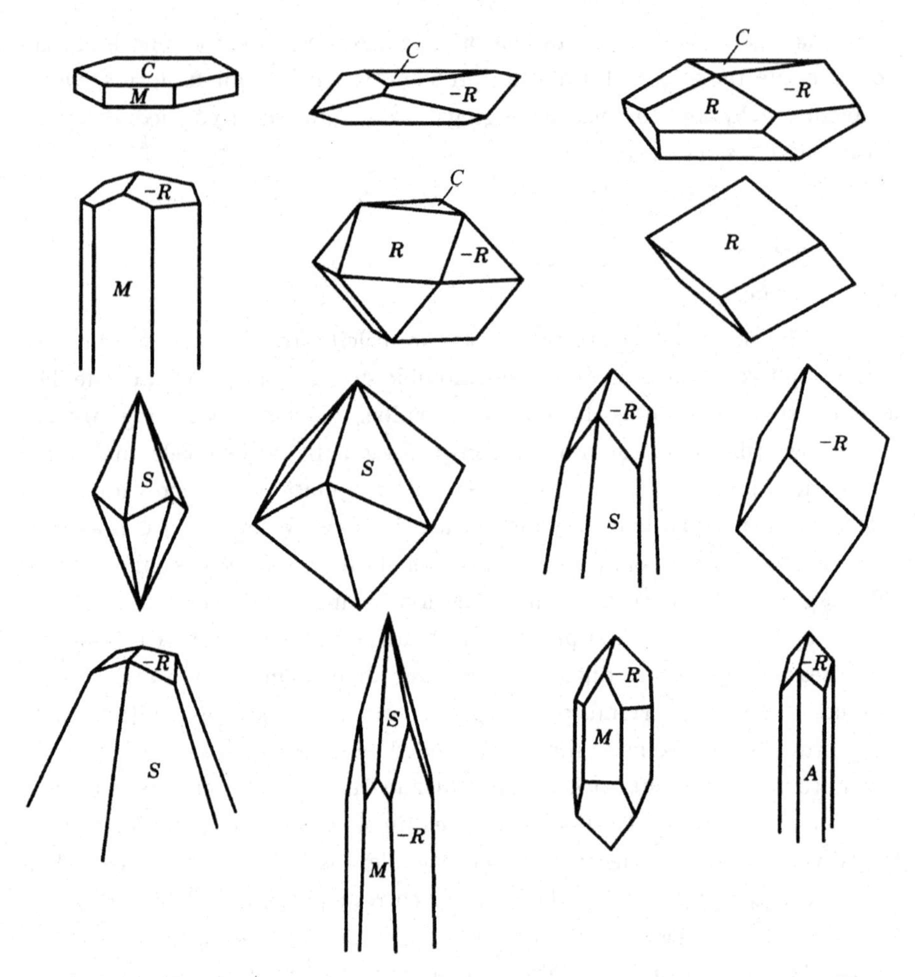

Figure 11.4. Various *Habitus* observed for natural calcite crystals: $C\{0001\}$, $M\{10\bar{1}0\}$, $R\{10\bar{1}1\}$, $-R\{0k\bar{k}l\}$, $S\{hk\bar{i}l\}$, and $A\{11\bar{2}0\}$ faces [3].

Figure 11.5(b) shows that the later-formed calcite crystal has a different *Habitus* from the earlier-formed prismatic crystal around which it grows, and Fig. 11.5(c) is an example of the growth of a crystal with dog-tooth *Habitus* on an earlier-formed crystal with rhombohedral *Habitus*. (See also Fig. 7.10.) By compiling these relations, it is possible to trace systematically how the *Habitus* of calcite changes from earlier to later stages or as the temperature decreases in the case of contact metasomatism [2], [3]. The *Habitus* variation with decreasing temperature summarized above is a general trend based on data of this type.

11.2.2 *Surface microtopography*

Although calcite crystals exhibit a great variety of *Habitus* change and many crystal faces are known, the crystal faces may be grouped as follows.

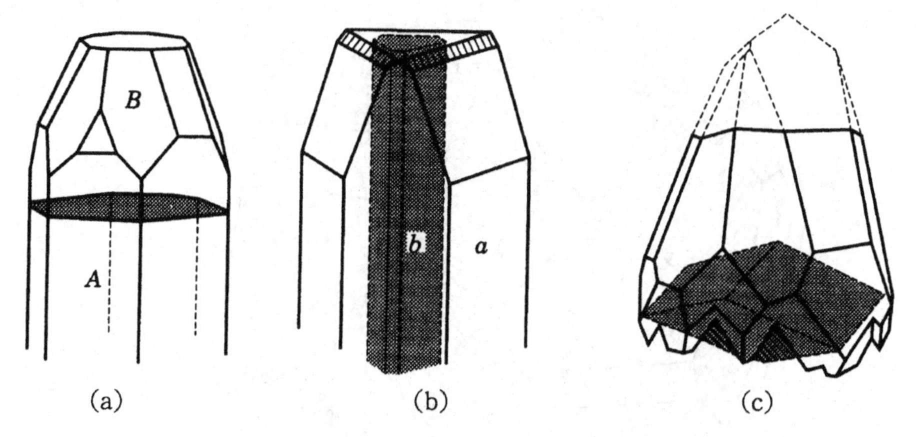

Figure 11.5. Three examples showing different *Habitus* of later-grown calcite crystals preferentially nucleated on the edges and corners of earlier-formed calcite crystal. Changes in *Habitus* depending on crystallization stages or growth temperatures are indicated [3]. (a) Earlier-formed hexagonal prism (*A*) and later-formed scalenohedral crystal (*B*). (b) Earlier-formed hexagonal prism (*b*) and later-formed thicker crystal (*a*). (c) The shaded area shows an earlier-formed rhombohedral crystal, and the remaining area represents later-formed scalenohedral crystals.

(1) $\{10\bar{1}1\}$. This is the unique F face from PBC analysis, and the face appearing through the spiral growth mechanism. Spiral growth hillocks of rhombic form are observed, and the step advancing rates have anisotropy following the symmetry element of the face. Since element partitioning will alter according to the growth step advancing rate, which affects optical properties, intra-sectorial sectors are formed (see Sections 3.14 and 6.2) [4], [5]. These were confirmed by atomic force microscopy (AFM) and X-ray fluorescence (XRF).

(2) $\{0k\bar{k}l\}$. These are minus rhombohedral faces ($-R$), and in most cases they appear as $\{01\bar{1}2\}$ faces with nail-head *Habitus*. Growth step patterns are not observed on these faces, which are characterized by the development of striations parallel to the edges with $\{10\bar{1}1\}$.

(3) $\{hk\bar{\imath}l\}$. This is a group of habit-controlling faces of dog-tooth *Habitus*, which commonly occur at low-temperature conditions. Many high-index faces, such as $\{21\bar{3}1\}$ and $\{65\bar{11}1\}$, are reported to grow as large as *Habitus*-controlling faces, but they are exclusively characterized by striations; step patterns due to layer growth are not observed.

(4) $\{10\bar{1}0\}$, $\{11\bar{2}0\}$. Among these prism faces, $\{11\bar{2}0\}$ appears as an extreme of dog-tooth *Habitus*, and is characterized similarly to $\{hk\bar{\imath}l\}$ by striations parallel to the edge with $\{10\bar{1}1\}$. No reliable observations have so far been

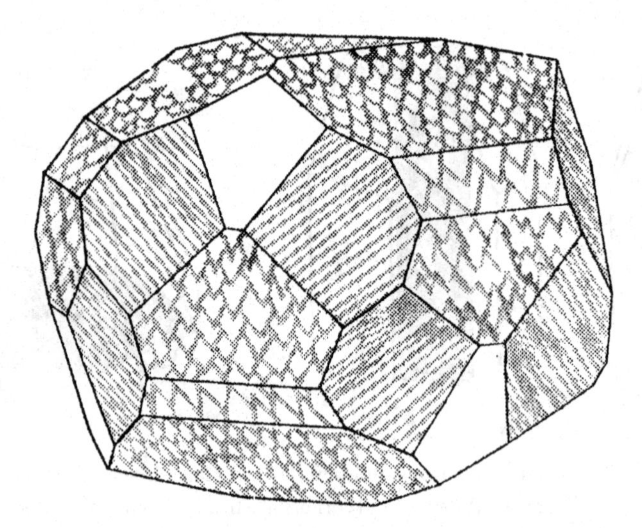

Figure 11.6. Characteristics of surface microtopography of crystal faces of calcite.

reported on the surface microtopography of $\{10\bar{1}0\}$, but step patterns are not reported.

(5) $\{0001\}$. This shows a rugged surface. No reports are available so far on step patterns on this face.

The characteristics of the surface microtopography of these faces are schematically illustrated in Fig. 11.6.

From the observations of the surface microtopographs of crystal faces of calcite, it is confirmed that the only face which behaves as a smooth interface under any conditions and grows by the spiral growth mechanism is $\{10\bar{1}1\}$. The $-R$ and $\{10\bar{1}0\}$ faces determining a nail-head or prismatic *Habitus* and the $\{hk\bar{i}l\}$ faces determining a dog-tooth *Habitus* are all characterized by striations parallel to the edge with $\{10\bar{1}1\}$, in spite of the fact that they develop as large as *Habitus*-controlling faces. These may be regarded to have appeared through the piling up of the edges of growth layers on the $\{10\bar{1}1\}$ face. Unless we assume that these faces become large in order to determine the *Habitus*, in spite of their nature as S faces, by stabilizing vicinal faces appearing due to the piling up of steps of growth layers on $\{10\bar{1}1\}$ faces, it is not possible to understand their appearance and development. It is possible that the vicinal faces are stabilized owing to the impurity adsorption along the steps forming these striations. It is necessary that the faces can adsorb the impurities, and that the temperature is set appropriately, so that adsorbed impurities can play such a role. The observations that calcite crystals formed under low-temperature conditions tend to take on a dog-tooth *Habitus*, and that dog-tooth *Habitus* can be produced by hydrothermal synthesis only when polymers are added as impurities, may reflect this relation. The spicules of sea urchins consist of calcite crystals with dog-tooth *Habitus*, as will be described in Chapter 14, which may be due to the same reason. On the other hand, the morphology of calcite crystals forming the exo-skeleton of coccoliths show an entirely different *Habitus*; see the discussion in Chapter 14.

References

1 Y. Endo and I. Sunagawa, Positive and negative striations in pyrite, *Am. Min.*, **58**, 1973, 930–5

2 I. Kostov and R. I. Kostov, *Crystal Habits of Minerals*, Sofia, Professor Martin Drinov Academic Publishing House and Pensoft, 1999

3 I. Sunagawa, Habit variation in calcite – with special reference to the relation with crystallization sequence, *Rep. Geol. Surv., Japan*, no. 156, 1953 (in Japanese with English abstract)

4 R. J. Reeder and J. C. Grams, Sector zoning in calcite cement: Implications for trace element distributions in carbonates, *Geochim. Cosmochim. Acta*, **51**, 1987, 187–94

5 W. J. Staudt, R. J. Reeder, and M. A. A. Schoonen, Surface structural control on compositional zoning of SO_4^{2-} and SeO_4^{2-} in synthetic calcite single crystals, *Geochim. Cosmochim. Acta.*, **58**, 1994, 2087–98

12

Minerals formed by vapor growth

Sulfur and hematite crystals formed around fumaroles during the final stages of volcanic activity, phlogopite and hematite occurring in small druses in igneous rocks, and large and highly perfect single crystals of quartz, beryl, topaz, tourmaline, and other minerals occurring in pegmatite are all formed in supercritical vapor phases concentrated in the final stage of magma solidification. These crystals are all grown in vapor phases in which a chemical reaction (such as oxidation) takes place, or in hydrothermal solution at elevated temperatures. How these crystals grow and how their perfection and homogeneity fluctuate in single crystals will be analyzed using beryl, hematite, and phlogopite as representative examples. The analysis will be made in relation to the size of free space in which the respective crystals grow.

12.1 Crystal growth in pegmatite

Since pegmatite* in granite produces large and highly perfect idiomorphic crystals, it is a treasure box of colored gemstones suitable for facet cutting, and also of rare mineral specimens containing rare earth elements. Free space on a large scale occurs due to the concentration of volatile components, and crystals can grow freely in such spaces. Since crystal growth proceeds either in a void in a solidifying magma or in a crack in the surrounding strata, at depths < 10 km, pressures < 30 kbar, and temperatures from 700 °C and lower, the conditions range from the supercritical vapor phase to hydrothermal solutions.

The morphology of single crystals, the surface microtopographs of crystal

* Pegmatite consists of much larger crystals than those in the mother rocks. It is also formed in rocks other than granite, but the scale is much smaller.

faces, and the inhomogeneities and imperfections in crystals have been investigated in detail for quartz, topaz, and beryl formed in pegmatite. We will select beryl as a representative example, and we will investigate crystal growth in pegmatite [1].

Among eight Be-containing minerals in the $BeO-Al_2O_3-SiO_2-H_2O$ system, beryl has the widest stable range (≤10kbar, 320~680 °C) [2], and exhibits a relatively stable *Habitus* bounded by {0001} and $\{10\bar{1}0\}$ faces: crystals showing spear-shaped tapered forms, or crystals containing channels parallel to the *c*-axis.

The first verification of the spiral growth theory, by Frank, was made from the observation of a horseshoe step pattern on the $\{10\bar{1}0\}$ face of a beryl crystal occurring in pegmatite, as already explained in Section 3.7 (ref. [18], Chapter 3). Elemental spiral step patterns are universally observed in beryl crystals formed in pegmatite that is only slightly etched: a hexagonal pattern on {0001} and a rectangular pattern, with the longer axis in the *a*-axis direction, or a rhombic pattern on $\{10\bar{1}0\}$ (Fig. 12.1). Composite spirals originating from a dislocation array or from a number of dislocations are also commonly observed.

The height of the elemental spiral steps is of unit cell size, and the step separation is in the order of 1~10 μm, corresponding to a separation versus height ratio of order 10^3–10^4. This indicates that growth occurs in a dilute ambient phase under a small driving force (estimated as $\sigma < 0.1\%$). Evidence to suggest precipitation of minute crystallites on the growing crystal surface, which affects the growth of the host crystal, is not generally detected.

There are many crystals that provide evidence of the fact that etching occurs once the growth has stopped, in addition to crystals exhibiting as-grown surface microtopography. Etch pits observed on {0001}, $\{10\bar{1}0\}$, and $\{11\bar{2}1\}$ are shown in Fig. 12.2. The etch pits show forms controlled by the symmetries of the respective faces. If etching proceeds further, it does so from outcrops of dislocations and from tube-like inclusions on the {0001} surface, in addition to corners and edges, and spear-like tapered forms (Fig. 12.3), or hollow-tube crystals appear. An X-ray topograph, a polarizing photomicrograph, and a reflection photomicrograph after etching of a single crystal plate cut perpendicularly to the *c*-axis are shown in Fig. 12.4. Figure 12.4(d) is a polarization photomicrograph of a similar section of another crystal.

Growth sectors, growth banding, inclusions, dislocations perpendicular (*A*), parallel (*B*), and inclined (*C*) to the *c*-axis, and the relation between dislocations and inclusions and optical anomalies may clearly be seen in Fig. 12.4. From the observation that the thicknesses of growth sectors of crystallographically equivalent faces are different in different directions, we see that there was an anisotropic flow of the mother liquid and that the flow directions varied during the growth process. From morphological changes associated with growth, as deduced from Fig. 12.4(a),

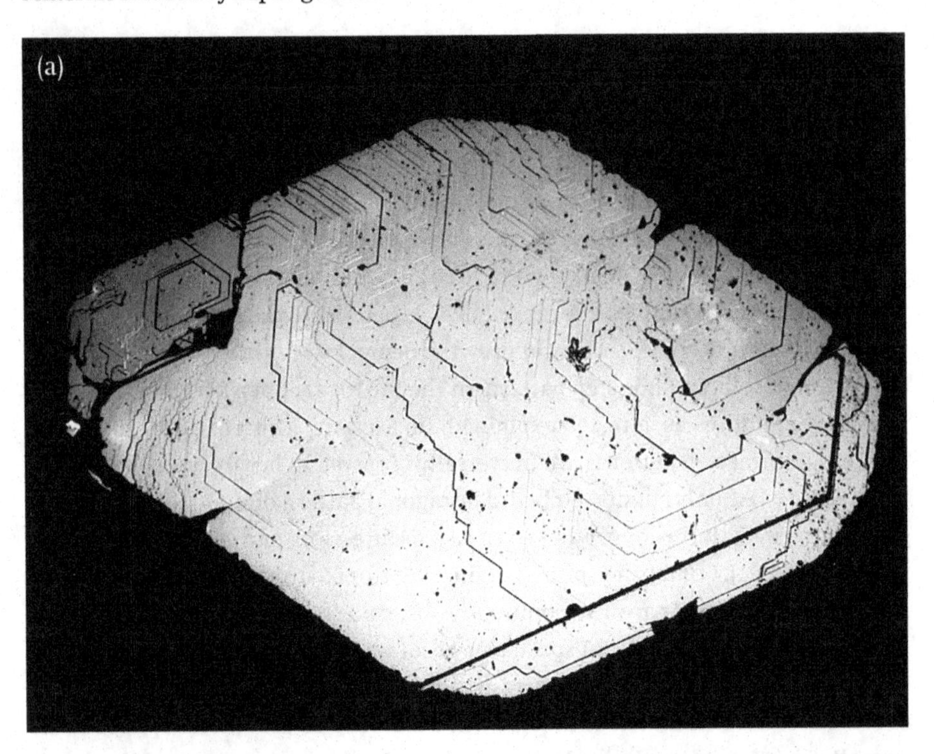

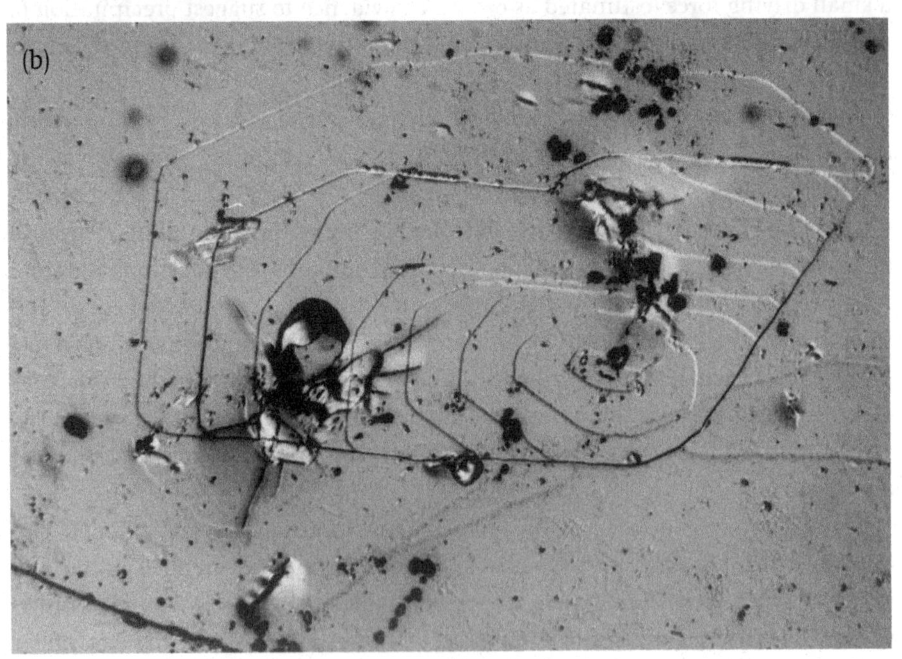

Figure 12.1. Elemental spiral step patterns observed on the {0001} face of beryl. (a) Low magnification, reflection; (b) high magnification, differential interference contrast photomicrograph.

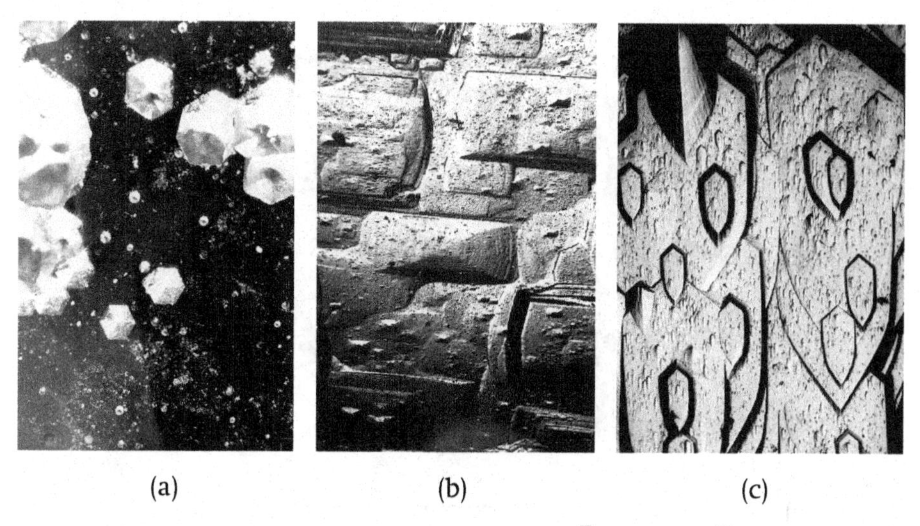

Figure 12.2. Etch pits observed on (a) {0001}, (b) $\{10\bar{1}0\}$, and (c) $\{11\bar{2}1\}$ faces of a beryl crystal [1].

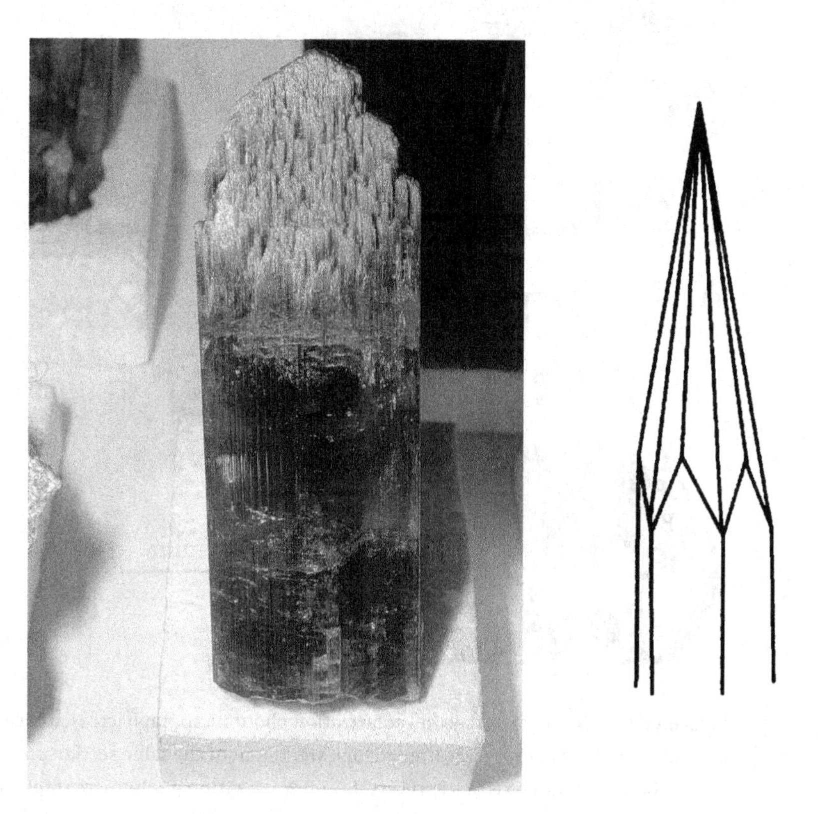

Figure 12.3. Spear-shaped beryl crystal etched by a natural process.

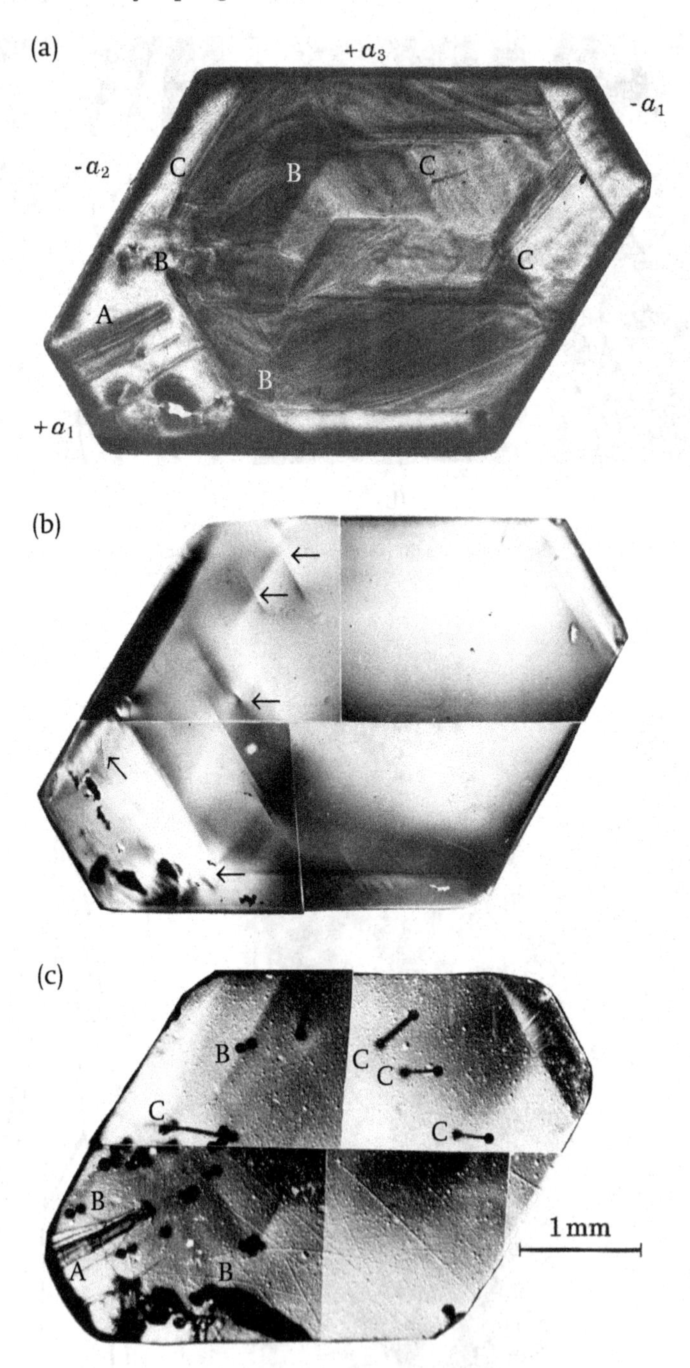

Figure 12.4. (a) X-ray topograph, (b) polarization photomicrograph (crossed Nicols), and (c) reflection photomicrograph after etching treatment of the same section prepared perpendicular to the *c*-axis of a beryl crystal. (d) Polarization photomicrograph of similar section of another sample [1].

(d)

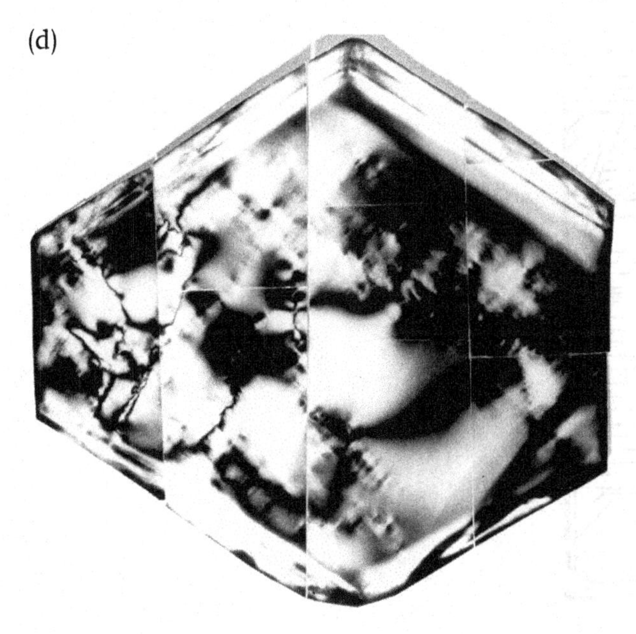

Figure 12.4 *(cont.)*

we see that there was an intermission in the growth followed by slight dissolution (contrast with rounded form) and trapping of inclusions, and an associated generation of dislocations. We also see that a large increment of growth rate of one of the crystallographically equivalent growth sectors is closely related to the generation of dislocations (compare the two crystallographically equivalent growth sectors in the lower left and upper right areas of Fig. 12.4(a)). Figures 12.4(b) and (d) show that the dislocations parallel to the *c*-axis cause optical anomalies due to strain birefringence observed in the optic axis direction under polarized light.

How growth and dissolution proceed during the whole process of the formation of beryl crystals in a pegmatite can be traced from the observation of a section prepared parallel to the *c*-axis. Figure 12.5 represents what sort of fluctuation took place during the formation of a hexagonal prismatic crystal, based on polarizing microscope observations and X-ray topographic data. It is seen that the conditions fluctuated at six stages, from A to F, dissolution started from the edges and corners of $\{0001\}$ and $\{10\bar{1}0\}$, and the morphology changed from hexagonal prismatic to tapered prismatic. Regrowth proceeds to recover the smooth interfaces, starting from a rounded rough interface, and during this process $\{h0\bar{h}l\}$ or $\{hk\bar{i}l\}$ faces may appear, but they eventually change to polygonal bounded by only $\{0001\}$ and $\{10\bar{1}0\}$ faces.

At the final stage, F, the trapping of vapor–liquid two-phase inclusions is particularly remarkable: a growth–dissolution–regrowth process of this type is repeated six times, and this is routinely observed in beryl crystals from pegmatite localities

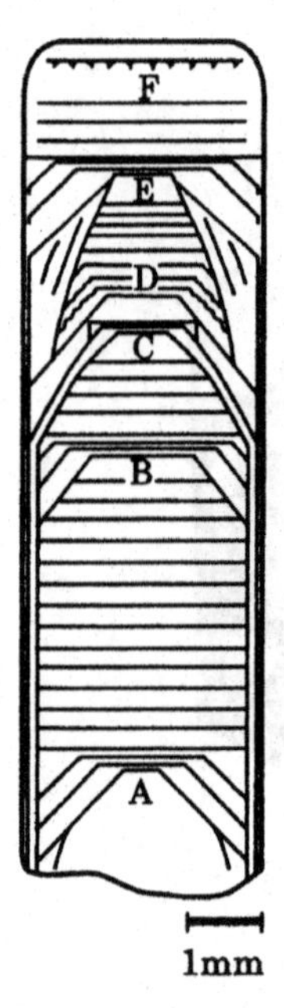

Figure 12.5. Schematic illustration of the fluctuation occurring during the growth process of a beryl crystal (growth history) based on the observations of a section prepared parallel to the *c*-axis of the crystal [1]. Please see text for an explanation of A–F.

all over the world, indicating a history of many growth–dissolution–regrowth periods. This is not due to the fluctuation in bulk conditions where the magma was solidified, but rather to the local fluctuations in supply and composition of the mother liquid during the process of magma solidification.

At the initial stage of regrowth, after the cessation of dissolution, occur the precipitation and adhesion of foreign mineral grains on the growing surface. These particles are trapped into the growing crystal as inclusions. In this process, tube-like liquid-phase inclusions are formed behind the particles, and dislocations are generated at the points where an inclusion is enclosed. In this way, tube-like two-phase inclusions and dislocations with a Burgers vector direction along the *c*-axis are formed, and spiral growth proceeds from the outcrop of these dislocations on the {0001} face. Figure 12.6 is a polarization photomicrograph showing this relation. Inclusions also generate dislocations that are perpendicular to or inclined to the *c*-axis.

These observations relating to beryl and the deduced growth history of the crystals in pegmatite are also commonly encountered for crystals of quartz, topaz, tourmaline, etc., which are grown in pegmatite. Although the formation of pegmatite, broadly speaking, occurs in a closed ambient phase, we have seen that in

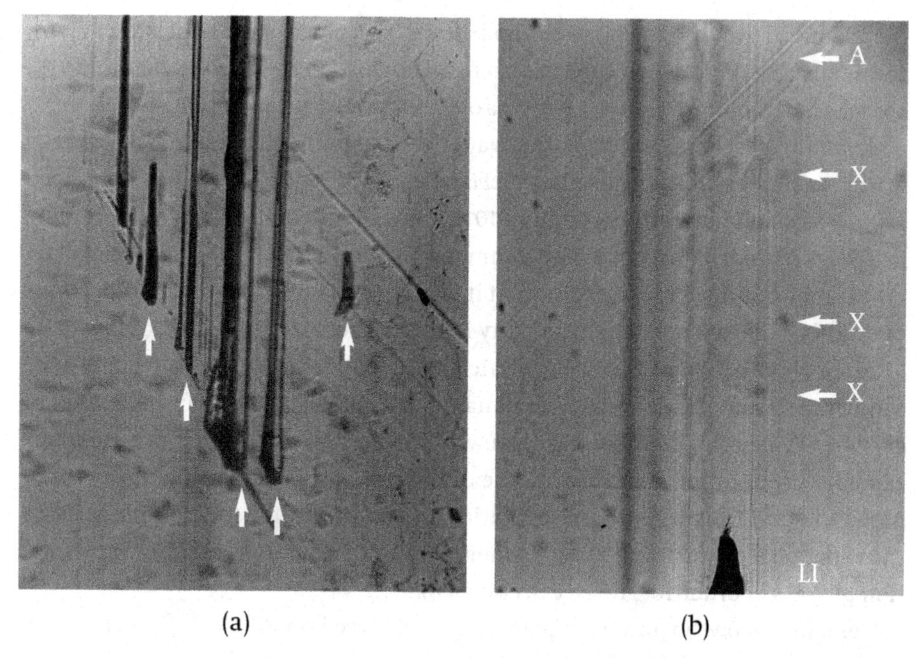

Figure 12.6. Polarization photomicrograph showing dislocation A with Burgers vector parallel to the *c*-axis, which is generated from a tube-like liquid inclusion (LI in (b)) formed behind foreign mineral grains (arrows in (a)), which were precipitated on a growing surface. The X symbols in (b) denote the banding of successive stages of formation of a negative crystal [1].

the real process growth conditions fluctuate many times in the composition or direction of supply of the supercritical vapor phase or hydrothermal solution, yet large crystals with high perfection are formed. However, the environmental conditions fluctuate less, and crystals grow in far less disruptive environmental conditions, as compared to crystals occurring in the druses of volcanic rocks (to be described in Section 12.3), in which a large number of minute crystals are formed in a small druse, in which they move around violently and agglutinate.

12.2 Hematite formed by post-volcanic action

At a later stage of volcanism, volatile components are supplied from the magma and crystals of sulfur or hematite crystallize around volcanic fumaroles or in fissures of surrounding rocks. Compared with crystallization in pegmatite, the environment is much more open, and the crystals of sulfur and hematite grow due to the chemical reaction occurring when the components supplied in the vapor phase oxidize at the Earth's surface. This crystallization therefore corresponds to the chemical vapor transport method using an open tube. Hematite crystals, whose surface microtopography has been investigated in the greatest detail, [3]–[6], are selected as representative examples of this type of crystallization.

Hematite crystals occurring in this way have been found in volcanoes all over the world, and all exhibit a thin platy *Habitus* bounded by the most well developed $\{0001\}$ face and narrow $\{10\bar{1}1\}$ and $\{10\bar{1}0\}$ faces. All hematite crystals occurring in this way show characteristically common thin platy *Habitus*; their surface microtopography is unique to the extent that it may be used to identify specific locations of origin of the crystals. The thin platy *Habitus* differs from the thick platy, nail-head, or rhombohedral *Habitus* exhibited by hematite crystals from vein-type or contact metasomatic deposits. The remarkable difference between the *Habitus* of crystals from vapor and solution growth is also noted in other minerals (for example, corundum), and is due to the difference in step separation of the spiral growth layers. The ratio of step separation to step height for crystals grown from the vapor phase is of order 10^3–10^4, whereas that for crystals grown from the solution phase is of order 10^2–10^3 (see Section 5.6).

Elemental growth spiral step patterns are observed on all $\{0001\}$, $\{10\bar{1}1\}$, and $\{10\bar{1}0\}$ faces of hematite crystals grown by post-volcanic action.

On $\{0001\}$ faces, growth spirals with step heights of 0.23 nm (corresponding to the height of one Fe_2O_3 molecule), 0.46 nm (two molecules), 0.7 nm (three molecules), and 1.4 nm (one unit cell height) and macro-steps corresponding to the height of a few Burgers vectors have been observed, and the ratio of step height to step separation is of order 10^4. The morphologies of elemental spirals vary from circular to regular triangular on $\{0001\}$, and spindle form containing one symmetry plane on $\{10\bar{1}1\}$ (Fig. 12.7). Circular spirals are observed on the $\{0001\}$ face on crystals assumed to have grown under slightly higher driving force conditions, whereas $\{10\bar{1}1\}$ and $\{10\bar{1}0\}$ faces show a hopper characteristic, from which the order of morphological importance is understood to be $\{0001\} > \{10\bar{1}1\} \geqslant \{10\bar{1}0\}$.

In many cases, several spiral centers are present on one $\{0001\}$ face, forming macro-steps by bunching of the spiral growth steps as they advance. The form of these macro-steps starts out as irregular, but, as bunching proceeds, a new rhythm is produced, forming macro-steps having a kind of regularity; for examples, see Fig. 12.8. When two crystals join incoherently, macroscopic spiral steps sometimes originate and spread from the boundary. Also, a different effect on the advancement of elementary spiral steps when minute crystallites precipitate on a growing $\{0001\}$ surface in the same or the twin orientation may be encountered. When precipitation occurs in the same orientation, the advancement of the elementary spiral steps is promoted around the precipitate, but it is retarded if it

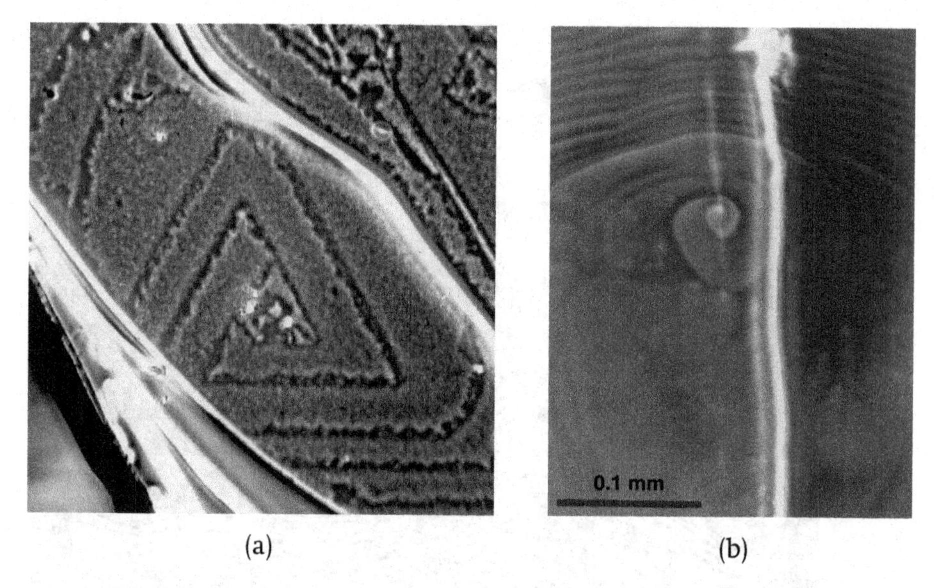

Figure 12.7. Spiral step patterns observed on (a) {0001} and (b) $\{10\bar{1}1\}$ faces of hematite. In (a), smooth growth steps are retarded two-dimensionally by etching, producing rough dissolution steps [3]–[6].

occurs in the twin orientation, forming a flow pattern reminiscent of that caused by water flowing around a rock in a stream. However, the proportion of precipitated crystallites is not high, unlike in the case of hematite to be described in Section 12.3.

Most of the hematite crystals formed in this way are single crystals, but there are places where most of the hematite crystals occur as contact twins with twin axis <0001> and composition plane $\{10\bar{1}0\}$. All twinned crystals take a thin ribbon-like *Habitus*, which is due to the pseudo re-entrant corner effect. The surface microtopographs of $\{10\bar{1}0\}$ and $\{10\bar{1}1\}$, which face the twin composition part, show different features from those observed on crystallographically equivalent faces appearing on the other part, and this indicates that the patterns are due to the rapid growth rate. Independent spiral centers are observed on two individuals of the common {0001} face on opposite sides of the twin boundary, respectively. Figure 12.9 is an example of the surface microtopographs showing this type of relation.

On the {0001} face of hematite originating from volcanoes having violent activities, such as Vesuvius in Italy, straight steps, originating from spiral centers and crossing as-grown spiral steps, are observed (Fig. 12.10). Since no displacement of the spiral steps is observed at the point where the straight step crosses the spiral step, it is understood that the straight steps appear due to dislocation movement

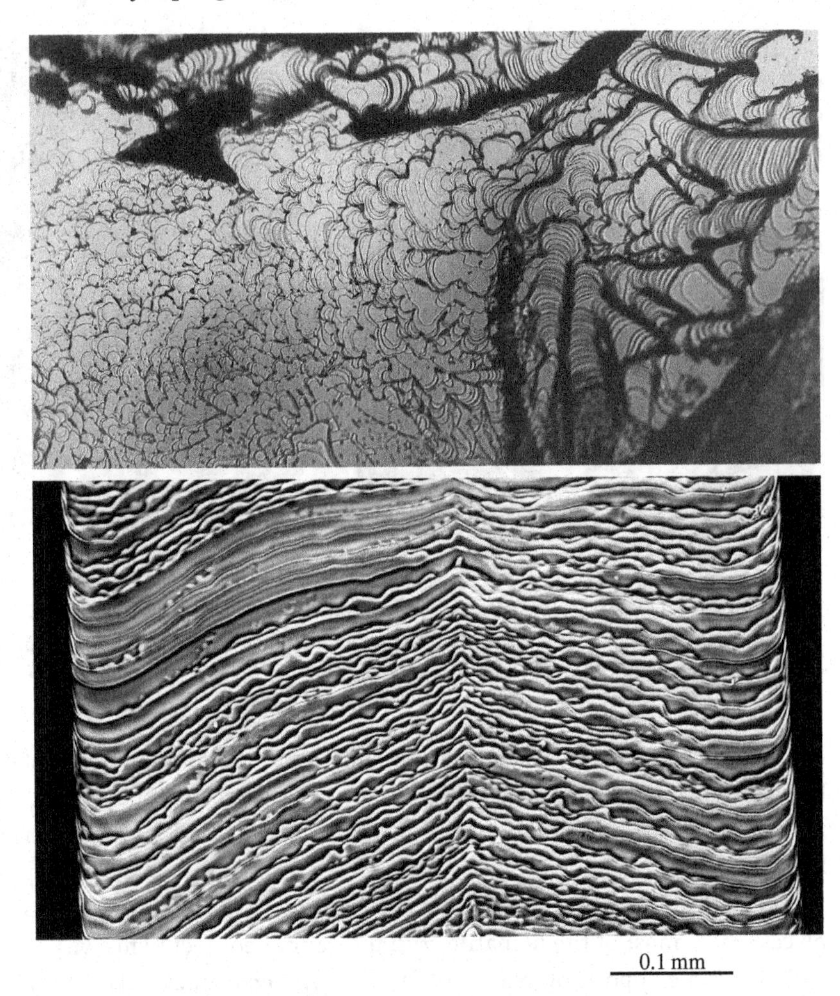

Figure 12.8. Regular pattern seen in macro-steps on a {0001} face of hematite.

on the glide plane after the cessation of growth. Straight steps of this type are observed only on hematite crystals from particular localities, and are related to the intensity of volcanic activity [8].

Etching occurring after the cessation of growth is well recorded in the form of step patterns on the crystal faces on hematite crystals from the Azores Islands. The {0001} face of hematite from this locality shows step patterns that resemble growth steps, but careful observations indicate that all these steps have a chopping waveform (see Section 5.10 and Fig. 5.18) and that the steps appear due to two-dimensional etching starting from growth steps. Figure 12.11 is a surface micro-topograph demonstrating this, and it should be noted that there is no elevated center of the step patterns. The circular step patterns seen on the photograph are all depressions.

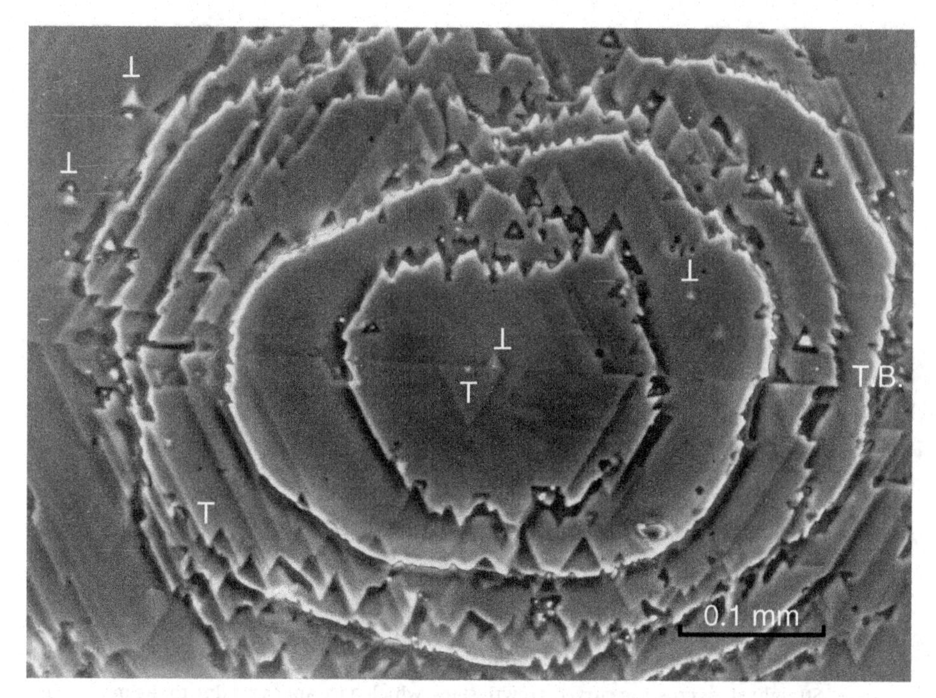

Figure 12.9. Spiral step pattern observed on both sides of a twin boundary on a {0001} face of a hematite crystal [7]. T.B. is the twin boundary, ⊥ and ⊤ indicate orientations. Phase contrast photomicrograph.

12.3 Hematite and phlogopite in druses of volcanic rocks

In volcanic rocks formed by the solidification of magma near the Earth's surface, small cavities of millimeter to centimeter order are often present, in which idiomorphic crystals of phlogopite and hematite occur. All these crystals grown from the vapor phase show typical spiral patterns. However, they grow in a much-reduced free space as compared to the case of pegmatite, and this characteristic is well represented on their surface microtopographs.

Figure 12.12 shows a step pattern on the {001} face of phlogopite; the step height is 1 nm, and the step separation is of order 10 μm. The characteristic form of the spiral is five-sided, i.e. hexagonal with one edge truncated [10]. This characteristic form is due to the fact that the stacking of the unit layer, which has a hexagonal form, is in one direction. Since an interlacing pattern is not observed in this step pattern, this crystal is identified as a 1M polytype. On the surface of the (001) face shown in Fig. 12.12, a five-sided step pattern rotated by 120° and 180° appears in an island form on one surface. This indicates that several crystals agglutinate in rotated orientation, and that crystals attaining a certain size move around in the cavity and come together to form the crystal [11].

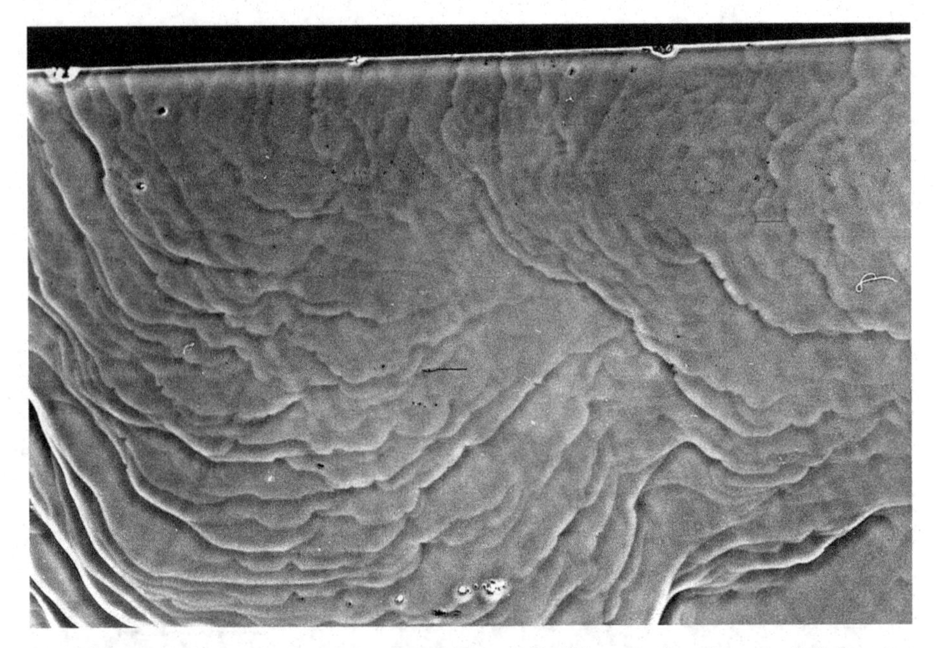

Figure 12.10. Phase contrast photomicrograph of hematite from Vesuvius in Italy. Straight steps crossing curved growth steps, which have appeared due to the movement of dislocations after the cessation of growth [8], are shown.

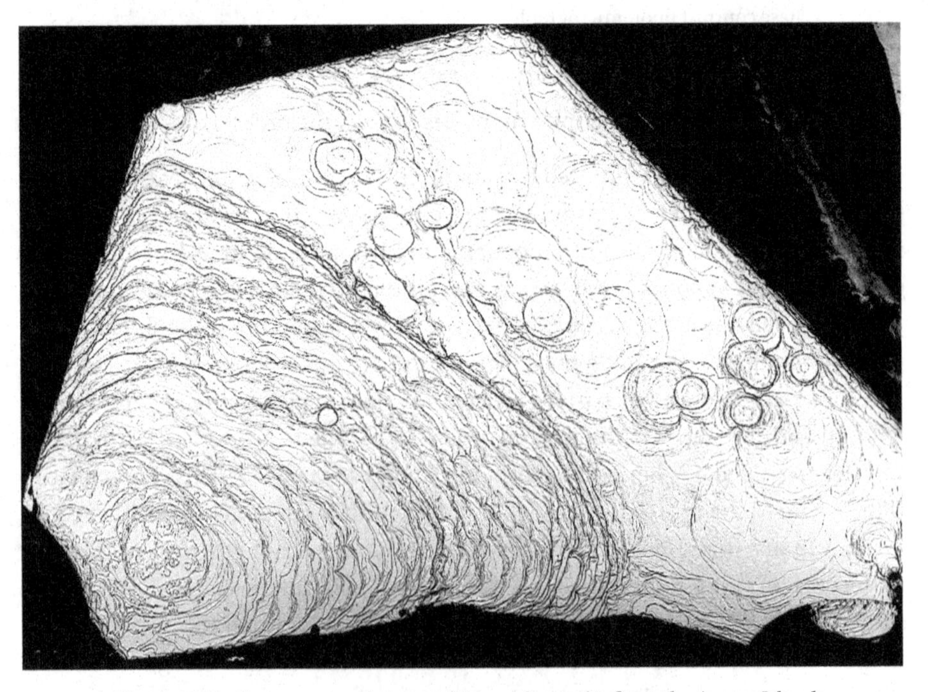

Figure 12.11. Step patterns due to etching on hematite from the Azores Islands. All circular step patterns are depressions. Note steps of chopping waveform [9].

Figure 12.12. Five-sided spiral steps observed on {001} face of phlogopite [10].

Another example showing the formation of many tiny crystallites in a small cavity, which move around and agglutinate, is found in hematite occurring in a similar way. Figure 12.13 is a surface microtopograph of the {0001} face of such a hematite crystal, and it indicates numerous hexagonal island-like portions on the terraces of step patterns. Around these islands, the step patterns of the host crystal are disturbed. This observation indicates that, as for phlogopite crystals, a large number of crystals with varying sizes are formed within a cavity, within which they move around and agglutinate.

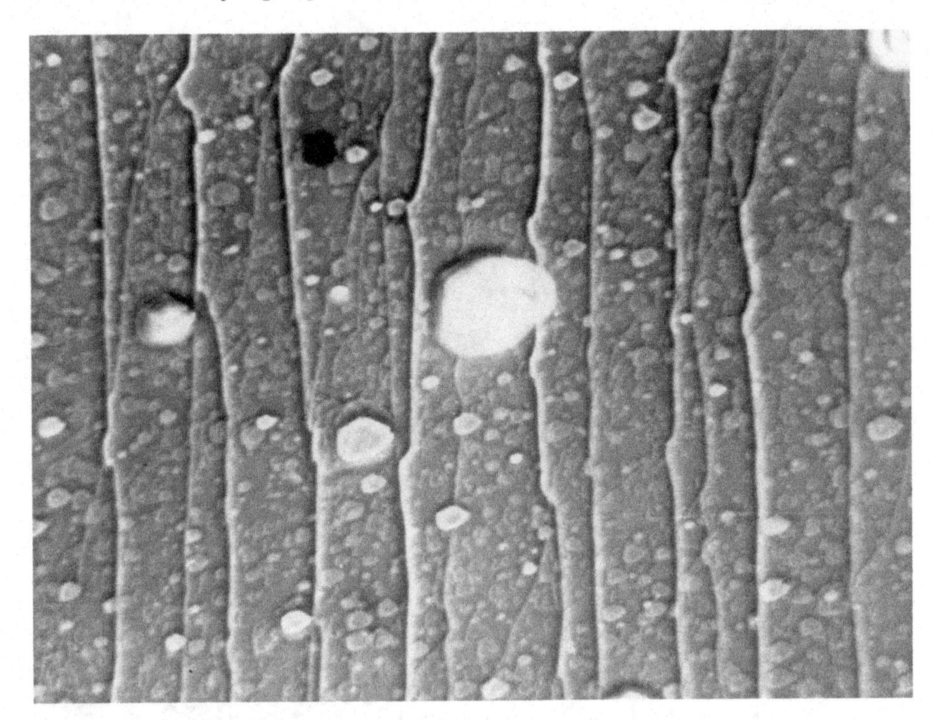

Figure 12.13. Surface microtopograph of hematite from a druse of ryolite. A large number of minute hematite crystals adhere and affect the advancement of the spiral growth layers on the host crystal.

References

1 I. Sunagawa and A. Urano, Beryl crystals from pegmatites: morphology and mechanism of crystal growth, *J. Gemmol.*, **26**, 1999, 521–33

2 M. D. Barton, Phase equilibria and thermodynamic properties of minerals in the BeO-Al_2O_3-SiO_2-H_2O system with petrologic applications, *Am. Min.*, **71**, 1986, 277–300

3 I. Sunagawa, Growth history of hematite, *Am. Min.*, **45**, 1959, 566–75

4 I. Sunagawa, Mechanism of crystal growth, etching and twin formation of hematite, *Mineral. J.*, **3**, 1960, 50–89

5 I. Sunagawa, Mechanism of growth of hematite, *Am. Min.*, **47**, 1962, 1138–55

6 I. Sunagawa, Mechanism of natural etching of hematite crystals, *Am. Min.*, **47**, 1962, 1332–45

7 Lu Taijing and I. Sunagawa, Origin of undulated growth steps on hematite crystals from Sasazawa, Japan, *Mineral. J.*, **13**, 1987, 409–23

8 A. F. Seager and I. Sunagawa, Movement of screw dislocations in hematite, *Min. Mag.*, **33**, 1962, 1–8

9 I. Sunagawa, Growth and etch features of hematite crystals from the Azores Islands, Portugal, *Surv. Geol. Portugal, Mem.*, **6**, 1960, 1–44

10 I. Sunagawa, Growth spirals on phlogopite crystals, *Am. Min*, **49**, 1964, 1427–34

11 I. Sunagawa and S. Tomura, Twinning in phlogopite, *Am. Min.*, **61**, 1976, 939–43

13

Crystals formed by metasomatism and metamorphism

When a hydrothermal solution permeates into a pre-existing rock, or heat and vapor components are supplied by the intrusion of magma, hydrothermal or contact metasomatism occurs, and mineral phases are newly formed. When deposited strata are placed at higher temperature and pressure conditions due to subduction owing to plate movement, metamorphism proceeds in a huge area, leading to the paragenesis of new minerals and regional metamorphic rocks. In this chapter, we shall analyze which aspects of crystal growth taking place in solid rocks are the same as or differ from those occurring in the ordinary vapor or solution phases. As representative examples, the kaolin group of minerals and trapiche emerald are selected for hydrothermal metasomatism; trapiche ruby is considered for contact metasomatism; and white mica is selected for regional metamorphism.

13.1 Kaolin group minerals formed by hydrothermal replacement (metasomatism)

Kaolin group minerals are sheet silicates, and they belong to a group of clay minerals, with 1:1 stacking of the unit layer with a chemical composition of $Al_2Si_2O_5(OH)_2$. Depending on the type of stacking, three polytypes, kaolinite, dickite, and nacrite, are distinguished. In addition, halloysite, containing (OH) and H_2O, belongs to this group, but this will not be discussed in this chapter.

Kaolin group minerals have been used since ancient times as raw materials for china and porcelain production. They occur by the weathering of igneous rocks, such as granite, that contain a quantity of feldspar, by hydrothermal replacement (metasomatism) of pre-existing rocks, or in hydrothermal veins. Both in weathering and hydrothermal metasomatic processes, kaolin group minerals grow from minerals constituting solid rocks, e.g. feldspar. How these processes proceed is analyzed based on the observation of surface microtopographs of these crystals. Crystals of kaolin group minerals are 0.5 ~ 1.0 μm in size, and their surface microtopographs may only be investigated using electron microscopy. If gold evaporation is applied in a vacuum on well dispersed samples, Au grains selectively nucleate along the steps; thus, after carbon coating and dissolving samples, it is possible to view replica samples showing the surface microtopograph under an electron microscope. This method is called the decoration method.

Step patterns due to elemental spiral growth steps, with a step height of 1 nm and a step separation of $10^2 \sim 10^3$ nm order, are universally observed on the surface of the {001} faces of kaolinite, dickite, and nacrite.

Commonly, one crystal face is covered by steps originating from one spiral center, and it is rare to see a case with a large number of growth centers on one surface. Interlaced step patterns (see Section 5.6) due to polytypes of kaolinite, dickite, and nacrite are discernible. This observation indicates that kaolin group minerals grow by the spiral growth mechanism in a dilute aqueous solution phase formed by the dissolution of pre-existing solid phases in hydrothermal metasomatism, and that the crystals grow in an environment in which growing crystals cannot move freely. Namely, hydrothermal metasomatism proceeds through the process of dissolution–precipitation and not through a direct phase transition from the solid state, and the dissolution front proceeds gradually in pre-existing rock, behind which crystallization of kaolin group minerals occurs. Two examples of surface microtopographs of kaolinite formed in this manner are shown in Fig. 13.1 [1].

In contrast to the above, aggregates of a few crystals of kaolinite, sericite (muscovite of micrometer size), or chlorite (a clay mineral) occurring in vein-type deposits are more frequently encountered than in hydrothermal metasomatic deposits. On individual crystal surfaces, growth spirals are observable, but one characteristic of this mode of occurrence is the aggregation of a number of crystals (Fig. 13.2), which suggests that the crystals moved freely during growth and grew in a moving environment [2].

13.2 Trapiche emerald and trapiche ruby

In the Muzo and Chivor mines, in Colombia, calcite veins, in which emerald crystals occur, develop, and they fill the cracks of sedimentary slate rock. Compared with emeralds occurring in basic metamorphic rocks, like those in the Urals, South Africa, or India, Colombian emerald crystals have a higher perfection, fewer inclusions, and attract higher evaluations as gemstones, since crystals grew freely in a hydrothermal solution.

In the Muzo and Chivor mines, emerald single crystals with a particular texture

Figure 13.1. Two examples of surface microtopographs of kaolinite using the decoration method [1].

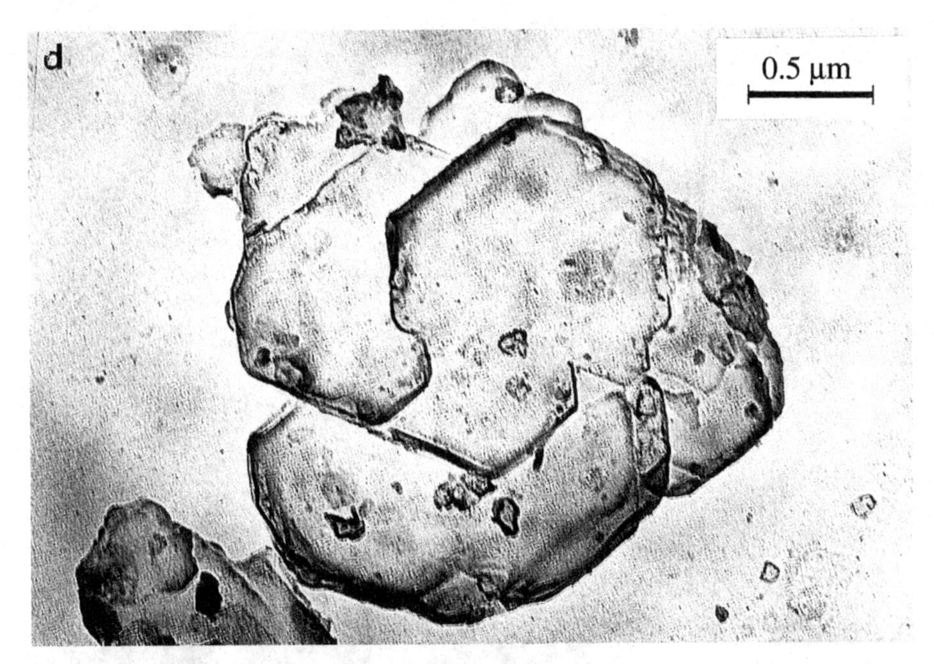

Figure 13.2. Characteristics of chlorite crystals in hydrothermal veins using the decoration method [3].

occur in sedimentary rocks around calcite veins. As schematically shown in Fig. 13.3(a), these crystals are hexagonal prismatic consisting of a hexagonal core and six growth sectors (arms) of trapezoid form. The boundaries between the hexagonal core and the trapezoid sectors, and those of the trapezoid, i.e. boundaries in six directions starting from the corners (edges) of the core hexagon, consist of two minerals, emerald and albite. Since the texture appears to resemble the gears (*trapiche* in Spanish) of a sugar cane compressor, they are named "trapiche emeralds" [4].

Ruby crystals with a similar texture are found at a contact metasomatic deposit in Myanmar, and are called trapiche rubies. Sapphires with similar textures are also known. Trapiche ruby crystals occur sporadically in marble formed by contact metasomatism due to the intrusion of a granitic magma into Mg-containing limestone. Trapiche sapphire is also formed by the contact metasomatism of Al-rich sedimentary rocks.

Since the terms used to describe the parts of the texture are different for trapiche emerald and trapiche ruby, the crystals are compared in Figs. 13.3(a) and (b). A photograph of trapiche ruby is shown in Fig. 13.3(c). The shared characteristics of the texture in both emerald and ruby are: the presence of a clear hexagonal or tapered-prismatic core portion; the dendritic portion starting from and stretching in six directions from the edges of the core prism; a similar dendritic portion

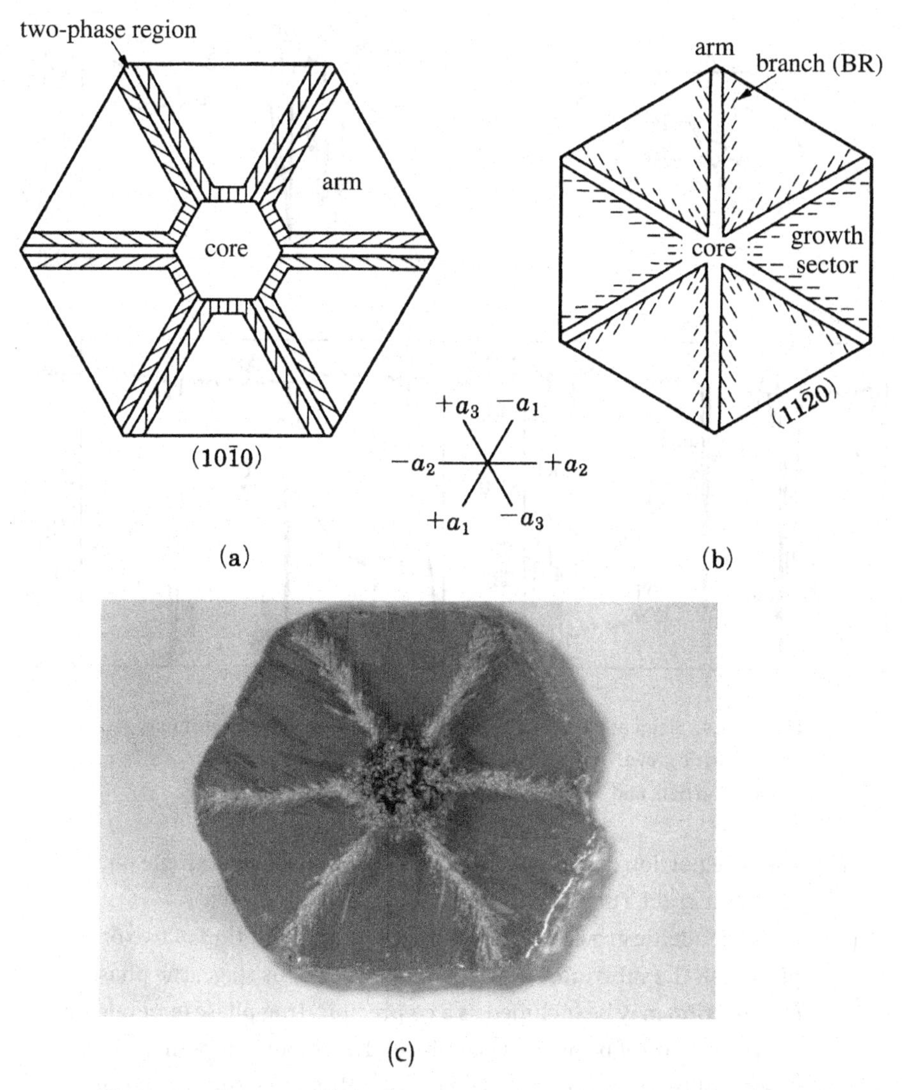

Figure 13.3. Textures of (a) trapiche emerald and (b) trapiche ruby [3]. (c) Photograph of trapiche ruby, taken by K. Schmetzer.

covering the surface of the core; and six clear trapezoid growth sectors surrounding the core and filling up dendrite interstices. In trapiche emerald, the dendritic portion consists of beryl and albite; in trapiche ruby, it consists of corundum, calcite, and a silicate mineral; and in trapiche sapphire, albite and rutile are present, without corundum [3], [4]. However, the core and the growth sector portions consist, respectively, of emerald, ruby, or sapphire, containing no other phases. From these textures, we understand that the growth process of these crystals should be as follows.

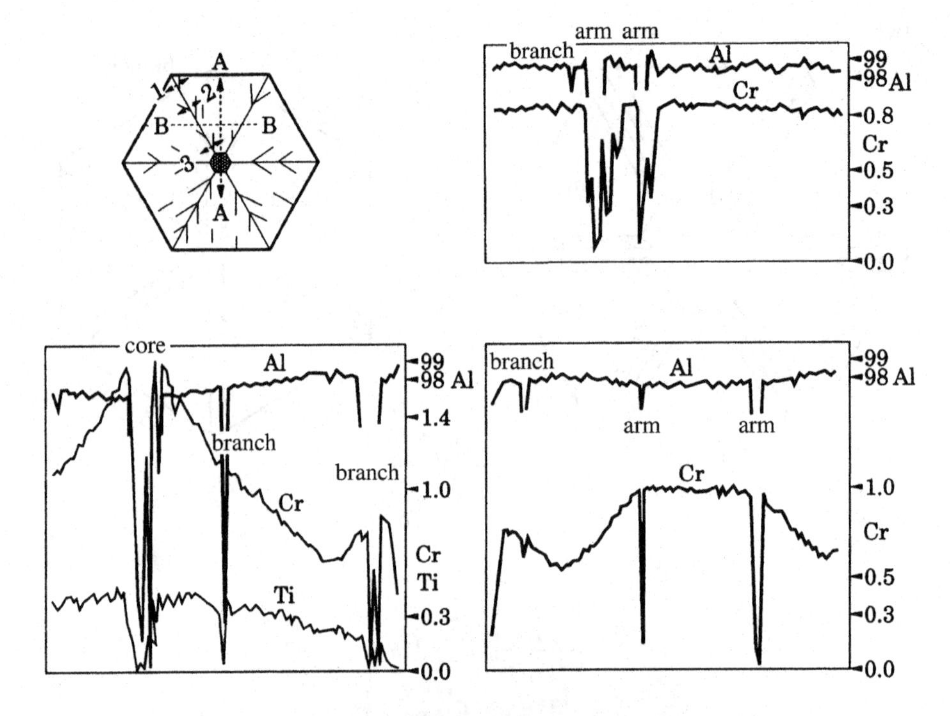

Figure 13.4. Element partitioning in trapiche ruby. Analyses are by micro-area XRF. The dendritic growth portions are denoted by "arm" and "branch"; "core" represents the core portion; and A, B, 1, 2, and 3 denote scanned lines [3].

(1) The core portion consists of an idiomorphic crystal of a single phase, either emerald, ruby, or sapphire.

(2) Rapid dendritic growth occurs through co-precipitation of two or more phases on the substrate of the core portion. In this stage, the phase of the core portion may be included as a co-precipitating phase (emerald or ruby), or may not be present (sapphire). The important point is that the growth of two or more phases is controlled by the structure of the core portion, and the dendritic growth proceeds very rapidly, so determining the size and the skeleton of the present crystal.

(3) The portions corresponding to the growth sectors are formed by filling up the interstices of the earlier-formed skeleton. The growth in this stage is single phase and slow.

The above process is well recorded in element partitioning in trapiche ruby. Figure 13.4 shows the summarized results of micro-area XRF (X-ray fluorescence spectroscopic analysis) [3]. The partitioning of Cr, Fe, and Ti in corundum in the core portion, the dendritic portion, and the growth sectors is indicated. The following points should be noted.

(1) The content of Cr in corundum in the dendritic portion is lower than that in both the core portion and the growth sectors, and is nearly constant from the root to the tip of the dendrite.
(2) Within the growth sector portion, the variation in the content of Cr and Ti is noticed alongside the external form, and the content is highest at the boundary with the dendritic portion, namely where the growth mechanism transforms from two-phase co-precipitation to the growth of a single phase, decreasing as growth proceeds, and gradually diminishing as a whole, although there is a slight fluctuation. At the final stage the contents increase slightly, and again decrease. The Ti content also shows a similar decreasing tendency as growth proceeds [3].

The characteristics of the element partitioning indicate that the dendritic portion was formed rapidly and over a short period; this element partitioning is principally controlled by the thermodynamics, i.e. the temperature and pressure, whereas the growth of the core portion and the growth sectors are by the layer growth mechanism, and the element partitioning is principally controlled by the kinetics.

When a crystal is formed and becomes large in the mother sedimentary rocks metamorphosed, by contact metasomatism, crystals with similar textures are often formed. Chiastolite, a form of andalusite (Al_2SiO_5) with a particular texture, may be such an example.

13.3 Muscovite formed by regional metamorphism

Muscovite, $KAl_2(AlSi_3O_{10})(OH)_2$, is a rock-forming mineral formed under a wide range of conditions. In granite, particularly in pegmatite, it occurs as a large idiomorphic crystal, and takes platy to tabular *Habitus* with well developed {001}. Step patterns of growth spirals are observed only on {001} surfaces, whereas prism faces are characterized by striations. Muscovite also occurs as hexagonal or lath-shape thin platy crystals of micrometer size in hydrothermal veins or hydrothermal metasomatic deposits. Muscovite crystals of this size are generally called sericite. On their {001} surfaces, typical growth spiral step patterns are revealed by the decoration method. In contrast, muscovite crystals grown from the supercritical vapor phase or hydrothermal solutions typically show growth spirals. We will investigate in this section the characteristics exhibited by muscovite crystals formed by regional metamorphism.

Regional metamorphism proceeds in a wide area, and characteristic mineral paragenesis appears in zones depending on the temperature and pressure conditions that were confronted by the original rock. These are called metamorphic zones (belts), and it is possible to distinguish between metamorphic facies, in successive zones, based on the grade of metamorphism, through geological and petrological investigations. Muscovite occurs as a major constituent mineral in all metamorphic facies, but the crystal size increases as the metamorphic grade progresses. In the zone with the lowest metamorphic grade, the size is of micrometer order, corresponding to sericite, but as the metamorphic grade increases, the size increases and the crystals are called white mica. On further increasing the metamorphic grade, the crystals become up to millimeter order across, and are called porphyroblastic (porphyritic large crystals formed by metamorphism) white mica.

Muscovite found in the metamorphic belt in Shikoku, Japan, whose metamorphic facies are well established by geological and petrological investigations, was investigated by the decoration method. The results indicated that spiral growth step patterns were not observed on these crystals, in contrast to the commonly observed spirals on muscovite crystals grown in pegmatite and by hydrothermal metasomatism. Instead of spiral patterns, the surface microtopographs of these muscovite crystals are characterized by closed-loop step patterns [5]. Also noticeable is the coexistence of sawtooth steps due to etching and smooth step patterns (Fig. 13.5); step patterns of this sort are most frequently encountered near the boundary of successive metamorphic facies.

The neighboring metamorphic facies correspond to the boundary from which the paragenesis of new minerals takes place; dehydration, which is associated with this reaction, is seen to occur when crystals of different sizes coexist under the same driving force conditions, the smaller crystals dissolve due to the effect of the surface energy, and the dissolved component is supplied to the neighboring larger crystals, which results in the larger crystals continuing to grow larger at the expense of smaller crystals.

This phenomenon, namely dissolution and growth occurring simultaneously under the same condition simply due to the size difference of crystals, with the larger crystals growing even larger, is called Ostwald ripening (see ref. [5], Chapter 3). This is one of the empirical rules which, together with Ostwald's step rule described in Section 3.4, is qualitative, yet pin-points the essentials of crystal growth. The Ostwald ripening rule has been observed in various crystal growth taking place on Earth, as well as in the synthesis of large single crystals. The origin of single crystals of NaCl with edge length attaining tens of centimeters occurring in salt deposits, or of a large single crystal of ice in the North and South Polar regions, may also be due to Ostwald ripening. As the metamorphic grade in regional metamorphic rocks is increased, the size of the muscovite increases, and the observed changes from sericite to white mica to porphyroblastic muscovite may also be due to Ostwald ripening [5]. The observation of the coexistence of step patterns due to growth and dissolution also support this argument.

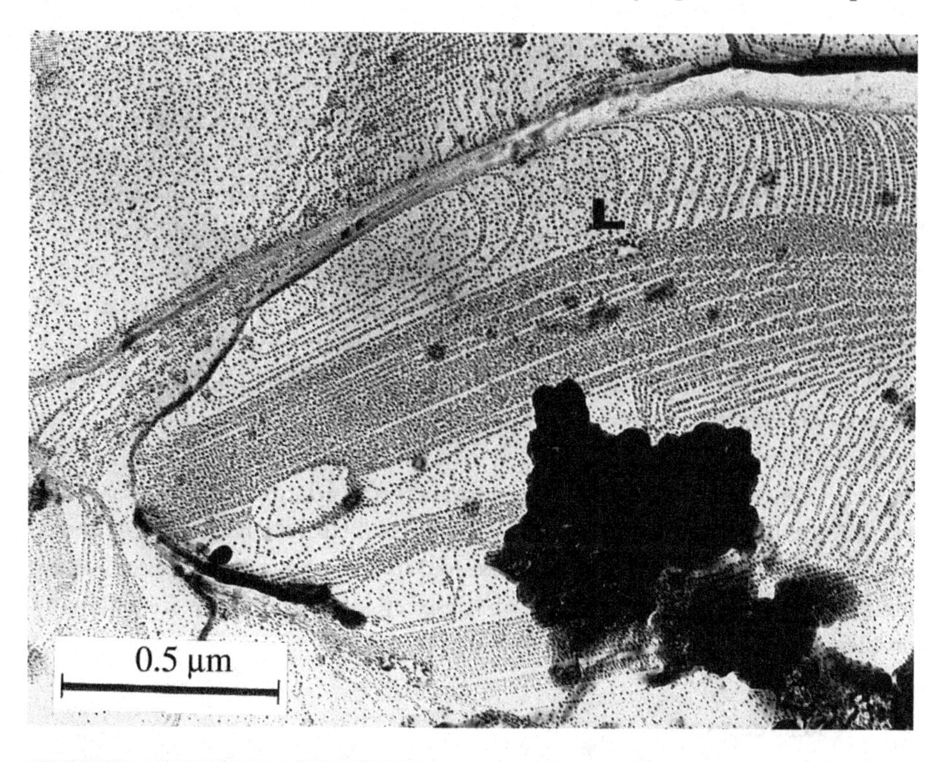

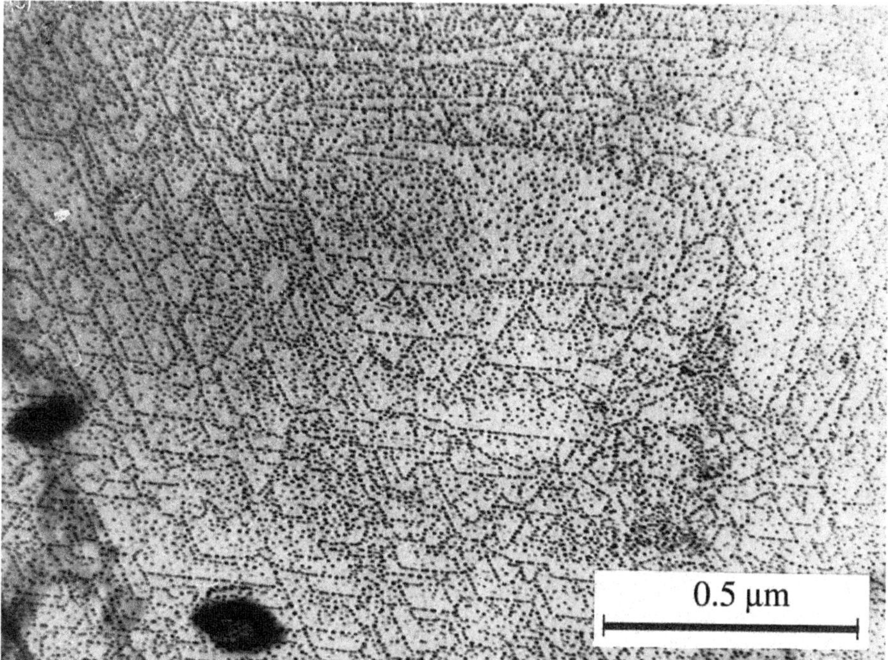

Figure 13.5. White mica in a regional metamorphic rock using the decoration method [5].

Compared with this, sericite (muscovite) crystals formed in hydrothermal solution and hydrothermal metasomatism exclusively show typical spiral step patterns; the {001} faces of muscovite in regional metamorphic rocks do not show spiral step patterns, but are characterized by step patterns of closed loop. The closed loop is not due to screw dislocations. The most likely step sources are contact points with other solid particles. We may understand this step pattern if we assume that the growth of muscovite takes place in thin interstices of other mineral grains. Observations of muscovite indicate that, even for crystal growth taking place in regional metamorphic rocks, the crystal growth is not a simple solid state crystallization but that growth from the solution phase plays an essential role. Similar step patterns are universally observed on crystals of a wide range of species formed by a sintering process with a solvent component.

Large idiomorphic porphyroblast of garnet crystals attaining sizes of more than 10 cm across occur in regional metamorphic rocks. Textures suggesting rotation of the crystals during growth are often observed in this type of garnet crystal. Many investigations have been carried out on these crystals [6], but there are still many problems left to be solved.

References

1 I. Sunagawa and Y. Koshino, Growth spirals on kaolin group minerals, *Am. Min.*, **60**, 1975, 407–12

2 I. Sunagawa, Y. Koshino, M. Asakura, and T. Yamamoto, Growth mechanism of some clay minerals, *Fortschr. Min.*, **52**, 1975, 515–20

3 I. Sunagawa, H-J. Berhardt, and K. Schmetzer, Texture formation and element partitioning in trapiche ruby, *J. Crystal Growth*, **206**, 1999, 322–30

4 K. Nassau and K. A. Jackson, Trapiche emeralds from Chivor and Muzo, Colombia, *Am. Min.*, **55**, 1970, 416–27

5 S. Tomura, M. Kitamura, and I. Sunagawa, Surface microtopography of metamorphic white micas, *Phys. Chem. Min.*, **5**, 1979, 65–81

6 T. Masuda, To rotate or not to rotate, this is the question; controversial opinions on the origin of snowball structures in garnet porphyroblasts, *J. Geograph.*, **108**, 1999, 94–109 (in Japanese)

14

Crystals formed through biological activity

Various inorganic and organic crystals are formed in living bodies through the biological activity in animals and plants. Some are indispensable, such as: hydroxyapatite, which constitutes teeth and bones; aragonite and calcite, which are the main constituents of shells and exo-skeletons; and those formed through the excretion of components that are either unnecessary or due to disease. The examples given respectively show the characteristic morphology of crystals and the textures of polycrystalline aggregates. In this chapter, we will summarize and discuss which features crystal growth in inorganic systems and in living bodies have in common and which are different. To do this we will consider the morphology of the crystals.

14.1 Crystal growth in living bodies

Many inorganic and organic crystals grow in living bodies and act as indispensable major components in the function of cells and organs. Teeth and bones consist of hydroxyapatite crystals, and shells, pearls, and the exo-skeletons of coral or coccolithophores are mainly composed of carbonate crystals, such as aragonite and calcite. There are magnetite crystals of single magnetic domain in size aligned in a rosary form in cells of magnetotactic bacterium, which acts as a direction sensor. Similar magnetite rosaries are found in the brain cells of pigeons, dolphins, and salmon, and it is suggested that they act as sensors for homing, wandering, or recurring instincts.

Crystals (or amorphous substances) are also formed in living bodies as a result of biological activities to act as a reservoir for necessary components (for example, amorphous silica in grass cells, and calcium oxalate or inuline crystals in dahlias and begonias), or crystals are formed by excretion processes or due to illness (for

Table 14.1 *Crystals formed by biomineralization*

[I] Bone, exo-skeleton, shell, carapace, spicule, etc.
Calcite: coccolith, foraminifera, mollusks
Mg-containing calcite: octocorals, spicules of sea urchin
Aragonite: corals, mollusks, gastropods, cephalopods, fish
Vaterite: gastropods, sea squirt
Amorphous $CaCO_3$: crustaceans, birds' eggs, plants
Hydroxyapatite: vertebrates, fish
Octacalcium phosphate: vertebrates
Celestite: acantharia
Amorphous SiO_2: diatoms, radiolarians

[II] Tooth
Hydroxyapatite: vertebrates, mammals, fish
Octacalcium phosphate: vertebrates
Amorphous SiO_2: limpets, chitons
Magnetite: chitons
Goethite: limpets
Phosphoferrite: chitons

[III] Sensors or receptors for orientation or gravity
Magnetite: magnetotactic bacteria, tuna, salmon, pigeons
Calcite: mammals, fish
Gypsum: jellyfish
Barite: charr

[IV] Storage
Whewellite $CaC_2O_4 \cdot H_2O$: various plants
Weddellite $CaC_2O_4 \cdot 2H_2O$: various plants
Amorphous silica: plants, grasses

[V] Excretion
Sodium uric acid: gout
Cholesterol: gallstones

[VI] Function not clarified[a]
Sulfides (pyrite, sphalerite, wurtzite, galena, greigite, mackinawite)
Oxides (magnetite, todorokite)
Carbonates (calcite, hydroxycalcite, siderite, aragonite)
Sulfates (barite, gypsum)
Phosphates (jarosite)

[a] All minerals in [VI] are found as selectively precipitated minerals on the surface of bacteria cells

example, calculus in various organs, or sodium uric acid in the case of gout). There remain some situations in which the reasons for crystal formation or function are still not well understood (for example, bacteria around which inorganic crystals precipitate selectively). Examples of these are summarized in Table 14.1.

There is no essential difference between crystal growth in living bodies and that of inorganic crystals in aqueous solution. In both cases, conditions in driving force higher than a critical value, such as a supersaturated state, should be achieved first, and then growth proceeds in the processes of nucleation and growth, through which the morphology of the crystals and their textures are determined. Since the textures formed in this manner play an essential role in organ function, the morphology and textures of the crystals are assumed to have been controlled purposely through biological activity. We may summarize the characteristic points of crystal growth in living bodies as follows.

(1) The environmental phase is limited, for example in cells or organs. In other words, this situation is comparable to crystal growth in a partly closed vessel.
(2) There are cases in which proteins or saccharoids, or organic sheets (protein foil) made up of these substances, play a cooperative role in crystal growth, and cases in which there is no relation. We may evaluate the degree of cooperation by evaluation of a coherency or misfit ratio between the crystal and the protein, or by the degree of epitaxial relation. The role of the protein may be variable. In crystal growth that is indispensable for biological activity, the role of the protein will be definitive, but for crystals formed in excretion processes, this cooperation will either not occur or will be very weak.
(3) All crystal growth takes place in low-temperature, low-pressure aqueous solution (at 1 atmospheric pressure and room temperature). This suggests a higher probability of formation of an amorphous state, phases of low crystallinity, and metastable phases as precursors, and therefore subsequent transformation to stable or metastable phases.
(4) Some crystals may remain in the cells, which act as vessels, and form textures consisting of crystals and protein films; in other cases, crystals protrude from the cell, where they may connect with others formed in neighboring cells or discharge from the cell and connect with other crystals to form higher-order textures, such as an exo-skeleton.

It is important to understand how these characteristics affect the morphology of the crystals and the textures of their aggregations. These specific characteristics allow crystals formed due to biological activity to exist as amorphous states or in metastable phases, and they bring about different characteristics in the morphology and textures of crystals from those present in uncontrolled inorganic systems.

The same crystal species may be formed either as an indispensable component for biological activity, or as an unnecessary (subsequently excreted) product; but the two may be different in their morphological features. It is the purpose of this chapter to investigate systematically crystals formed through biological activity, based on the morphology and texture exhibited by these crystals. It is intended to classify crystals formed by biological activity based on the fundamental images of crystal growth and morphology discussed in Part I. Therefore, observations on actual examples are based on the results achieved so far in the field of biomineralization. For details, readers should refer to refs. [1]–[6]. We will classify the crystal growth of indispensable solids for biological activity under the control of proteins as "normal" growth and that formed by excretion processes of unnecessary compounds as "abnormal" growth, and we shall discuss the characteristics of the morphology and textures shown by respective crystal species. The growth of inorganic crystals constituting teeth, bones, shells, and exo-skeletons has been investigated over many years as calcification or biomineralization [1], [2], which constitute a scientific discipline now called "biomineralization." International conferences have been regularly organized on this topic [3]. Owing to the increasing demand in recent years for protein single crystals suitable in size and perfection for structural analysis, investigations have been extended to include the study of single crystal synthesis of protein crystals, a subject known as "biocrystallization." At present, no satisfactorily close cooperation between the two disciplines has been achieved; however, such cooperation is expected in the future. In contrast, investigations on crystals formed by abnormal growth have been mainly centered on description and classification, but a new understanding has recently been achieved for gallstones from the standpoint of crystal growth [7].

14.2 Inorganic crystals formed as indispensable components in biological activity

As summarized in Table 14.1, teeth, bones, shells, etc. are indispensable components, consisting of inorganic mineral crystals and protein film, with sizes, morphologies, and textures suitable to fulfil the function of the particular organs involved. In this section we will look at hydroxyapatite, aragonite and calcite (two polymorphs of $CaCO_3$), and magnetite in greater detail.

14.2.1 Hydroxyapatite

The constructions of the teeth of an adult human and herbivorous animals are illustrated in Fig. 14.1. Both types of teeth consist of an enamel portion, a dentine portion, and a cementum portion [4]. The enamel portion consists of 97% hydroxyapatite, 2% protein, and a small amount of carbonate; the dentine consists of 30% hydroxyapatite, and 70% protein and water; and the cementum contains an inorganic part below 30%, around 10–20%. The enamel is compact and hard, and assists in the cutting and grinding of food, whereas the dentine and cement portions are softer, are more permeable, and support the enamel region. The propor-

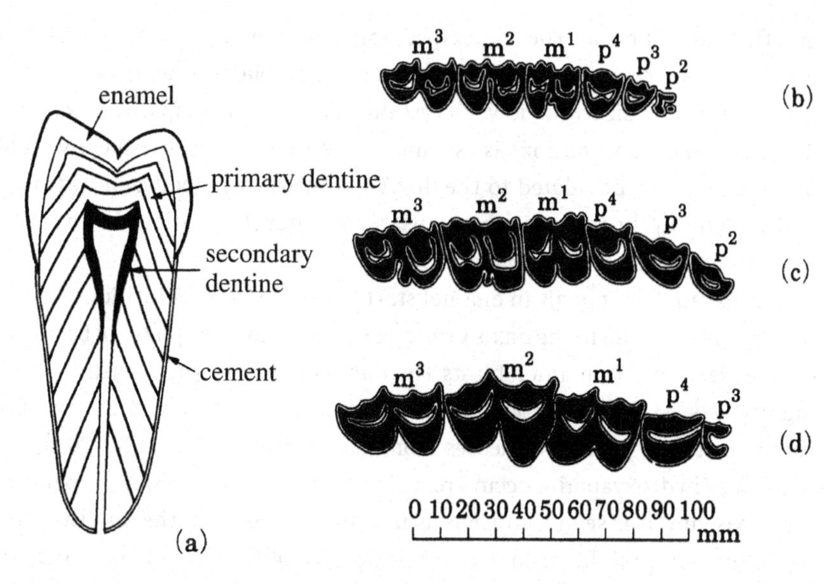

Figure 14.1. (a) Tooth of healthy adult human. (b)–(d) The teeth of herbivorous animals: (b) oryx; (c) bison; (d) camel. Taken from ref. [4].

tion of the respective components is nearly the same for all mammals; the difference in textures results in different functions. As can be seen in Fig. 14.1, the differences seen in different species appear at the higher-order textural level; this is due to the difference in eating habits between the species.

Although the main components of teeth are protein and hydroxyapatite, $Ca_{10}(PO_4)_6(OH)_2$, the protein species and the size, form, and crystallinity of the hydroxyapatite are different in the enamel, dentine, and cement portions. The enamel part consists of proteins of amelogenines, enamelins, etc., and the dentine portion comprises 90% collagen and 10% non-collagen protein, and the sizes of the respective cells are different. They are called ameloblasts (enamel cells), odontblasts (dentinal cells), and cement cells, respectively. The size and form of hydroxyapatite crystals in the enamel part of a tooth of an adult human (and also in the teeth of other mammals) are 70 nm in thickness, $40 \sim 50\,\mu m$ in length (although sometimes they attain $100\,\mu m$), and are of a flattened hexagonal prismatic form elongated along the *c*-axis; the texture consists of crystals aligned along the *c*-axis and interstitial protein. This is the reason for the extreme hardness of the enamel portion. In contrast, the hydroxyapatite comprising the dentine is 50 nm in thickness and 30 nm in length, much smaller than that in enamel; the form is irregular short prismatic, and it is poorer in crystallinity and alignment than in enamel [4]. Apatite crystals formed in an inorganic environment usually take hexagonal thick tabular or prismatic *Habitus*. In the fact that the crystals are bounded by $\{0001\}$ and $\{10\bar{1}0\}$, they are similar to the crystals in the enamel portion, but they differ greatly in aspect ratios. The large aspect ratio corresponding to the enamel portion occurs because crystal growth takes place in ameloblasts (enamel cells). The reason for the smaller size and lower crystallinity of hydroxyapatite crystals in the dentine and cement potions is assumed to be related to the respective cells. This difference might be related to the degree of misfit ratio between hydroxyapatite and proteins in the respective portions: the smaller the misfit ratio, the denser the well oriented texture is expected to be.

Hydroxyapatite crystals in enamel start to grow as very thin, elongated, ribbon-like crystals parallel to the *c*-axis in a "cheese-like" ambient phase of base protein of enamel secreted from ameloblasts The *c*-axes are well aligned with the neighboring crystals. As growth proceeds, the thickness increases, and the ratio of inorganic crystal to protein increases from the starting value of 30:70 up to 97:3. Growth of hydroxyapatite occurs principally to increase the size perpendicular to the *c*-axis, and it is seen that there is not much change in the length. The length along the *c*-axis might be controlled by the size of the ameloblasts. Although the dentine portion consists of much smaller crystals elongated along the *c*-axis, the orientations of the neighboring crystals are less regular than in the enamel portion [4].

14.2.2 *Polymorphic minerals of $CaCO_3$*

There are two stable polymorphs of $CaCO_3$: a low-pressure, high-temperature polymorph, calcite, which belongs to the rhombohedral system $\bar{3}m$; and a high-pressure, low-temperature polymorph, aragonite, which belongs to the orthorhombic system mmm. There is also a metastable phase called vaterite, which belongs to the hexagonal system. Amorphous $CaCO_3$ is also known to exist.

These polymorphic minerals play an important role in protecting the soft organs in a wide range of living things; examples include the shells of mollusks, the exo-skeletons of coccoliths, foraminifera, and coral, the spicules of sea urchins, the scales of fish, the carapaces of crustaceans, and eggshells. The most interesting point is the fact that aragonite, which is in a thermodynamically stable phase above 5000 Pa, is formed under atmospheric pressure as if it were a stable phase when it is formed as a result of biological activity. However, when a living body dies and fossilizes, it is seen that metastable aragonite transforms to the stable phase, calcite, over a timescale of about 1000 years. It is interesting to note that in the organs which require molting or breaking, the amorphous or metastable $CaCO_3$ phase is present, whereas for exo-skeletons, which require strength, such as in a shell or coccolith, the stable phase, calcite or aragonite, is present.

The growth mechanism, form, and texture of aragonite crystals in the shells of bivalves and snails or in pearls have been extensively investigated over many years in order to control the quality of cultured pearls [6].

The shells of bivalves consist of periostracum, consisting of hard protein tissue, and ostracum, consisting of $CaCO_3$, which is further divided into outer, middle, and inner shell layers. The shell structure of bivalves is classified into the following seven layers based on the textures and structures.

(1) Prismatic structure layer: the layer consisting of polygonal rhombic prismatic aragonite crystals perpendicular to the inner surface of the shell.
(2) Nacreous structure layer: a layer consisting of stacked platy aragonite crystals in brick-like layers and an interstitial protein layer.
(3) Foliated structure: a layer formed by the stacking of narrow elongated platy crystals of calcite, like a slate roof.
(4) Crossed lamellar structure: aragonite needle crystals align in the *c*-axis direction forming rectangular blocks. The *c*-axes are inclined at 45 ~ 55° to the inner surface of the shell, and the neighboring blocks are oppositely inclined.
(5) Complex crossed lamellar structure: a layer consisting of aragonite needle crystals with orientations inclined at 45° to the inner portion of the cell. These needle crystals aggregate in a radial manner, forming a conical body, which is arranged nearly perpendicularly to the inner surface of the shell.
(6) Composite prismatic structure: the *c*-axis of rhombic prisms align parallel to the shell. The rhombic prism is an aggregation of aragonite needle crystals aligned in conical form in its elongation direction.
(7) Homogeneous structure: consists of polycrystalline granular aggregate of aragonite crystals of varying sizes.

Representative structures are schematically shown in Fig. 14.2. Different species of bivalves have shells composed of aragonite alone, or just calcite, or both, and the compositional structures of the shell also vary. The sizes and forms of the aragonite or calcite crystals, and the resulting textures, are also different. However, it should be noted that crystals of the respective structures are uniform in size and form. In those cases in which the structure is composed of idiomorphic crystals or polycrystalline aggregates of sheaf or spherical forms, the sizes and arrangement are regular. This indicates that the proteins forming the respective structure are different, and that inorganic crystals grow under their control. We shall discuss how crystallization proceeds and how the morphology and textures are controlled, using as an example the nacreous layers present in both shell and pearl, which has been most extensively investigated [6].

In the formation of pearls and the nacreous layer in shells, an organic sheet is formed 0.2 ~ 0.3 μm away from and between the surfaces of the substrate pearl

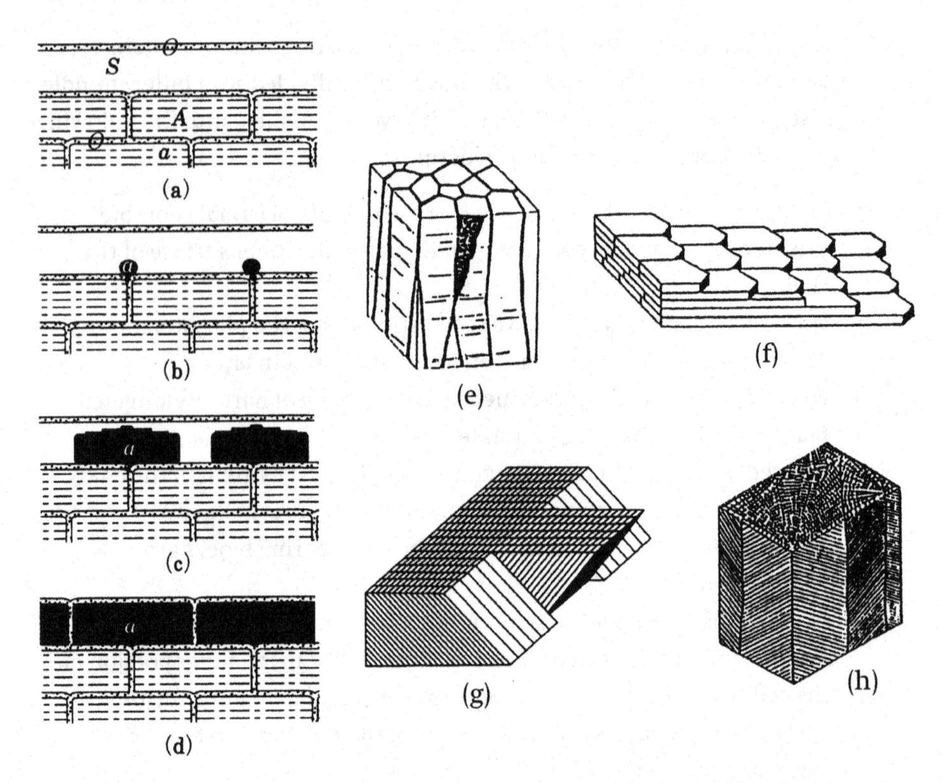

Figure 14.2. Structures forming the shells of bivalves [6]. Parts (a)–(d) indicate how the nacreous layer of pearl is formed. *A* = platy aragonite crystals; *O* = organic sheet; *S* = solution. (e) Prismatic structure; (f) foliated structure; (g) crossed lamellar structure; (h) complex crossed lamellar structure.

layer and the pearl sac epithelium, and this space acts as a compartment in which nucleation and growth of aragonite crystals can occur. Nucleation occurs either at the outcrops of aragonite or at a protrusion on the surface of the organic sheet. The thickness of the aragonite crystals is controlled by the thickness of the compartment, and the size is determined by the separation between neighboring crystals. As a result, aragonite crystals take platy form with uniform thickness, and the texture formed by aggregation resembles a stack of bricks. The {001} face grows large on the platy crystals, and takes either a rhombic form bounded by {110} or a hexagonal form bounded by {100} and {010}. The organic sheet is sandwiched between neighboring crystals. Since the organic sheet covering the surface of a pearl is porous, nucleation and growth are highly likely to be controlled by the substrate crystal that is outcropped there, but the important point is that the misfit ratio between the organic sheet and the aragonite crystal is small and that they are in an epitaxial relation. As a result, the *c*-axes of the aragonite crystals are all aligned perpendicularly to the pearl surface. The *a*- and *b*-axes align with the same

Figure 14.3. Step pattern observed on the surface of a pearl. Photographed by K. Wada [6].

orientation owing to the same reason, and have an epitaxial relation with the substrate. As a result, step patterns with heights corresponding to the thickness of individual aragonite crystals appear. The step patterns are observable under an optical microscope. In many cases, the step patterns exhibit closed loops, but spiral patterns are also observed (Fig. 14.3). This is due to the introduction of twist boundaries, owing to a slight misorientation between neighboring crystals. Spiral growth layers are also observed on the {001} face of individual aragonite crystals, and a spiral growth mechanism is suggested as the growth mechanism of individual aragonite crystals [6].

In the case of bivalves, only one organic sheet is formed at the front, whereas many sheets are formed in snail shells. As a result, unlike bivalves, snail shells grow as a tower-like pyramidal stacking structure, as illustrated in Fig. 14.4(b).

In the nacreous layer (aragonite) and the foliated structure (calcite), both the aragonite and calcite have uniform platy form. Orientations of crystals in prismatic structures and crossed lamellar structures are orderly, whereas in complex crossed lamellar, composite prismatic, and homogeneous structures, calcite and aragonite crystals form polycrystalline aggregates showing sheaf-like form composed of needle crystals or granular (spherulitic) structures. However, the sizes and alignment of sheaf-like and granular aggregates are regular, suggesting cooperation between the protein and the inorganic crystals.

Among the organs of living things consisting of calcite, the spicules of sea urchins and the exo-skeleton of coccolithophores attract particular interest. A sea

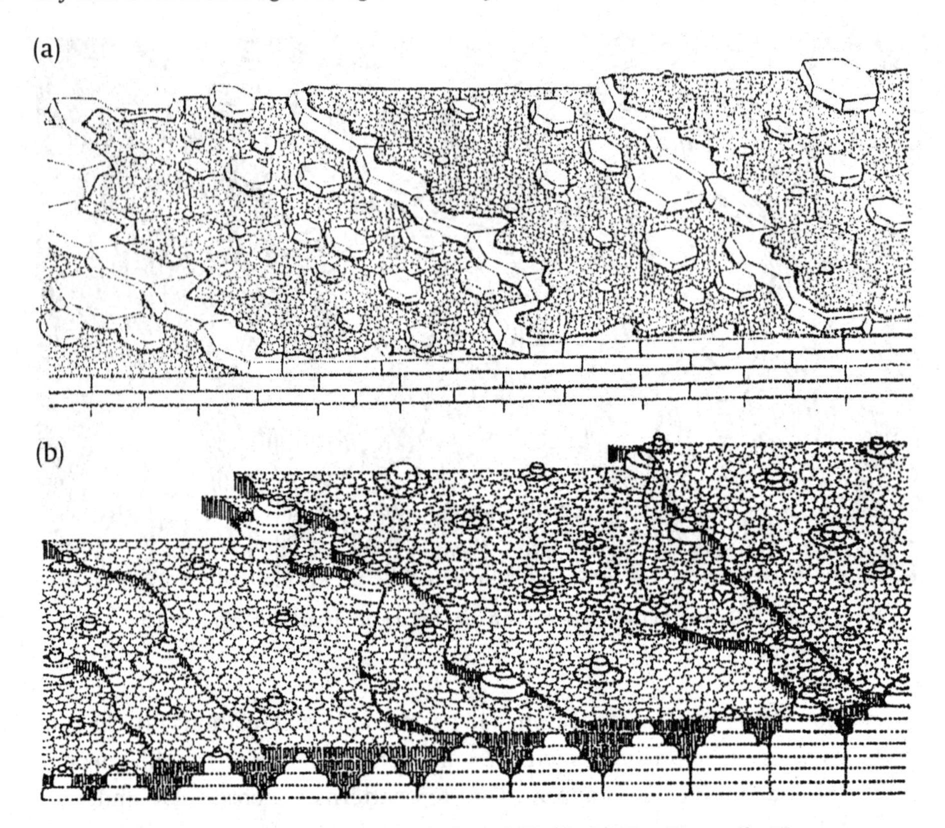

Figure 14.4. Formation of nacreous layer [6] in (a) a bivalve; (b) a snail.

urchin's spicule consists of many single crystals with scalenohedral forms bounded by $\{hk\bar{\imath}l\}$ faces, all aligned with a common *c*-axis. Although one spicule consists of a large number of single crystals, because they have a common crystallographic axis, the individual spicules can be regarded as single crystalline. However, at the root of a spicule, there is a radiating group of semi-spherical spherulites consisting of radially grown calcite crystals of dog-tooth (scalenohedral) morphology. Out of many crystals forming this semi-spherical aggregate, only one crystal becomes a spicule; the others seem to act as part of a bearing to move the central spicule.

Coccolith, an exo-skeleton of coccolithophores, consists of calcite crystals of uniform size showing a most unusual morphology, which resembles a trug. The structure of the exo-skeleton consists of about thirty calcite crystals of equal size, which are regularly aligned and conjugated. This unusual form of calcite crystals (shown in Fig. 14.5) has stimulated particular interest, and many studies have been conducted on this structure [8].

Calcite crystals with this unusual morphology are formed in the cell called the Golgi complex, and are expelled continuously from the cell to form a circular

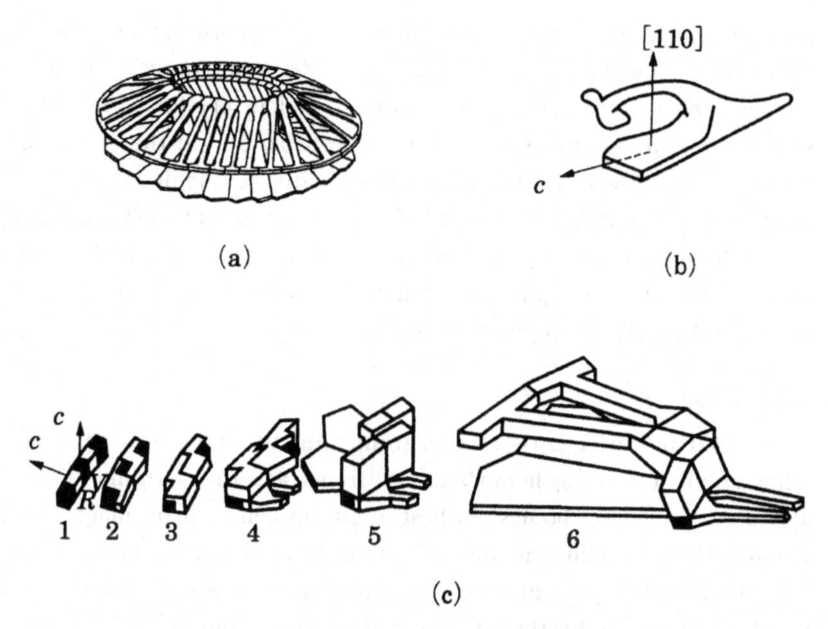

Figure 14.5. Crystal orientation and mechanism of formation of calcite constituting the segment of coccolithophores. *Emiliania huxleyi* is presented as an example [8]. (a) One coccolith (diameter approximately 4 μm) consists of approximately thirty segments. There exist two opinions as to the construction of calcite crystals. Westbroek *et al.* [10] and Didymus *et al.* [10] assume that all crystals of segmental calcites are single crystalline (b), whereas Young *et al.* [11] assume that calcite crystals consist of R crystals, whose *c*-axis is parallel to the cell surface, and V crystals, whose *c*-axis is perpendicular to the cell surface. In the latter model, only R crystals become large [11]. The growth process is illustrated in (c).

arrangement by conjugating outside the cell. The strength of this circular exoskeleton is due to the arrangement of trug forms. It is clear that the growth of calcite crystals with unusual trug forms is controlled by protein. In earlier investigations, calcite crystals showing this trug form were believed to be single crystalline, but recent research has shown that both R and V crystals were formed on the surface of the cell and that only the R crystals became larger. This unusual morphology and the relation between R and V crystals are factors that are clearly controlled by the Golgi cell. The structural form of calcite is rhombohedral $\{10\bar{1}1\}$, an F face, whereas $\{hk\bar{\imath}l\}$ faces are S faces. Therefore, an $\{hk\bar{\imath}l\}$ face cannot become as large as the controlling *Habitus*. However, natural calcite crystals show extremely variable *Habitus* (see Chapter 11), and particularly those crystals formed at lower temperatures tend to show scalenohedral (dog-tooth) *Habitus* bounded by $\{hk\bar{\imath}l\}$ faces. On $\{hk\bar{\imath}l\}$ faces, only striations parallel to the edges with $\{10\bar{1}1\}$ faces are observable, which implies that impurity adsorption on the steps of growth layers on $\{10\bar{1}1\}$ is the main reason why $\{hk\bar{\imath}l\}$ faces become large. There is a paper reporting that synthetically grown calcite crystals exclusively show rhombohedral *Habitus*, but, when an organic compound is added as an impurity, a scalenohedral *Habitus* is exclusively formed. The reason why calcite crystals forming sea urchin spicules have a scalenohedral *Habitus* is probably due to the cooperation with protein in their growth. This is in agreement with the spicule shape.

There is no doubt that the *Habitus* of calcite crystals forming the exo-skeleton of coccoliths is designed naturally to match the function of spicules, but further investigation may be required to explain the growth mechanism of this peculiar form and the relations between R and V crystals.

14.2.3 *Magnetite*

Magnetite (Fe_3O_4) is a spinel group mineral belonging to the cubic system m3m, and it shows a wide mode of occurrence as a rock-forming mineral, but it also occurs in living bodies. Earliest attention centered on magnetite crystals found in the cells of magnetotactic bacteria. Later, magnetite crystals were found in the brain cells of pigeons, dolphins, and salmon, suggesting that the crystals are possibly responsible for the homing instincts of pigeons and the behavior shown by fish that return to their spawning grounds. The adult teeth of chiton (marine mollusks) also consist of magnetite (the immature teeth are composed of goethite, α-FeOOH, and lepidocrosite, γ-FeOOH, which transform into magnetite as the chiton matures).

The sizes of magnetite crystals formed in the cells of magnetotactic bacteria, fish, and pigeons are 700 Å across, corresponding to one magnetic domain size, and they form a rosary chain. The alignment of individual crystals forming the chain is characteristic of biological magnetite. Individual magnetite crystals are idiomorphic and polyhedral crystals; skeletal or dendritic morphology, which represents crystal formation under a high driving force, has not been observed. However, the morphology of polyhedral crystals is markedly different from that of magnetite crystals formed as rock-forming minerals in inorganic growth environments. For magnetite crystals formed by biological activities, simple octahedral *Tracht* is seldom observed; they are mostly cubo-octahedral with {100} or cubo-dodecahedral with {110}, but rarely take one directionally malformed conical weight-like form. These crystals align to form a single rosary chain. Magnetite is a mineral belonging to the spinel group, and has a reverse spinel structure. The {111} face is the close-packed plane of oxygen, and the structural form should be simple octahedral bounded by {111} only, not only from the BF law and the DH expanded law, but also by HP PBC analysis. Magnetite crystals formed in an inorganic environment as a rock-forming mineral almost exclusively take this form. The {100} and {110} faces are encountered only exceptionally. There is a report that {100} faces occur only when crystals are smaller than 0.5 μm.

The observations that the sizes of magnetite crystals are uniform, corresponding to the size of a single magnetic domain when they grow as a result of biological activity, and that crystals connect to form a single rosary-like chain, indicate that the crystal growth of magnetite proceeds through cooperation with protein, and also that the presence of magnetite is indispensable for biological activity. The commonly observed *Habitus* and *Tracht* of magnetite, which deviate from the structural form deduced from the crystal structure only, also suggest that the growth of inorganic crystals proceeds under the control of protein. The size and form of the container (cell) appears to play a role in determining the size and form of magnetite crystals. In particular, the form of the cell is highly likely to determine conical-weight-like form.

14.3 Crystals formed through excretion processes

There is no doubt that calculus formed in organs, such as gallstones, kidney stones, and urethra stones, are not formed as indispensable factors in biological activities, unlike teeth, bones, and shells. Instead, these are products of illness, formed by precipitation in the respective organs during the process of excretion of components formed in unnecessary amounts. Clearly, cooperation with protein is not expected in their formation. A wide variety of crystals are known to exist in living bodies due to these processes.

Gout is caused by precipitation, at the knuckle of the big toe, of needle crystals of sodium uric acid, which are formed in blood corpuscles and transported via blood flow. When the crystals precipitate, a polycrystalline aggregate results from the intermingling of the needle crystals, around which radiating growth of the needles occurs, resulting in precipitation of polycrystalline aggregates with a close-cropped chestnut form. The size and form of needle crystals of sodium uric acid, and their spherical precipitates, are not uniformly controlled, and they change as the illness progresses, suggesting that their growth proceeds outside the control of protein.

In earlier times, when dietary habits were poor, gallstones consisted mainly of pigment precipitation, but, in recent years, they mostly consist of cholesterol crystals. Thin platy crystals of cholesterol are formed in the gallbladder, and coagulate to form spherical aggregates. The growth of thin platy crystals of cholesterol was confirmed to be by the spiral growth mechanism [7], [12]. Thin platy crystals formed in the gallbladder move and agglutinate by movement of the environmental field around something that acts as a heterogeneous nucleation site, so forming gallstones. Although hydroxyapatite was suggested as a possible material that could act as the nucleus, this has not yet been confirmed. The sizes and forms of single crystals of cholesterol are uneven. When a small gallstone is discharged through the gallbladder tube, the patient experiences a sharp pain, whereas when gallstones grow larger than a certain size in the gallbladder, the stone does not cause pain.

Both kidney stones and urethra stones are spherical polycrystalline aggregates of calcium oxalate. They are not uniform in size and form, although they are referred to as spherical aggregates. It is known that hydroxyapatite, which is the main component of teeth and bone, is sometimes formed in muscle due to disease. In this case, hydroxyapatite crystals form a spherulitic polycrystalline aggregate. It is interesting to note that the same hydroxyapatite crystals in kidney and urethra stones are different in form and size from those that exist in teeth and bones. It seems that crystals formed in living bodies through the excretion of unnecessary components do not show characteristics indicating cooperation with protein in their formation.

14.4 Crystals acting as possible reservoirs for necessary components

Many examples have been reported of organic or inorganic crystals in plant cells. Crystals are formed within cells, on cell walls, or at cell boundaries. Calcium oxalate, weddellite, and whewellite crystals are observed in dahlia and spinach, and needle crystals of inuline are seen in begonias. Crystals are mostly sheaf-like aggregates (Fig. 14.6). Amorphous silica (opal, $SiO_2 \cdot nH_2O$) is found in the cells of grasses, and their sizes and forms are variable depending on the species. In addition, stalactic aggregates of calcite, attaining sizes of a few 100 μm (vaterite, amorphous $CaCO_3$, tartaric acid, and citric acid) are known to grow in plants.

Judging from the sizes and forms of the crystals and aggregates, crystals formed in plants are mostly poor in uniformity, and thus it is not clear what sort of role these crystals play in the biological life cycle. If these crystals act as a reservoir for essential components of the plant, such as Ca and Si, we may expect to see characteristics showing growth and/or dissolution in crystal morphology. There have been no reports of investigations that might substantiate this claim; this would be a suitable subject for further discussion. Investigations into the decalcification process have been made with respect to hydroxyapatite forming teeth and bones, or polymorphs of $CaCO_3$ constituting shells.

14.5 Crystals whose functions are still unknown

Various bacteria, such as spherical bacteria, bacillus, and fibrous bacteria, selectively precipitate inorganic crystals around the surface of the bacterium, though the reasons for this are unclear [13]. Although many studies have been performed on the precipitation of inorganic crystals around bacteria in relation to

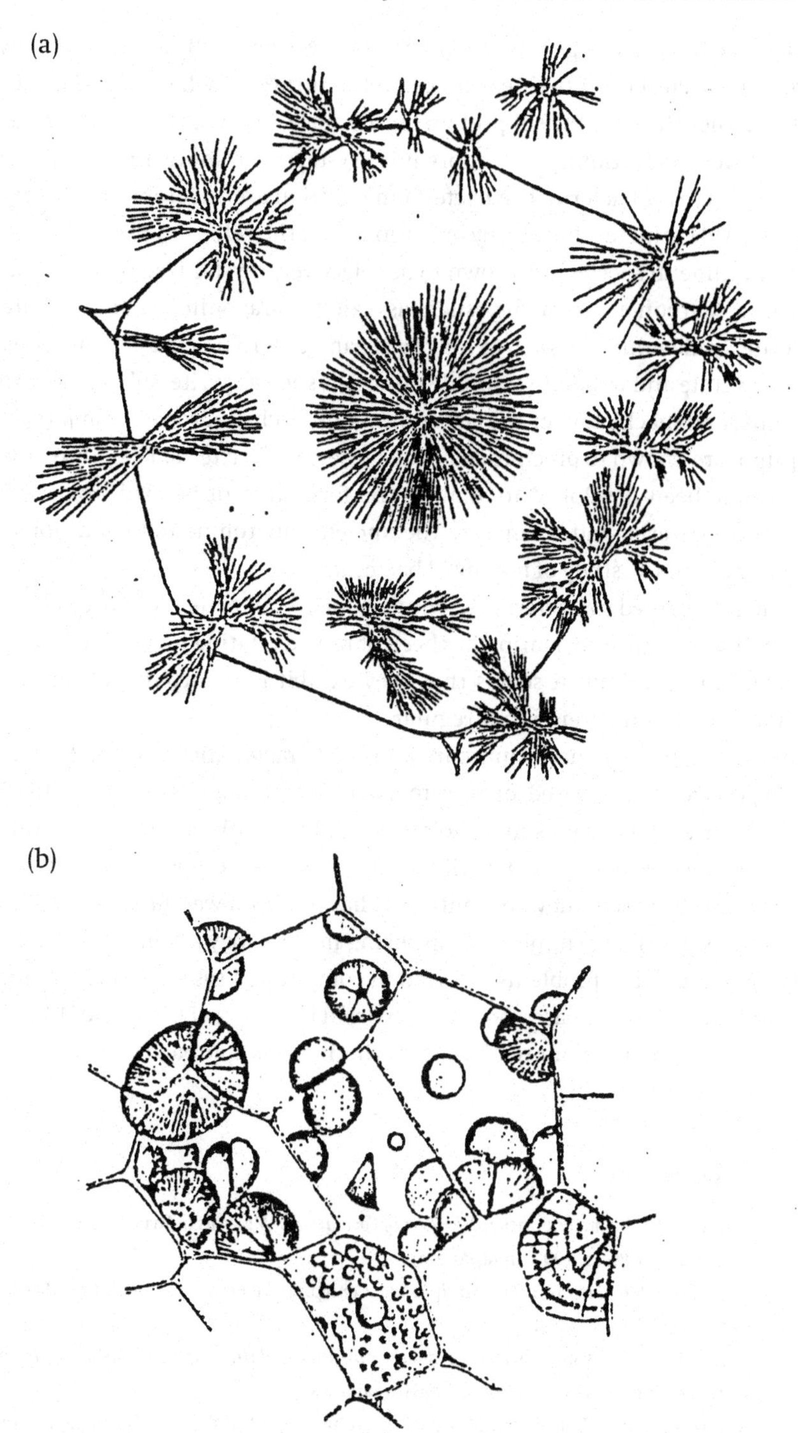

Figure 14.6. Crystals formed in plant cells. (a) Calcium oxalate in begonias. (b) Inuline in dahlias.

environmental or ore genesis problems, it is yet not well understood how they affect the bacteria and their existence. In the future, it will be necessary to establish whether the structure of protein on the surface of a bacteria cell acts as a heterogeneous nucleation site for inorganic crystals, or if inorganic crystals play an essential role for bacteria. The bacteria in question are anaerobic, and so they discharge oxide minerals formed by oxidation from the cell.

Many mineral species are known to be selectively crystallized by the presence of bacteria. Carbonate minerals, such as calcite, aragonite, hydroxycalcite, and siderite; oxide minerals, such as magnetite and todorokite; oxalate minerals, such as whewellite and weddellite; sulfide minerals, such as pyrite, sphalerite, wurtzite, greigite, and mackinawite; and other minerals, such as jarosite, iron-jarosite, and gypsum, are known to precipitate in the presence of bacteria. Therefore, investigations have been developed to analyze the formation of banded iron ore by the action of bacteria, and to analyze the ancient environmental conditions of the Earth through the study of fossilized bacteria.

Minerals formed by bacterial activity are formed outside the cell. Since they are so small, very little information has been obtained relating to the characteristics of their morphology, but it seems that they exhibit idiomorphic polyhedral form, although their alignment is not regular.

In this chapter, we have summarized the characteristics of crystals formed by biological activities, based on the morphology of single crystals and polycrystalline aggregates. Various morphologies exhibited by biominerals are well documented, and we have tried to illustrate these as best we can; however, the mechanism by which they are controlled has not been well understood. Based on the understanding of simple systems, using morphology as our key, we may properly understand the problems involved in complicated systems, such as biomineralization. The data are based on those reported in refs. [1]–[8], [12], and [13]. We shall await the judgment of our readers as to whether this approach was appropriate or not.

References

1 T. Watabe, *Biomineralization, Wonders of Formation of Minerals in Living Things*, Tokyo, Tokai University Press, 1997 (in Japanese)

2 S. Mann, *Biomineralization, Principles and Concepts in Bioinorganic Materials*, Oxford, Oxford University Press, 2001

3 K. Wada and I. Kobayashi (eds.), *Biomineralization and Hard Tissue of Marine Organisms*, Tokyo, Tokai University Press, 1996 (in Japanese)

4 S. Hilton, *Teeth*, Cambridge Manuals in Archaeology, Cambridge, Cambridge University Press, 1986

5 M. Iizima and Y. Moriwaki, In vitro study of the formation mechanism of tooth enamel apatite crystals – Effects of organic matrices on crystal growth of octacalcium phosphate (OCP), *J. Japan. Assoc. Crystal Growth*, **26**, 1999, 175–83 (in Japanese with English abstract)

6 K. Wada, *Science of Pearl – Mechanism of Formation and Method to Distinguish*, Tokyo, Shinjyu Shuppan Co. (in Japanese)

7 H. Komatsu, Crystallization of gallstones, *J. Japan. Assoc. Crystal Growth*, **12**, 1985, 23–41 (in Japanese with English abstract)

8 Y. Shiraiwa, Calcification by photosynthetic organisms and global CO_2 environment, *J. Japan Assoc. Crystal Growth*, **28**, 2001, 53–60 (in Japanese with English abstract)

9 R. Westbroek, F. W. de Jong, P. van der Wal, A. H. Borman, J. P. M. de Vrind, and D. Kok, Mechanism of calcification in the marine alga *Emiliania huxleyi*, *Phil. Trans. Roy. Soc. London*, **B304**, 1984, 435–44

10 J. M. Didymus, J. R. Young, and S. Mann, Construction and morphogenesis of the chiral ultrastructure of coccoliths from the marine alga *Emiliania huxleyi*, *Proc. Roy. Soc. London*, **B258**, 1994, 237–45

11 J. R. Young, J. M. Didymus, P. R. Bown, B. Prins, and S. Mann, Crystal assembly and phylogenetic evolution in heterococcoliths, *Nature*, **356**, 1992, 516–18

12 H. Komatsu, Coincidence site conjugation of cholesterol monohydrate crystals, in *Morphology and Growth Unit of Crystals*, ed. I. Sunagawa, Tokyo, Terra Scientific Publications, 1989, pp. 761–75

13 K. Tazaki, T. Mori, and T. Nonaka, Microbial jarosite and gypsum from corrosion of Portland cement concrete, *Can. Min.*, **30**, 1992, 431–44

Appendixes

A.1 Setting of crystallographic axes

Refer to Figs. A.1.1–A.1.3.

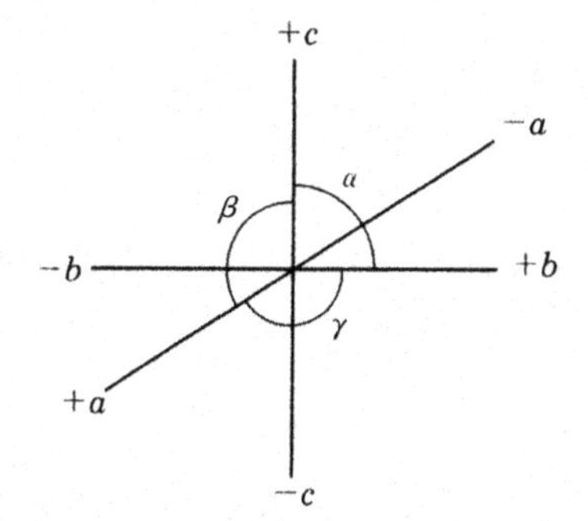

Figure A.1.1. Miller axes applied to crystal systems other than hexagonal (including rhombohedral = trigonal) system.

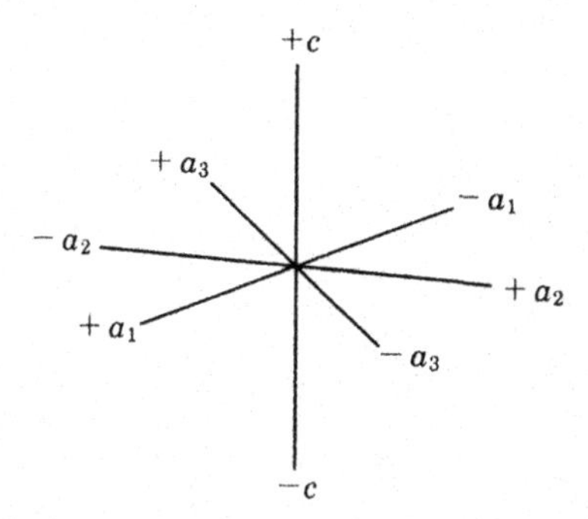

Figure A.1.2. Miller–Bravais axes in the hexagonal system.

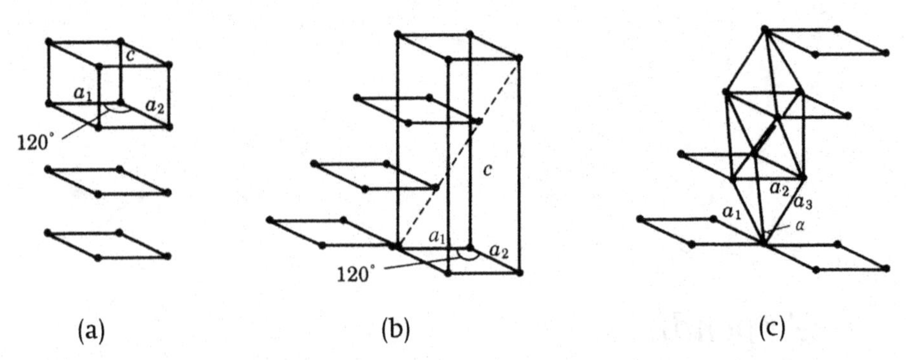

Figure A.1.3. Relation between hexagonal and rhombohedral (trigonal) systems. (a) Hexagonal *P* lattice. (b) Relation between rhomobohedral *R* lattice and hexagonal *P* lattice. (c) Rhombohedral *R* lattice.

A.2 The fourteen Bravais lattices and seven crystal systems

Refer to Figs. A.2.1 and Table A.2.1.

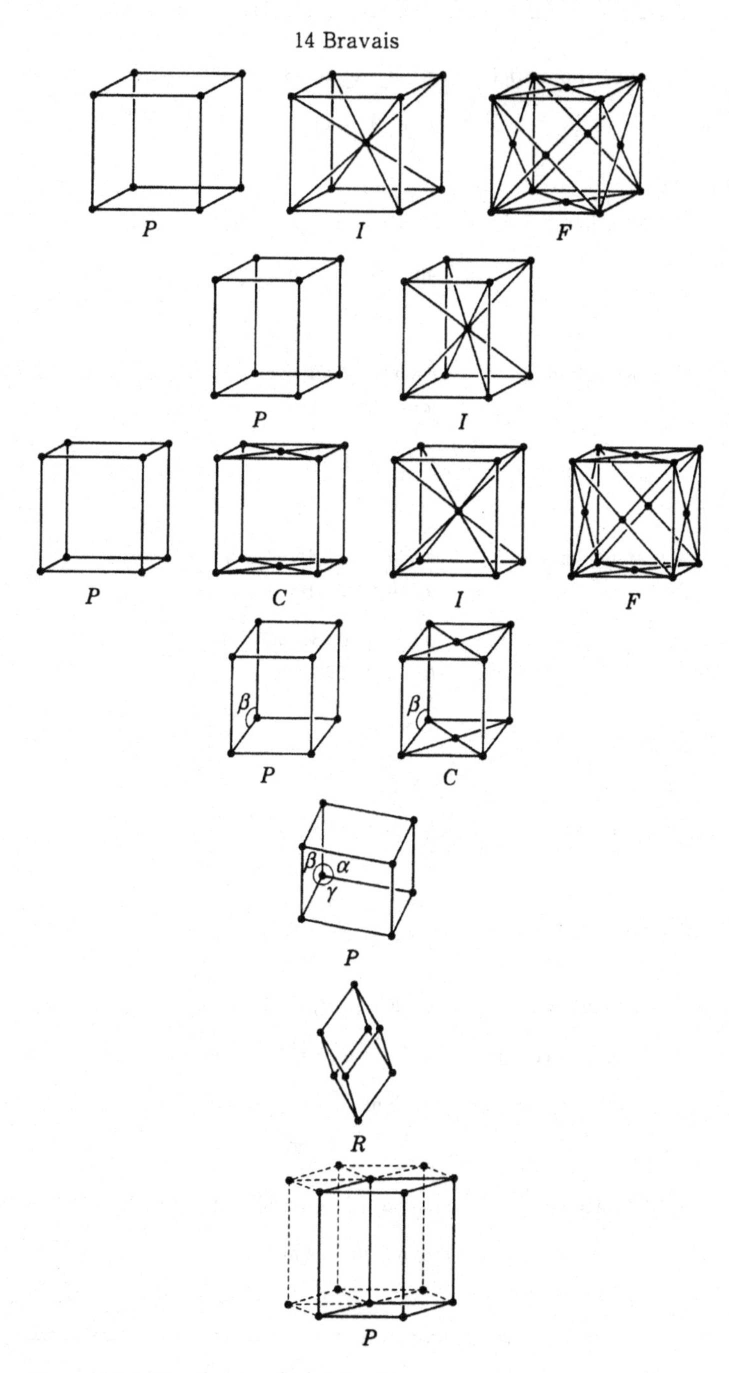

Figure A.2.1. The fourteen Bravais lattices.

Table A.2.1 *The seven crystal systems*

Crystal system	*Axial length, angles, and lattice symmetry*
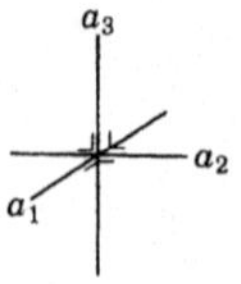 Cubic (isometric)	$a_1 = a_2 = a_3$, $\alpha = \beta = \gamma = 90°$ 4/m $\bar{3}$ 2/m
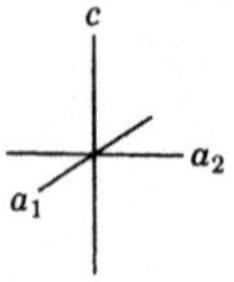 Tetragonal	$a_1 = a_2 \neq c$, $\alpha = \beta = \gamma = 90°$ 4/m 2/m 2/m
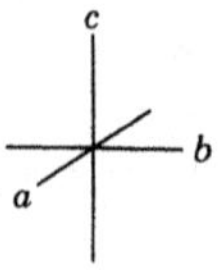 Orthorhombic	$a \neq b \neq c$, $\alpha = \beta = \gamma = 90°$ $(c < a < b)$ 2/m 2/m 2/m
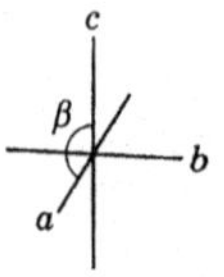 Monoclinic	$a \neq b \neq c$, $\alpha = \beta = 90° \neq \gamma$ $(a < b, \gamma \geq 90°)$ $a \neq b \neq c$, $\alpha = \gamma = 90° \neq \beta$ $(c < a, \beta \geq 90°)$ 2/m
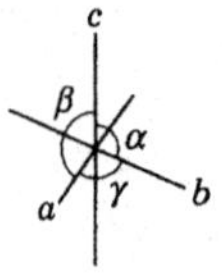 Triclinic	$a \neq b \neq c$, $\alpha \neq \beta \neq \gamma \neq 90°$ $(c < a < b)$ $\bar{1}$
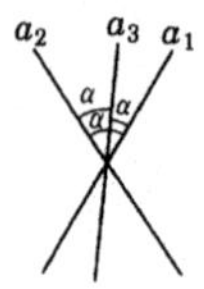 Trigonal (rhombohedral)	$a = b \neq c$, $\alpha = \beta = 90°$, $\gamma = 120°$ $a_1 = a_2 = a_3$, $\alpha = \beta = \gamma \neq 90°$ $\bar{3}$ 2/m 1
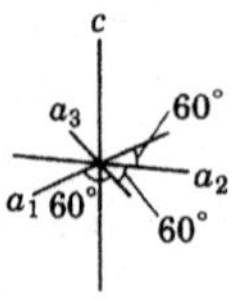 Hexagonal	$a_1 = a_2 = a_3 \neq c$, $\alpha = \beta = 90°$, $\gamma = 120°$ 6/m 2/m 2/m

A.3 Indexing of crystal faces and zones

The indices of crystal faces are traditionally expressed by the points at which the lattice plane parallel to the face cuts the *a*-, *b*-, and *c*-axes, respectively. Indices are always expressed as integers. A particular face is expressed by parentheses ( ), and crystallographically equivalent faces are denoted by curly brackets { }. Miller indices use the reciprocal number of the unit length at which the respective axes are cut, and are widely adopted.

For example, a crystal face that is parallel to the *a*- and *b*-axes but intersects the *c*-axis may be expressed in the form $(\infty\infty 1)$ (this is called the Weiss index), but this can be expressed in Miller index form as $(1/\infty, 1/\infty, 1/1) \Rightarrow (001)$, and the positive and negative orientation of each respective axis is distinguished by adding a bar above the index: (001) and $(00\bar{1})$. In the cubic system, (100), $(\bar{1}00)$, (010), $(0\bar{1}0)$, (001), and $(00\bar{1})$ are all crystallographically equivalent faces, and so they can be denoted by {100}.

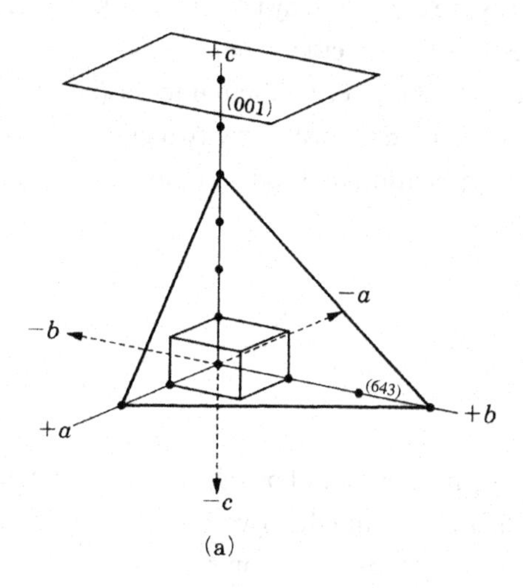

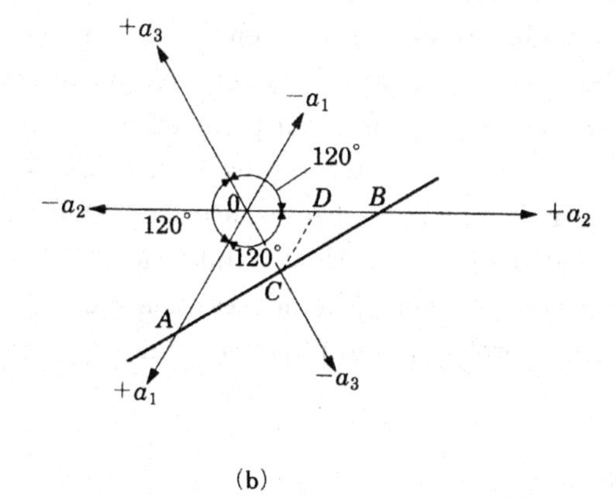

Figure A.3.1.

In the tetragonal system, (100), ($\bar{1}$00), (010), and (0$\bar{1}$0) are crystallographically equivalent and can be expressed by {100}, but (001) and (00$\bar{1}$) are not included in this system. A crystal face cutting the *a*-, *b*- and *c*-axes by 2, 3, and 4 units is indexed as (1/2, 1/3, 1/4)⇒(6/12, 4/12, 3/12)⇒(643). See Fig. A.3.1(a).

Since a subsidiary axis a_3 is assumed in addition to the a_1 and a_2 axes in the hexagonal system (and in the hexagonal expression of the trigonal system), the index is expressed by four indices for a general face ($hk\bar{i}l$). This is called the Miller–Bravais index. For example, in Fig. A.3.1(b) a face *ACB* cuts the $+a_1$ and $+a_2$ axes at 1 ($=OA, OB$), and the $-a_3$ axis at $OC=OD=OB/2$. Therefore this face is indexed as (11$\bar{2}$0). From geometry, $h(1)+k(1)=i(2)$.

Crystal zones are crystallographic directions, which express the direction of an edge formed by two crystallographic faces, or the direction perpendicular to a crystal face, or the direction of a crystallographic axis. This corresponds to a point row in a crystal lattice. Their indices are expressed by integer numbers arranged in the order of the *a*-, *b*-, and *c*-axes [*uvw*] (a specific zone), <*uvw*> (general zone), similar to the case of the face index. The zone perpendicular to the (001) face (*c*-axis) is expressed as <001>. The index of an edge formed by two crystal faces can be easily obtained by the crossing multiplication method, as follows.

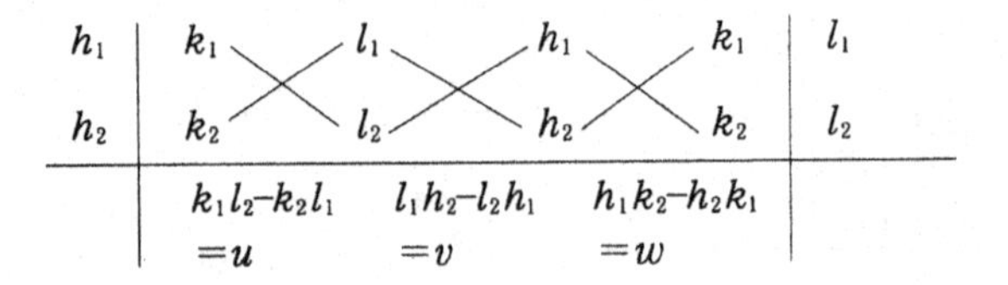

When a face index has a minus sign, this should be taken into consideration in the multiplication. A crystal face (*hkl*) determined by two zones [$u_1v_1w_1$] and [$u_2v_2w_2$] can be similarly obtained by the crossing multiplication method.

The indexing of crystal faces and zones was achieved using morphologically determined axial ratios and angles before the commencement of X-ray structural analyses. In most cases, morphologically determined axial ratios and angles were in accordance with structurally determined unit cell parameters, but in some cases the two were not the same. In such cases, structurally deduced indexes have become more commonly used. For example, the cleavage face of $CaCO_3$ is [10$\bar{1}$1] in morphological indexing, but the structurally determined index [10$\bar{1}$4] has become more generally used. However, since morphological indexes are more straightforward in dealing with morphological problems, morphological indexing is adopted in this book.

A.4 Symmetry elements and their symbols

In the upper row of Fig. A.4.1, rotation axes 1, 2, 3, 4, and 6, and in the lower row, rotation-inversion axes $\bar{1}$, $\bar{2}$, $\bar{3}$, $\bar{4}$, and $\bar{6}$ are illustrated by stereographic projections. Symbols × and ○ denote the positions of a general face projected onto the northern and southern hemispheres, respectively. Numbers 0, 1, 2, 3, 4, and 5 are orders appearing by rotation or rotation inversion. A center of symmetry is expressed by ∘, and a symmetry plane is expressed by thick lines; $\bar{1}$ has a center of symmetry, $\bar{2}$ corresponds to the symmetry plane perpendicular to 1, and so can be expressed as $\bar{2} = 1/\mathrm{m}$. Similarly, $\bar{3} = 3 \times \bar{1}$, $\bar{6} = 3/\mathrm{m}$.

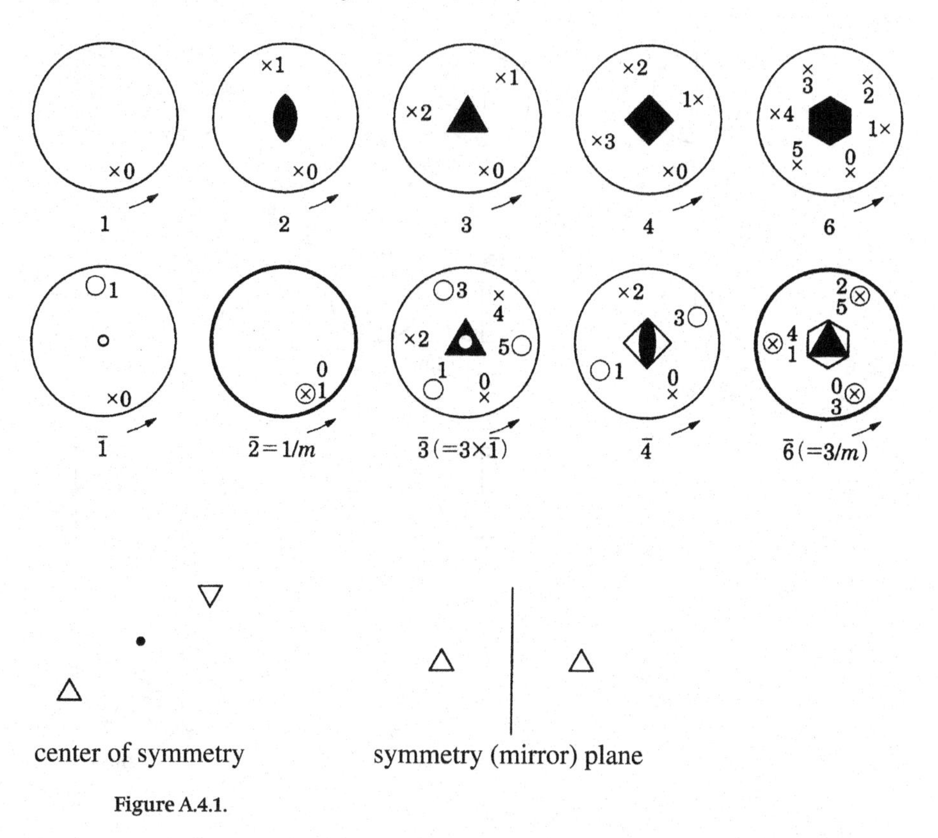

Figure A.4.1.

A.5 Stereographic projections of the symmetry involved in the thirty-two crystal classes (point groups)

Please refer to Table A.5.1. In each row a general face is shown on the left, and the symmetry elements appear on the right; Hermann–Mauguin symbols are shown beneath. Points on the general face are distinguished by • for the northern hemisphere and ○ for the southern hemisphere. For symmetry element symbols, refer to Appendix A.4.

Table A.5.1

	Triclinic	*Monoclinic/ orthorhombic*	*Tetragonal*
X	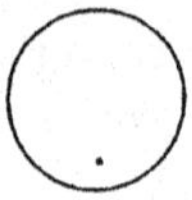 1	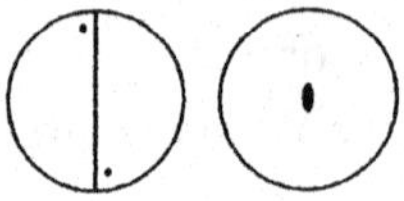 2	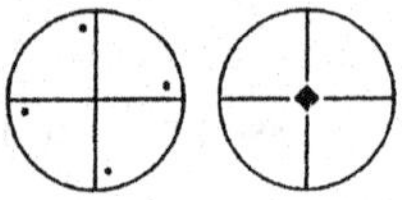 4
$\bar{X}$	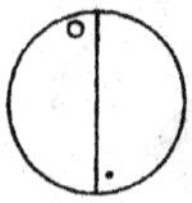 $\bar{1}$	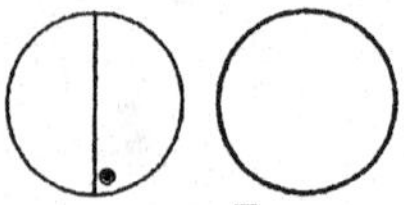 m (= $\bar{2}$)	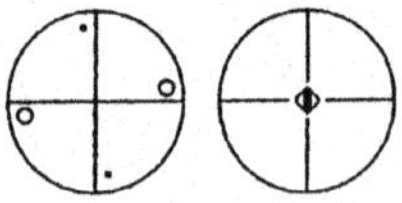 $\bar{4}$
X/m	1/m = $\bar{2}$	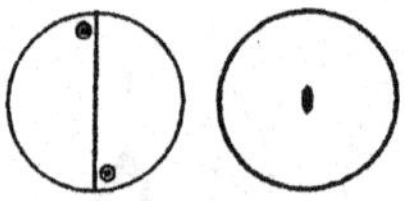 2/m	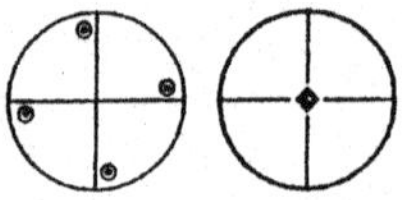 4/m
Xm	1/m = $\bar{2}$	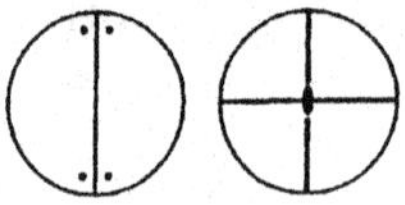 mm2	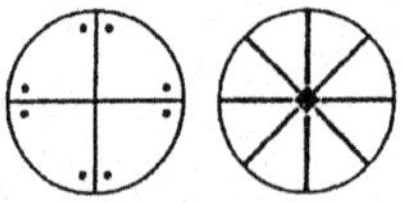 4mm
$\bar{X}$m	$\bar{1}$m = 2/m	$\bar{2}$m = mm2	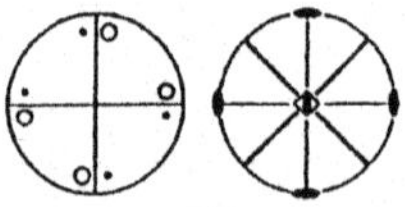 $\bar{4}$2m
X2	12 = 2	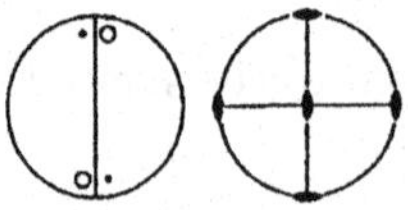 222	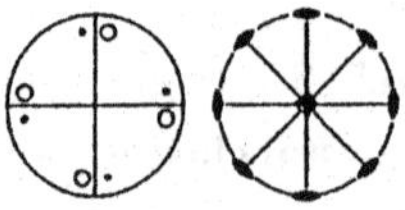 422
X/mm	1/mm = mm2	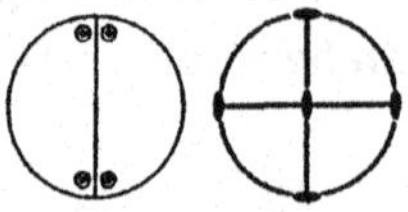 mmm	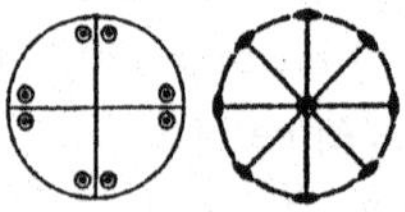 4/mmm

Trigonal/ rhombohedral	*Hexagonal*	*Cubic*
3	6	23
$\bar{3}$	$\bar{6}$	$\overline{23} = 2/m3$
$3/m = \bar{6}$	6/m	m3
3m	6mm	2m3 = 2/m3
$\bar{3}$m	$\bar{6}$m2	$\overline{43}$m
32	622	432
$3/mm = \bar{6}m2$	6/mmm	m3m

Materials index

Subject index

Printed in the United States
By Bookmasters